ROUTLEDGE HANDBOOK OF WILDLIFE CRIME

The *Routledge Handbook of Wildlife Crime* provides a comprehensive and state-of-the-art overview of wildlife crime in its various forms.

The effects of wildlife crime and overexploitation are contributing to the possible extinction of one million species. These activities also harm human and nonhuman animals, ecosystems, and communities. To understand and tackle these issues, this handbook presents critical approaches to the study of wildlife crime grounded in empirical, methodological, and conceptual perspectives. Curated for an international audience of researchers, practitioners, and policymakers, its contributors are drawn from diverse disciplines, backgrounds, and geographies. The handbook addresses recognised challenges associated with wildlife crime, including transnational security, the role of corporations, legislative frameworks, and enforcement strategies, as well as broader concerns related to conservation, sustainable development, socio-environmental harm, and well-being. Importantly, it also delves into emerging areas, such as gender dynamics, digital markets and social media, social inequality and the marginalisation of vulnerable groups, and moral philosophy and ethics. This handbook equips readers to understand and respond to the multifaceted challenges of wildlife crime in the 21st century.

The *Routledge Handbook of Wildlife Crime* will be of great interest to students and scholars of wildlife crime, wildlife management and conservation, environmental crime, and green criminology more widely. The book will also be of use to practitioners and policymakers involved in developing and implementing strategies to reduce wildlife crime.

Francis Massé is an Assistant Professor in the Department of Geography at Durham University, UK. He takes a political ecology approach to examine the intersection of wildlife trade economies, conservation, and development.

Annette Hübschle is Chief Research Officer and Co-Director of the Global Risk Governance Programme at the University of Cape Town, South Africa. Her research explores illicit economies, natural resource extraction, contested regulation, and the emergence of new harmscapes.

Laura Gutiérrez is a Lecturer in Criminology at the University of Liverpool, UK, focusing on green criminology and socio-environmental harm.

Rebecca W.Y. Wong is an Associate Professor in the Department of Social and Behavioural Sciences at the City University of Hong Kong. She is the author of *The Illegal Wildlife Trade in China: Understanding Distribution Networks* (2019).

Tanya Wyatt is the Lead Researcher on Crimes that Affect the Environment for the United Nations Office on Drugs and Crime's Research and Trend Analysis Branch. Prior to this role, she was Professor of Criminology in the Department of Social Sciences at Northumbria University, UK.

ROUTLEDGE ENVIRONMENT AND SUSTAINABILITY HANDBOOKS

ROUTLEDGE HANDBOOK OF FOREST ECOLOGY
Edited by Kelvin S.-H. Peh, Richard T. Corlett, and Yves Bergeron

ROUTLEDGE HANDBOOK OF CLIMATE LAW AND GOVERNANCE
Courage, Contributions and Compliance
Edited by Marie-Claire Cordonier Segger and Christina Voigt

ROUTLEDGE HANDBOOK ON CLIMATE CRISIS COMMUNICATION
Edited by Alison Anderson and Candice Howarth

ROUTLEDGE HANDBOOK OF DEGROWTH
Edited by Anitra Nelson with Editorial Advisor Vincent Liegey

ROUTLEDGE HANDBOOK OF WETLANDS
Edited by Alan Dixon and Ian Maddock

ROUTLEDGE HANDBOOK OF WATER DIPLOMACY
Edited by Shafiqul Islam, Kevin Smith, Martina Klimes, and Aaron Salzberg

ROUTLEDGE HANDBOOK OF WILDLIFE CRIME
Edited by Francis Massé, Annette Hübschle, Laura Gutierrez, Rebecca W.Y. Wong, and Tanya Wyatt

ROUTLEDGE HANDBOOK OF GRASSLANDS
Edited by Heather Hager, David Gibson, and Jonathan Newman

For more information about this series, please visit: www.routledge.com/Routledge-Environment-and-Sustainability-Handbooks/book-series/RESH

ROUTLEDGE HANDBOOK OF WILDLIFE CRIME

Edited by
Francis Massé, Annette Hübschle, Laura Gutiérrez, Rebecca W.Y. Wong, and Tanya Wyatt

Routledge
Taylor & Francis Group
LONDON AND NEW YORK

earthscan
from Routledge

Designed cover image: Photo taken by Annette Hübschle during fieldwork on illegal succulent harvesting and alternative livelihood opportunities for local communities.

First published 2026
by Routledge
4 Park Square, Milton Park, Abingdon, Oxon OX14 4RN

and by Routledge
605 Third Avenue, New York, NY 10158

Routledge is an imprint of the Taylor & Francis Group, an informa business

British Library Cataloguing-in-Publication Data
A catalogue record for this book is available from the British Library

ISBN: 978-1-032-36814-6 (hbk)
ISBN: 978-1-032-36815-3 (pbk)
ISBN: 978-1-003-33390-6 (ebk)

DOI: 10.4324/9781003333906

Typeset in Optima
by codeMantra

CONTENTS

ABOUT THE EDITORS

Francis Massé is is an Assistant Professor in the Department of Geography at Durham University, United Kingdom. Focusing on Southern Africa, he bridges political ecology and political geography to critically examine the intersections between poaching and the legal/illegal wildlife trade and concerns about development, security, and public health.

Annette Hübschle is Chief Research Officer and Co-Director of the Global Risk Governance Programme in the Public Law Department at the University of Cape Town. Her research focuses on il/legal economies, emerging harmscapes, and alternative governance models. She approaches these issues through a sociological and criminological lens, drawing on critical criminology, economic sociology, and studies of formal and informal governance.

Laura Gutiérrez is a green criminologist based at the University of Liverpool. She specialises on the study of socio-environmental harm stemming from legitimate industries and wildlife crime, paying particular attention to how global trends materialise in non-Western contexts, and the complex relationship between the Global North and the Global South.

Rebecca W.Y. Wong is an Associate Professor in the Department of Social and Behavioural Sciences at the City University of Hong Kong, Hong Kong. Her research interests concern the intersections of green criminology, illegal trade in protected and endangered wildlife, animal abuse, criminal networks, and issues of trust in the underworld.

Tanya Wyatt is a green criminologist with nearly 20 years of experience researching Crimes that Affect the Environment. She joined the United Nations Office on Drugs and Crime's Research and Trend Analysis Branch in November 2022. There, she is the Lead Researcher on Crimes that Affect the Environment. Before joining United Nations Office on Drugs and Crime, she was a Professor at Northumbria University in Newcastle, United Kingdom.

During this time, she published nearly 100 books, journal articles, book chapters, and reports on Crimes that Affect the Environment. Prior to that, she was a US Peace Corps Volunteer in Ukraine and a police officer for nearly five years in the United States. She holds a BA in Biology from Mills College in Oakland, California, USA and a Master's in Criminology and Criminal Justice from Eastern Michigan University in Ypsilanti, Michigan, USA. She earned her Doctorate in Criminology from the University of Kent in Canterbury, United Kingdom.

CONTRIBUTORS

Astrid A. Andersson is a Post-Doctoral Researcher at the University of Hong Kong, Hong Kong. Her research focuses on wildlife crime and urban biodiversity. She is a National Geographic Explorer, a member of the Hong Kong Wildlife Trade Working Group and the International Union for the Conservation of Nature (IUCN) Wild Parrot Species Specialist Group. She has published papers on wildlife forensics and multiple traded species, including pangolins, parrots, sharks, hippo teeth, and more.

Bupe Banda is working in the Zambian CRB Association, Lusaka, Zambia. She is a Community-Based Natural Resource Management (CBNRM) practitioner and community rights advocate with over nine years of hands-on experience, dedicated to empowering local communities to sustainably manage their natural resources for the improvement of their livelihoods. With a Master of Philosophy in Conservation Leadership from Cambridge University and a Bachelor's degree in Environmental Education from the University of Zambia, Bupe embraces the philosophy that conservation's success hinges on empowering those whose livelihoods depend on natural resources – the front lines in the fight against poaching and IWT.

Cecilia Banda is part of the Zambia CBNRM Forum, Lusaka, Zambia. She is an experienced Project Manager with eight years of expertise in Natural Resources, Climate Change, and Community Engagement, underpinned by a Degree in Agroforestry. Her skills encompass project design, implementation, and monitoring, with a focus on strengthening community-based governance structures for sustainable resource management.

Duan Biggs is an Associate Professor and the Olajos-Goslow Chair of Environmental Science and Policy at Northern Arizona University, United States. He focuses on developing partnerships between researchers and NGOs, governments, and the private sector to conduct science that informs the development of conservation actions and policies for the complex challenges of the Anthropocene.

José Luis Carpio-Domínguez is a Research Professor in Criminology at the Faculty of Law of the Univerisdad Autónoma de San Luis Potosí, Mexico and Member of the National System of Researchers (CONAHCyT-SNII-I).

Jesús Ignacio Castro-Salazar is a Research Professor at the Instituto Tecnológico Superior de Abasolo (TecNM-ITESA), Guanajuato, Mexico and a Member of the National System of Researchers (CONAHCyT-SNII-I).

Hubert Cheung is an Assistant Professor at Ritsumeikan Asia Pacific University, Japan. His research interests are in wildlife trade governance, conservation geopolitics, and inclusion in conservation and science.

Peter Coals is a conservation biologist with a background in practical law enforcement as an anti-poaching officer and police detective. He is an Honorary Researcher at the University of the Witwatersrand, focusing on international and domestic trade in African lion body parts.

Aubrey Collins serves as the EU Wild Animal Policy Coordinator at FOUR PAWS, where she specialises in advocating for stronger EU legislation on wildlife trafficking, with an emphasis on improving policies related to exotic pet trade and the protection of non-CITES species.

Hannah Dickinson is a Simon Research Fellow in the Department of Geography, University of Manchester, United Kingdom. Her research focuses on commodification and (illegal) trade in marine species and the role of ocean governance in shaping marine biodiversity conservation in Europe and the United States.

Nick Fromherz is an independent consultant and attorney focusing on wildlife and environmental issues. He is also an Adjunct Professor at Lewis & Clark Law School, where he previously served as the Latin American Program Director for the Global Law Alliance for Animals and the Environment.

Nicia Givá is Professor of Rural Development at the Department of Economics and Agrarian Development in the Faculty of Agronomy and Forestry Engineering, at Universidade Eduardo Mondlane in Maputo, Mozambique. Her research focuses on livelihoods, socio-ecological resilience, and environmental justice in contexts of wildlife conservation.

Chloe E.R. Hatten is a Post-Doctoral Researcher at the University of Hong Kong, Hong Kong. She is a wildlife forensic and conservation biology researcher, using multi-disciplinary scientific tools to tackle questions about endangered species threatened by illegal and unsustainable wildlife trade and other anthropogenic impacts. Her research has focused on a range of taxa including sharks, hornbills, and groupers, and she is a member of the Hong Kong Wildlife Trade Working Group and the IUCN SSC helmeted hornbill working group.

Janine Heim is a PhD Researcher at Maastricht University researching the exotic pet trade between Germany and South Africa. With a background in communication and security studies, her interests also include the trade structures and networks of other niche sectors.

Amy Hinsley is a Research Fellow at the University of Oxford, United Kingdom, working on the trade of many different species of plants, animals, and fungi. Her research uses a range of methods to understand the people and processes involved in trade, how and why supply chains change, and how trade can be made more sustainable.

Dang Vu Hoai Nam is a Postdoctoral Researcher from the University of Copenhagen and an expert with the Global Initiative Network of Experts. His research focuses on non-market economic valuation, wildlife trade, public health, consumer behaviour, and social marketing.

Harriet Ibbett is a researcher working in the Bangor University. She is an interdisciplinary conservation scientist with a specific interest in understanding human behaviours that drive biodiversity loss. To date, much of her research has involved using novel and experimental approaches to answer complex questions regarding wildlife crime and conservation rule breaking.

George Iordăchescu is a political ecologist working as a Senior Researcher in the Forest and Nature Conservation Policy group at Wageningen University, The Netherlands. Here, he leads the ERC-funded project Green Frontier: Politics of Conservation and Unequal Ecological Exchange in European Peripheries (2024–2029). George has a background in social anthropology, and his work spans across the disciplines of human geography, green criminology, and conservation social sciences. His research interests include the governance of nature protection in the European Union, environmental crime, and environmental politics in Europe.

Freya A.V. St. John is a Reader in Conservation Science at Bangor University. Her research focuses on understanding the links between human behaviour, well-being, and conservation, with a specific focus on conservation conflict as well as unlawful resource extraction.

Julia P.G. Jones is a Professor in Conservation Science at Bangor University. She is a conservation scientist interested in conservation impact evaluation and the impacts of conservation interventions. She has a particular focus on the social dimensions of conservation, with extensive experience working in Madagascar.

Dener José Giovanini is a journalist and the founder and current coordinator-general for RENCTAS, the Brazilian Network to Fight the Wildlife Trafficking.

Teresa Lappe-Osthege is a Senior Consultant in the Environment and Society team at Risk & Policy Analysts Ltd. Teresa's expertise spans the fields of EU internal and external policy processes, multi-level governance, biodiversity conservation, environmental crime and security, and environmental peacebuilding. She holds a PhD in Politics from the University of Sheffield, an MA in Conflict, Security and Development from King's College London, and a BA in European Studies from Maastricht University.

Elizabeth (Libby) Lunstrum is Professor of Environmental and Global Studies at Boise State University in the United States. Her research interests include the human dimensions of biodiversity decline and ecological restoration, in particular green militarisation, restoration in post-war contexts, and Indigenous-led approaches to restoration.

Rodgers Lubilo is working in the Zambia CBNRM Forum, Lusaka, Zambia. He is a seasoned expert in CBNRM and Rural Development, with over 25 years of experience. His expertise encompasses policy analysis, traditional governance, and research, resulting in substantial publications in scientific journals. He specialises in environmental and natural resources policy formulation, particularly in protected areas.

Erica Lyman is a practising international environmental lawyer who specialises in international wildlife and wildlife trade law and is Professor of Law at Lewis & Clark Law School, where she directs the Global Law Alliance for Animals and the Environment.

Jared Margulies is Lecturer in Environmental Governance in the Department of Geography, University College London, UK. His first book, *The Cactus Hunters: Desire and Extinction in the Illicit Succulent Trade*, was published in 2023 by the University of Minnesota Press.

Lorraine Mazerolle is an Australian Research Council Laureate Fellow (2010–2015) and a Professorial Research Fellow in the School of Social Science at the University of Queensland, Australia. Her research interests are in experimental criminology, policing, drug law enforcement, regulatory crime control, and crime prevention.

Nolwazi S. Mbongwa is a Social Ecological Systems Scientist at SANParks (Cape Region), whereby she is tasked with leading research, collaborations and engagements with Indigenous and local communities, private landowners, traditional leaders and healers to foster adaptive and co-management of biodiversity and cultural heritage. She is a Visiting Researcher at Wits University and is completing her PhD at the University of Cape Town looking at the cultural value and sustainability of wildlife trade among traditional healers and muthi traders in South Africa.

Ophélia Miralles is a statistician at the Centre for Climate Systems Modelling (C2SM) at ETH Zurich, Switzerland. She specialises in generative deep learning for nowcasting in partnership with MeteoSwiss. She earned her PhD in Mathematical Statistics from EPFL, Switzerland, focusing on environmental hazards and assessing human activities' impact on global climate change. Additionally, she actively volunteers as a scientific collaborator with RENCTAS, a Brazilian NGO dedicated to combatting wildlife trafficking.

Shylock Muyengwa is working in the Resource Africa, South Africa. He has more than 15 years of experience as a CBNRM Practitioner in Southern Africa. He currently works as the Program Director for Resource Africa.

Vincent Naude is the Science Interface Manager at African Parks, as well as an Extraordinary Lecturer at Stellenbosch University and a Visiting Researcher at the University of the Witwatersrand. His work focuses primarily on large carnivore ecology and wildlife trade, with a particular interest in reintroduction science and monitoring technology. His passion has become the science–management interface, ensuring that research is addressing the needs of communities and management on the ground, while translating complex international policy and cutting-edge science into practical value for local stakeholders. He is particularly interested in challenging norms around cultural use of wildlife and how to sustainably and equitably harness natural resources towards securing traditional rights of access and enforcement mechanisms.

Thelma Nkandu is working in the Zambia CBNRM Forum, Lusaka, Zambia. She is a wildlife biologist who is passionate about the conservation of natural resources and promoting community engagement in the management of their natural resources. She believes that community involvement in the management of the natural resources is the best way to go in achieving conservation goals and making the communities realise the benefit of taking part in conserving the resources around them.

Angus Nurse is Professor of Law and Environmental Justice at Anglia Ruskin University, United Kingdom. He was previously the Head of Criminology and Criminal Justice at Nottingham Trent University. He has research interests in green criminology, criminality, critical criminal justice, environmental law, and animal rights and human rights law. His books include *Policing Wildlife* (Palgrave Macmillan, 2015); *Animal Harm: Perspectives on Why People Harm and Kill Animals* (Ashgate, 2013); *Miscarriages of Justice: Causes, Consequences and Remedies* (Policy Press, 2018), co-authored with Sam Poyser and Rebecca Milne; *The Citizen and the State* (Emerald, 2020); and *Wildlife Criminology* (Bristol University Press, 2020), co-authored with Tanya Wyatt.

Alice Pasqualato is a Global Policy Specialist at the Wildlife Conservation Society, working to strengthen national, regional, and international responses to wildlife trafficking and other nature crimes.

Mònica Pons Hernández is a Postdoctoral Researcher at the Universitat Rovira i Virgili (Spain) MSCA SEAS Postdoctoral Research Fellow at the University of Bergen (Norway). Her research focuses on green criminology, especially the legal and illegal trade of marine wildlife.

Hugh P. Possingham is the Biodiversity Council's Co-Chief Councillor, as well as a Professor of Mathematics and Professor of Ecology at the University of Queensland, Australia. His research interests are in conservation research, operations research, and ecology.

Michael 't Sas-Rolfes is a Research Fellow at the African Wildlife Economy Institute at Stellenbosch University, South Africa. He has an academic background spanning commerce, environmental resource economics, and conservation science and four decades of varied experience in the wildlife conservation sector as a practitioner, researcher, and consultant.

Dina Siegel is Professor of Criminology at Willem Pompe Institute for Criminal Law and Criminology at Utrecht University, The Netherlands. She conducted ethnographic research and published on migration, crime of mobility, human trafficking and smuggling, transnational organised crime, and green and cultural criminology.

Ragnhild Sollund is a Professor in the Department of Criminology and Sociology of Law at the University of Oslo, Norway. She has published widely within green criminology, particularly about wildlife crimes and harms, migration, with a focus on women, police racial profiling, and violence. She was a partner to the EFFACE project (2012–2016) and led the research project Criminal Justice, Wildlife Conservation and Animal Rights (2019–2024).

Hannah Bethany Tilley is a biologist currently working with Asian elephants. She primarily examines animal behavioural questions but is also investigating measures for tackling trade issues concerning this species. She is a member of the Hong Kong Wildlife Trade Working Group, collaborating on wildlife trade research in Hong Kong and in Asia and is a Post-doctoral Associate in animal behaviour and applied conservation at the University of Hong Kong.

Daan van Uhm is Professor of Environmental Crime and Associate Professor of Criminology, specializing in green criminology (Open University/Utrecht University). He has conducted research on various forms of environmental crime, including illegal mining, wildlife trafficking, deforestation, and the criminalization of ecocide.

Thiago Vargas is an Anthropologist and Political Scientist and currently works as the institutional coordinator for RENCTAS, the Brazilian Network to Fight the Wildlife Trafficking.

Valeria Vegh Weis is a Senior Researcher at the University of Konstanz in Germany and a faculty member at Buenos Aires University in Argentina. Her last co-authored book is *Lawfare: The Criminalisation of Democratic Politics in the Global South* (Brill 2023).

Amanda S. Whitfort is an Associate Professor in the Faculty of Law, The University of Hong Kong and a member of the Hong Kong Bar. Her research is focused on animal-related crimes in greater China.

Vivienne Williams is a "freelance academic" and a Senior Visiting Researcher at the School of Animal, Plant & Environmental Sciences at the University of the Witwatersrand, South Africa. Her research focuses mainly on wildlife trade, particularly the informal trade in animals and plants for zootherapeutic use, the monitoring of traditional medicine markets for threatened vertebrates, and the domestic and international trade in lion body parts and bones.

Raulff Zanferrari Lima is a Brazilian biologist and Executive Director of RENCTAS with over 20 years' experience in wildlife conservation.

1

INTRODUCTION - BEYOND BOUNDARIES

Foundations and frameworks in critical wildlife crime studies

Francis Massé, Annette Hübschle, Laura Gutiérrez, Rebecca W. Y. Wong, and Tanya Wyatt

Introduction

Wildlife crime has grown exponentially as an area of research, policy, and practitioner interest in recent years. As an umbrella term, wildlife crime encompasses a range of practices, including illegal hunting, capturing, harvesting, and trafficking of plants, animals, fungi, and their parts for consumptive, collector, and pet trades. While charismatic megafauna have received the bulk of attention, wildlife crime also includes lesser-known and supposedly less charismatic species (Hutchinson et al., 2021; Wyatt, 2021). Wildlife crime is also not confined to trafficking and related practices but includes crimes such as blood sports (e.g. badger baiting) (Nurse & Wyatt, 2020), harms perpetrated directly against specific nonhuman animals (Sollund, 2016), or indirectly via the exploitation of environments and related resources that negatively impact wildlife (Zabyelina & van Uhm, 2020). While conservation outcomes and animal-centric harms remain core concerns of wildlife crime research, what has driven a wide range of researchers from a variety of disciplinary backgrounds and orientations to the study of wildlife crime is its intersections with myriad other concerns from money laundering and corruption (van Uhm & Moreto, 2018), zoonotic-related public health concerns (Ibanez Alonso & South, 2024), instability and security risks linked to organised crime (Wyatt, van Uhm & Nurse, 2020), and issues concerning livelihoods, sustainable development, and broader ecological well-being and justice (Massarella et al., 2022).

Wildlife crime is front-page news and at the top of policy agendas nationally and internationally, with geopolitical ramifications (Massé & Margulies, 2020). It has also experienced a meteoric rise in academia and scholarship, featuring prominently in the disciplines of criminology, law, and policing, as well as in disciplinary journals across the conservation sciences, geography, the humanities, development studies, and international relations, among others. As of early 2025, a search for "wildlife crime" as a topic in Web of Science yields 326 articles, and in Google Scholar, for example, yields over 10,000 results. Five hundred and eighty-one results have "wildlife crime" in the title. However, the term "wildlife crime" does not account for the many acts and issues that individually constitute wildlife crime, like trafficking, poaching, captivity, abuse and harm, and related crimes like fraud, corruption, and

DOI: 10.4324/9781003333906-1

money laundering. A search for these would yield an unknown number of results. The overlap between a range of concerns, disciplines, and challenges highlights the urgency of researching wildlife crime in a way that takes seriously its broader social, economic, and ecological dynamics, intersections, and impacts on both human and nonhuman species.

To address wildlife crime, there is also a need to go beyond traditional legal frameworks and criminal justice responses to tackle the full range of harms associated with wildlife crime. As such, this handbook aims to provide an overview of critical approaches and frameworks grounded in rigorous empirical, methodological, and conceptual perspectives. We work with an expanded scope of what constitutes wildlife crime to include broader socioecological harms and the impacts on human and nonhuman communities. We argue that researching, understanding, and addressing wildlife crime requires a multidisciplinary approach that integrates insights from criminology, political ecology, political science, sociology, and the humanities, among others. It must also incorporate a broader lens that includes the role of communities in mitigating wildlife crime, and the potential of environmental and restorative justice frameworks as an alternative to punitive responses. This handbook is thus designed not only for those researching wildlife crime but also for practitioners, policymakers, and other stakeholders working at the interface of wildlife crime and its related concerns.

Before we introduce the structure of this handbook and proceed to the different chapter contributions, we foreground some of the pillars of what we understand as a critical approach to wildlife crime. This includes (1) an overview of key concepts of crime, wildlife, and wildlife crime; (2) understanding what critical approaches offer and how they might, and do, differ from other perspectives and approaches to wildlife crime; and (3) outlining some of the different disciplinary and theoretical perspectives that inform our understanding of critical approaches to wildlife crime and that weave through the chapters that follow.

What is wildlife? What is crime? What is wildlife crime?

The collection of studies from this handbook draw from and highlight a variety of disciplinary and conceptual perspectives, including various approaches of critical criminology from green (Eliason, 2012, 2020; Ruggiero & South, 2013; Brisman & South, 2019; South & Brisman, 2020; White & Heckenberg, 2014), feminist (Chesney-Lind, 2006; Chesney-Lind & Morash, 2013; DeKeseredy, 2015), and southern and postcolonial orientations (Carrington, Hogg, & Sozo, 2016; Carrington et al., 2019; Goyes, 2019; Travers, 2019). These all reflect on and offer alternative paths to researching and understanding wildlife crime that depart from, yet complement, existing traditional criminological and crime science perspectives. This includes a focus on restorative justice, community engagement, and the multispecies and socioecological dimensions of wildlife crime (Hübschle & Margulies, 2024; Massé et al., 2017; Roe & Booker, 2019). This handbook, however, is not meant to be a criminology handbook. While drawing on these key critical criminological traditions, this handbook moves beyond criminology to capture a broader interdisciplinary audience, highlighting innovative work and the diversity of empirical, methodological, and theoretical research from scholars working across the disciplines of geography, political ecology, conservation science, anthropology, international relations, development studies, and the humanities. This handbook also includes contributions from practitioners working in the field of wildlife crime to critically engage with and reflect on the state of practice and policy across a range of spheres from biodiversity conservation, international development, and the policing and law enforcement sector.

More specifically, the objective of this handbook is to introduce some of the approaches, theories, and methods being used to critically examine wildlife crime. The hope is to provide a foundation for those seeking to begin research on the topic, and a reference for more experienced scholars and practitioners looking to expand their toolkit to include a diversity of critical theorisations and praxis. We aim to offer a set of chapters developed with the level of depth and breadth necessary to holistically understand wildlife crime, its related concerns, and various consequences. Below, we provide an overview of the constituent parts of wildlife crime, namely 'crime' and 'wildlife', and how a critical approach engages with each.

What is crime?

'Crime' was long the purview and subject matter of the discipline of law, taking a fairly positivistic approach that focuses on the use of laws to prevent unwanted behaviours and the subsequent breaking of said laws. Criminology countered some of these positivist tendencies, establishing itself as a discipline that was much more comprehensive in thinking about crime, drawing from a number of social science disciplines like sociology, economics, politics, and further afield like psychology, among others (Brisman, 2019; Michalowski, 1996). These disciplines were brought together to explore 'all things crime', including law breaking, but also motivations for criminal acts and deviant behaviour, responses to criminal behaviour, and broader interactions between society and crime. However, for a long time, criminology itself remained constrained by the law, studying only that which was legally defined as criminal.

Critical criminology emerged in the 1960s–1970s as a needed response to thinking about crime and criminality outside of the bounds of the law (Brisman, 2019; Michalowski, 1996). One of the most important observations that is credited to critical criminology is that crime has no ontological reality – crime is a behaviour that has been labelled as such (Becker, 1963; Michalowski, 1996). Critical criminology does not take the legal, judicial, and enforcement systems and ideas about crime and policing as apolitical or natural. Rather, critical criminology questions, deconstructs, and interrogates the laws, legal structures, ideas of (il)legality, and the knowledge and values underpinning them (Brisman, 2019; Eliason, 2012, 2020; Michalowski, 1996). It followed from this observation that critical criminologists – and indeed all who study damaging practices – should expand their gaze to include the study of all things *harm*, given that by focusing only on unlawful practices, a swathe of destructive, harmful behaviours would remain unchecked and underexamined (Schwendinger & Schwendinger, 1975). This critical turn is particularly important in the context of wildlife crime, where many harmful acts, such as environmental degradation and corporate exploitation, fall outside traditional legal definitions but still cause significant harm to both wildlife and human communities (e.g. excessive consumerism, irresponsible production or extraction practices, and denial of climate change) (Lynch & Stretsky, 2003; Ruggiero & South, 2013; White 2011). Key here is that "what matters is the harm caused (in this case to wildlife) rather than whether that harm is classed by legal systems as 'crime'" (Nurse & Wyatt, 2020, p. 2).

The approach to crime in this handbook thus draws from the critical criminological tradition and its acknowledgement of multiple human and nonhuman harms. Yet, we do not discount the value of examining those harmful acts that fall within the narrow definition of 'crime', and some contributions in this handbook do indeed focus on behaviours and activities that are violations of relevant wildlife or environmental law. As such, this handbook examines wildlife harm through legal, socio-legal, and zemiological lenses. More importantly, the wildlife harm discourse presented in this volume departs from a human-centred understanding of crime to acknowledge and recognise that nonhuman beings and the environment can also be subject to direct and indirect harm and be victims of crime.

Finally, it is worth noting that the works presented in this volume do not understand crimes and harms in a vacuum, but as the product of multi-layered interactions. Useful here is the concept of harmscapes. The term harmscapes refers to complex, interconnected landscapes where harm – whether criminal, ecological, or social – is perpetuated by broader socio-political and economic structures (Berg & Shearing, 2024). This framework challenges traditional criminological views that treat harm as an isolated, event-driven phenomenon. Instead, harmscapes situate harm within larger governance failures, systemic inequalities, and environmental degradation, all of which contribute to both human and nonhuman suffering (Hübschle, forthcoming). This lens is especially useful for addressing wildlife crime, which often occurs in these multifaceted spaces where exploitation and inequality intersect, shaping the broader context in which harm is experienced by both ecosystems and communities.

What is wildlife?

Others have done the hard work of charting various definitions of wildlife and how they do and do not include different forms of nonhuman life like plants, fish, and fungi (Wyatt et al., 2021; Wyatt & Hutchinson, 2023). Importantly, all countries legally define 'wildlife' in their own ways, as do international conventions like CITES (the Convention on the International Trade in Endangered Species of Wild Fauna and Flora) and other global agreements. While there are differences in defining wildlife across these and what gets included and excluded, there are some commonalities. First, wildlife always refers to nonhuman animals and species, although the extent to which it includes plants, fish, and fungi varies. Second, wildlife almost always refers to species that are not domesticated or companion animals (Beirne, 2018). Livestock, for example, is excluded. Of course, wildlife is sometimes kept and traded as companion animals, and sometimes bred in breeding facilities for a variety of reasons and not always captured or killed from the 'wild'. Yet these nonhuman animals are still considered as wildlife because they have not been domesticated in the same sense as cows or dogs.

While this definition is useful, it often excludes plants and other organisms, such as fungi, that are integral to ecosystems and biodiversity. This oversight is compounded by the phenomenon of "plant blindness", where plants are undervalued in wildlife crime discussions despite being subject to similar overexploitation (Margulies et al., 2019). Similarly, non-traditional wildlife forms, such as fungi and certain plant species, face increasing threats from illegal harvesting and trafficking in markets that are often poorly regulated or subject to irregular regulation (Hübschle & Margulies, 2024). These markets are frequently inadequately governed, allowing for criminal exploitation, whether for medicinal purposes, ornamental use, or in response to the growing demand for rare plants in international trade. In this context, we recognise the need to expand the definition of wildlife to include not only nonhuman animals but also plants, fungi, and other life forms, especially in the face of irregularly regulated markets where these entities are often traded in legal grey areas.

Hence, we adopt a definition of wildlife already established in wildlife criminology that encompasses the spirit of the above. Here, wildlife is defined as "All non-human animals, plants, and fungi which form part of a country's natural environment, or which are visitors in a wild or captured state" (Nurse & Wyatt, 2020, p. 4). While this definition is useful, we also recognise the importance of expanding it to include nonhuman animals, plants, and fungi that do not fall under the jurisdiction of a 'country' and its laws, such as marine species in international waters. In the context of wildlife crime, this expanded definition allows for a more comprehensive understanding of how wildlife trafficking, habitat destruction, and exploitation can occur across borders, impacting ecosystems and communities globally.

What is wildlife crime?

The critical nature of our understanding of crime and wildlife leads us to a definition of wildlife crime that is expansive, inclusive, and reflective of contested illegality, necessitating critical engagement. In keeping with critical criminology, we move beyond the legalistic definitions of wildlife crime often adopted in crime science, namely the "illegal taking, trading, exploiting, possessing, or killing of animals or plants in contravention of national or international laws" (Kurland et al., 2017, p. 1), to include broader notions of crime that encompass harm involving and against wildlife (Eliason, 2020; Nurse & Wyatt, 2020; Ruggiero & South, 2013; Sollund, 2019; Wyatt, 2021). The contested nature of illegality in wildlife crime often arises from the grey areas in law where certain actions are legal in some jurisdictions but illegal in others (Dickinson, 2022), and where cultural practices may be deemed criminal in one context but legitimate in another (Hübschle, 2017a 2017b, 2022). More-than-human social harm caused by actions against and involving wildlife is thus integral to our understanding of wildlife crime. This includes expanding our understanding beyond anthropocentric harm to consider direct and indirect harm done to nonhuman life (Flynn & Hall, 2017).

Additionally, much has been said about how wildlife crime, inclusive of harms, is often too narrowly focused on illegal hunting, poaching, and trafficking of wildlife and wildlife derivatives (Nurse & Wyatt, 2020). While we, and the chapters in this handbook, do not shy away from these and related harms, it is important to recognise the myriad other harms and illegal activities that constitute wildlife crime. Some of these are included in this book, such as the use of nonhuman animals in sport (Chapter 6), state-corporate crime like the harm perpetuated against wildlife by corporations (Chapter 5), and broader notions of ecocide (Chapter 12). There are, however, many more, including but not limited to the use of nonhuman animals in bloodsport (Nurse & Wyatt, 2020) and sexual assault of nonhuman animals (Beirne et al., 2017).

No one approach: a multidisciplinary tradition and foundation

To reiterate, critical criminology is central to this handbook, due to both its approach to the conceptualisation of crime-related issues and the empirical areas of research that it has opened up. In terms of conceptual contributions, this volume builds on three key ideas proposed by critical criminology. First, critical criminology highlights the role of the powerful in the shaping of what is regarded by the law and in the zeitgeist as harmful, undesirable behaviour (Pearce, 1976); an issue that some of the chapters of this volume will zoom in on. Second, and to repeat from above, we draw from green criminology, which alerts us to the fact that the destruction of the environment (including harm to wildlife) is often the product of lawful practices and that, therefore, we must look beyond the law to study wildlife harm

(White, 2011; South & Brisman, 2020). We use this logic to set the thematic boundaries of this volume, which encompasses both legal and illegal harms to wildlife. Third, the interface between legality and illegality is a central theme of this volume: critical criminology suggests that when it comes to certain types of crime (e.g. 'organised' crime), there is a very grey area. This includes the concept of contested illegality, where the same actions may be both legal and illegal depending on the jurisdiction, actors involved, and the broader cultural and political context (Hübschle, 2017b). Legal actors can engage in illegal practices, and those causing harm can operate in the legal world. Some of our contributors explore this grey area and interface (Chapters 2 and 13).

While critical and green criminology have been very good at setting the foundations to think about wildlife harm, focus on the topic in itself has been lacking in the discipline. Wildlife criminology itself is a project to push the boundaries of the discipline to further bring in non-domesticated nonhuman animals, plants, and fungi into the fold (Nurse & Wyatt, 2020). Additionally, cultural criminology offers a framework for exploring the interplay between culture and crime, emphasising how social contexts, meanings, and representations shape criminal behaviours and perceptions (Ferrell, 2013). In the context of wildlife crime, this perspective is increasingly relevant as it highlights how cultural attitudes towards and presentations of nature, conservation, and wildlife exploitation influence illegal activities such as poaching, trafficking, and habitat destruction and responses to them (Brisman & South, 2014; also see Chapters 4, 9, and 16). Also relevant to this volume is the work on southern and decolonial (green) criminology that highlight how harm and crime, including those related to wildlife, are not experienced the same way nor do they necessarily have the same origins, as crime in the West or North (Carrington, Hogg, & Sozo, 2016; Carrington et al., 2019; Goyes, 2019). Furthermore, the West and North have dominated scholarship and narratives about (wildlife) crime, whereas there are perspectives and lessons to be learned from the South (Carrington et al., 2019; Goyes, 2019; Travers, 2019). Some such perspectives are found in this handbook (Chapters 15 and 18).

As wildlife crime is fundamentally about the human–environment nexus, and specifically human–wildlife relations, our framing and approach and some of the contributions in this handbook also draw heavily on political ecology (Robbins, 2019). Political ecology is concerned with how processes of material and discursive power shape human–environment relations, how processes of environmental and socioecological change are experienced, understood, and engaged with, and the subsequent social and ecological outcomes. For example, political ecology analyses are increasingly being mobilised to examine "how processes of (post)colonial, neoliberal, racial, and capitalist power and state-making shape the production of, and response to, environmental crimes and their socio-ecological outcomes" (Margulies, Massé, & VandeBerg, 2023, p. 25; see also McSweeney, 2023). These outcomes can have real, material impacts for humans and nonhumans. On the discursive front, political ecology complements green criminology in understanding how discourses about wildlife harm and crime frame the problem, who/what is responsible, and therefore the policy and related actions that follow (Massé et al., 2020). Political ecology similarly moves away from thinking about the legal vs. illegal and licit vs. illicit in terms of actors and actions to think more critically about their blurring, and how certain forms of social and ecological harm are sanctioned and sustained, while others are demonised (Dickinson, 2022; Vasile & Iordăchescu, 2022; Margulies, Massé, & VandeBerg, 2023). Bringing together political ecology and critical, especially green, criminology, provides a comprehensive and productive approach to researching and theorising wildlife crime and the various power dynamics that shape and contribute to its multifaceted harms and how they are (or are not) addressed (Massé, Giva, & Lunstrum, 2021; Duffy & Brockington, 2022; Gutierrez & Duffy, 2024; Iordăchescu, 2023). Several chapters in this handbook specifically bring political ecology insights to wildlife crime (Chapters 2 and 11).

A multidisciplinary framework for wildlife crime includes other disciplines and subdisciplines that we and the contributing chapters draw on. Sociology, the study of social behaviour, provides valuable insights into wildlife crimes by examining the social contexts and cultural factors that contribute to such activities. It helps explain how socioeconomic conditions, community norms, and globalisation influence illegal practices like poaching and trafficking (Hübschle, 2017a). Critical legal studies (CLS) examine the dynamics between law and power, emphasising how legal frameworks can perpetuate social injustices (Tushnet, 2008). In the context of wildlife crime, CLS deconstructs, questions, and critiques existing environmental laws and regulations, highlighting how they may favour corporate interests over ecological protection and marginalised human and nonhuman communities (Eliason, 2012, 2020; Sollund, 2019, 2022). As much wildlife crime occurs within spaces of conservation, like national parks and other protected areas, and has profound implications for conservation outcomes, practice, and policy (Challender et al., 2014), conservation studies and related disciplines of conservation biology, ecology, and other sciences also analyse the causes and effects of wildlife crime. By integrating socioeconomic factors, community engagement, and law enforcement, conservation studies can inform policies that mitigate wildlife crime. Conservation criminology (Gibbs et al., 2010; Gore, 2017) is a recently established field that bridges green criminology, conservation crime science approaches (Moreto, 2018; Moreto & Pires, 2018), and conservation studies and natural resource management to understand wildlife crime.

Throughout this volume, the reader will encounter chapters that seamlessly weave the disciplines and perspectives mentioned above, deploying the full range of analytical tools, theories, and methods to present robust understandings of and approaches to wildlife crime. All of the perspectives, approaches, disciplines, and subdisciplines mentioned have something to offer when it comes to researching and understanding wildlife crime, whether that be through theoretical insights, empirical methods, analytical frameworks, or academic debates. They are also united in their concern for harm that extends beyond legal boundaries; the defence of well-being for both human and nonhuman entities; the use of critical thinking as a motor for change; and a deep interest in the project of ecological, species, and justice-oriented futures.

Responding to wildlife crime through alternative, justice-oriented mechanisms

A core thread of this handbook is the need to develop and understand new and already existing alternative, multispecies, justice-oriented mechanisms to respond to and address wildlife crime. While critical approaches do not offer a single 'solution', they provide a foundation for thinking beyond legalistic, juridical, and policing standpoints. For example, community-based approaches to poaching (Booker & Roe, 2017; Cooney et al., 2017; Massé et al., 2017) and socioecological harm reduction approaches to the illegal wildlife trade (Hübschle & Margulies, 2024) provide useful, practice-oriented insights and frameworks that seek to work with rather than alienate people in and around conservation areas or areas where wildlife is sourced.

Alternative mechanisms also require re-thinking existing laws, legal frameworks, and enforcement or policing efforts to take account of wildlife crime's various human and non-human harms and intersections. Development and livelihood-oriented approaches, for example, acknowledge the socioeconomic harms caused by wildlife crime *and* law enforcement, security-oriented strategies to address it (Annecke & Masubelele, 2016; Hübschle & Shearing, 2018; Lunstrum et al., 2023; Massé, Lunstrum, & Giva, 2021). The ongoing and further criminalising of the wildlife-based livelihoods of some of the most vulnerable people can often have unintended consequences, including the further marginalisation of people as they lose access to land and resources for subsistence and livelihood purposes, thus deepening their hardship and poverty (Hübschle, 2025; Iordăchescu & Vasile, 2023; Witter, 2021). Critics also point to the disproportionate use of violence by law enforcement against marginalised people in and around conservation areas (Lunstrum, 2014; Witter & Satterfield, 2019; Weldemichel, 2020; Ramutsindela, Matose & Mushonga, 2022; Lunstrum et al., 2025). Both are examples of social harms stemming from efforts to address wildlife crime, which can also lead to a negative impact on wildlife. For one, poaching or retaliatory killing of wildlife can sometimes emerge as a response and resistance to these conservation or law enforcement-induced social harms (Hübschle, 2017a; Witter, 2021). Additionally, a law enforcement or security-first approach can redirect resources away from wildlife conservation and care to militarised policing approaches (Annecke & Masubelele, 2016; Duffy et al., 2019).

One set of alternatives to an enforcement-first and criminalising approach falls under the umbrella term of 'sustainable use'. Rather than criminalising the use of wildlife for food or livelihood purposes, these approaches seek to understand how the use of wildlife by marginalised communities for consumptive food security purposes and for livelihoods through the development of wildlife-based economies can yield socioeconomic and ecological benefits (Hübschle & Shearing, forthcoming; Roe & Booker, 2019). The rationale is that by extending marginalised, rural populations the ability to use wildlife and therefore benefit from it in a sustainable way, these same people will work to conserve and protect wildlife. Contestations over the definitions of sustainable are, of course, a challenge, as are ethical and philosophical differences over whether people ought to be able to use wildlife for their own benefit.

Another area of critical scholarship argues that we should move beyond human-centred ways of knowing and understanding wildlife crime. One such example is Earth jurisprudence (Cullinan, 2011; Lampkin & Wyatt, 2020). Earth jurisprudence is a radical approach to legal systems that would require completely new laws – not simply adjusting the existing ones or implementing more environmental legislation. The goal of Earth jurisprudence is to remake the system to be Earth-centred rather than human-centred.

According to Cullinan (2011, p. 13), Earth jurisprudence contains five principles. The first principle hinges on the idea that the laws by which our planetary ecosystem operates are given by the universe, not by human legal systems. Second, all beings on Earth have the rights to exist, to have a place to live in, and to be part of the evolution of the Earth community. Third, the rights of beings must be balanced against one another in order to protect the health of all communities. Fourth, human-made laws that contravene the above rights are in violation of the 'Earth Jurisprudence' (the underlying principles that rule relationships and rights in the Earth community) and are therefore illegitimate and unlawful. Finally, humans should modify the laws and principles that currently inform their practices to align them with Earth jurisprudence. This would require that all governance systems focus on restorative justice (restoring relationships rather than punishing) and respect the principles outlined above to preserve life, foster well-being and ultimately meet the rights of the whole Earth community (Cullinan, 2011, p. 13).

While some of the approaches in this handbook and their underpinnings might seem at odds, we focus on what brings them together, namely, an acknowledgement that the existing way of responding to and addressing wildlife crime needs re-thinking for more just, sustainable futures.

Structure of the handbook

While this handbook is organised into different sections, readers should not regard these as mutually exclusive categories. There is much overlap between all sections, with chapters in one often making contributions to another. The chapters in Part 1 of this handbook examine the myriads of ways in which wildlife crime interfaces and intersects with a host of actors, harms, and power dynamics from the cultural (Chapters 4 and 6) and patriarchal (Chapter 3), to those concerning organised crime (Chapter 7), corporate harm (Chapter 5), and broader interfaces of legal and illegal (Chapter 2).

Researching wildlife crime is often fraught with difficulties and challenges given the sensitivities around it and its illegal or illicit nature. Yet, such research is important for understanding the motivations of those engaging in harmful practices, the effects of such practices and efforts to address them, and how to craft more sustainable, just approaches to curbing wildlife crime. The contributions in Part 2 make contributions in this regard. Coming from a variety of disciplinary and methodological backgrounds, these chapters offer innovative methods to study and research the practices and actors contributing to and encompassing wildlife crime and its multifaceted harms (Chapters 8, 9, and 10). Other chapters take a step back to offer reflections on how to study wildlife crime not necessarily through the lens of crime or harm, but broader lenses, such as a form of environmental conflict (Chapter 11).

The contributions in Part 3 highlight some of the alternative and emerging responses to wildlife crime. These include more policy and practice-oriented strategies (Chapters 16, 17, and 21), and broader thinking about the perspectives that underpin, shape, and inform government and society's approaches to addressing and responding to wildlife crime (Chapters 12, 13, 15, and 18). Often pushing against legalistic and enforcement-first approaches, together these contributions highlight the many policy and ontological possibilities for holistic, just, and sustainable pathways to addressing the myriads of harms involving and against wildlife.

As wildlife crime and its associated harms, dynamics, and connections are always evolving and changing, the chapters in Part 4 provide an overview of some of the emerging and ongoing issues in wildlife crime and responses to it. This includes forms of wildlife crime that have always existed but are garnering increasing attention given the scale of their harms (Chapters 19, 23, and 24), as well as exploring ongoing discussions pertaining to how to address such harms and the challenges in doing so (Chapters 14, 20, and 22).

Conclusion

Wildlife crime has become an increasingly important area of academic research and policy interest with a range of wildlife crimes including hunting, capturing, harvesting, and trafficking of protected plants and animals. While much attention has focused on charismatic species, wildlife crime also involves less recognised species and includes activities arising from corporate practices and environmental exploitation that harm wildlife. This growing interest is driven by wildlife crime's contribution to the imminent biodiversity and extinction crisis as well as its links to human and nonhuman harms, organised crime, public health issues, and sustainable development concerns, drawing interests from diverse disciplinary backgrounds.

The objective of this chapter was to highlight the complexity of wildlife crimes and to emphasise the need for interdisciplinary collaboration to better understand the threats facing wildlife and the environment. In the collection of work presented in this handbook, we have brought together different perspectives from both practitioners and scholars to offer a comprehensive understanding of crimes and harms against wildlife. However, the field is continuously evolving, and this handbook does not provide a definitive picture of wildlife crimes. Instead, it serves as a resource for current knowledge and to encourage further exploration. Indeed, further academic and other research is urgently needed to understand wildlife crimes and how to address them in sustainable, just ways. We hope this handbook prompts more critical thinking and research in this regard. Lastly, we extend our appreciation to all contributors for their effort and time. We hope that readers will gain valuable insights from the knowledge they have generously shared in this work.

References

Annecke, W., & Masubelele, M. (2016). A review of the impact of militarisation: The case of rhino poaching in Kruger National Park, South Africa. *Conservation and Society, 14*(3), 195–204.

Beirne, P. (2018). *Murdering Animals: Writings on Theriocide, Homicide and Nonspeciesist Criminology*. London: Palgrave Macmillan.

Beirne, P., Maher, J., & Pierpoint, H. (2017). Animal sexual assault. In Maher, J., Pierpoint, H. and Beinre, P. (eds). *The Palgrave International Handbook on Animal Abuse Studies*. London: Palgrave Macmillan: 59–86.

Becker, H. S. (1963). *Outsiders*. New York: Free Press.

Berg, J., & Shearing, C. (2024). 'Everything-old-is-new-again': Private urban security governance responses to new harmscapes. *Urban Studies, 61*(14): 2703–2718.

Booker, F., & Roe, D. (2017). *First Line of Defence? A Review of Evidence on the Effectiveness of Engaging Communities to Tackle Illegal Wildlife Trade*. London: IIED.

Brisman, A. (2019). Editor's introduction to the special issue: "crucial critical criminologies—revisited and extended". *Critical Criminology, 27*(1), 1–4.

Brisman, A., & South, N. (2014). *Green Cultural Criminology: Constructions of Environmental Harm, Consumerism, and Resistance to Ecocide*. London: Routledge.

Brisman, A., & South, N. (2019). Green criminology and environmental crimes and harms. *Sociology Compass, 13*(1), e12650.

Carrington, K., Dixon, B., Fonseca, D., Goyes, D. R., Liu, J., & Zysman, D. (2019). Criminologies of the global south: Critical reflections. *Critical Criminology, 27*, 163–189.

Carrington, K., Hogg, R., & Sozzo, M. (2016). Southern criminology. *The British Journal of Criminology, 56*(1), 1–20.Challender, D. W., & MacMillan, D. C. (2014). Poaching is more than an enforcement problem. *Conservation Letters, 7*(4), 277–278.

Chesney-Lind, M. (2006). Patriarchy, crime, and justice: Feminist criminology in an era of backlash. *Feminist Criminology, 1*(1), 6–26.

Chesney-Lind, M., & Morash, M. (2013). Transformative feminist criminology: A critical re-thinking of a discipline. *Critical Criminology, 21*(3), 287–304.

Cooney, R., Roe, D., Dublin, H., Phelps, J., Wilkie, D., Keane, A., … & Biggs, D. (2017). From poachers to protectors: Engaging local communities in solutions to illegal wildlife trade. *Conservation Letters, 10*(3), 367–374.

Cullinan, C. (2011). A history of wild law. In Burdon, P. (ed). *Exploring Wild Law: The Philosophy of Earth Jurisprudence*. Mile End: Wakefield Press: 12–23.

DeKeseredy, W. S. (2015). New directions in feminist understandings of rural crime. *Journal of Rural Studies, 39*, 180–187.
Dickinson, H. (2022). Caviar matter (s): The material politics of the European caviar grey market. *Political Geography, 99*, 102737.
Duffy, R., Massé, F., Smidt, E., Marijnen, E., Büscher, B., Verweijen, J., & Lunstrum, E. (2019). Why we must question the militarisation of conservation. *Biological Conservation, 232*, 66–73.
Duffy, R. V., & Brockington, D. (2022). Political ecology of security: Tackling the illegal wildlife trade. *Journal of Political Ecology, 29*(1), 21–35.
Eliason, S. (2012). From the King's deer to a capitalist commodity: A social historical analysis of the poaching law. *International Journal of Comparative and Applied Criminal Justice, 36*(2), 133–148.
Eliason, S. L. (2020). Poaching, social conflict, and the public trust: Some critical observations on wildlife crime. *Capitalism Nature Socialism, 31*(2), 110–126.
Ferrell, J. (2013). Cultural criminology and the politics of meaning. *Critical Criminology, 21*, 257–271.
Flynn, M., & Hall, M. (2017). The case for a victimology of nonhuman animal harms. *Contemporary Justice Review, 20*(3), 299–318.
Gibbs, C., Gore, M. L., McGarrell, E. F., & Rivers III, L. (2010). Introducing conservation criminology: Towards interdisciplinary scholarship on environmental crimes and risks. *The British Journal of Criminology, 50*(1), 124–144.
Gore, M. L. (Ed.). (2017). *Conservation Criminology*. New Jersey: Wiley-Blackwell.
Goyes, D. R. (2019). *Southern Green Criminology: A Science to End Ecological Discrimination*. Bingley: Emerald Publishing Limited.
Gutierrez, L., & Duffy, R. (2024). Harms and the illegal wildlife trade: Political ecology, green criminology and the European eel. *Critical Criminology, 32*(1), 61–76.
Hübschle, A. (2017a). The social economy of rhino poaching: Of economic freedom fighters, professional hunters and marginalized local people. *Current Sociology, 65*(3), 427–447.
Hübschle, A. (2017b). Contested illegality: Processing the trade prohibition of Rhino Horn. In *The Architecture of Illegal Markets: Towards an Economic Sociology of Illegality in the Economy*. Oxford University Press: 177–197.
Hübschle, A., & Shearing, C. (2018). *Ending wildlife trafficking: Local communities as change agents*. Switzerland: The Global Initiative. Ending wildlife trafficking: Local communities as change agents [accessed September 17 2025] https://globalinitiative.net/wp-content/uploads/2018/08/TGIATOC-Wildlife-Trafficking-Report-WEB-4.pdf
Hübschle, A., (2022). Community responses to wildlife, forest and fisheries crime. UNODC Education for Justice (E4J) Wildlife, Forest, and Fisheries Crime. Vienna: UNODC. 114 pages. https://www.unodc.org/e4j/en/wildlife-crime/module-5/index.html
Hübschle, A., & Margulies, J. (2024). The need for a socioecological harm reduction approach to reduce illegal wildlife trade. *Conservation Biology, 38*(5), e14335.
Hübschle, A. (2025). Community Responses to Illegal Wildlife Trade. In White, R. (ed). *Elgar Encyclopaedia of Environmental Crime*. Cheltenham, UK: Edward Elgar Publishing: 285–292.
Hübschle, A., & Shearing, C. (forthcoming). *Governing Wildlife Security: Towards Pragmatic Conservation*. London: Routledge.
Hübschle, A. (forthcoming). *Harmscapes and Security Governance: A Critical Perspective. Encyclopedia of Social Justice in Education – Restorative Justice*. Oxford: Bloomsbury.
Hutchinson, A., Stephens-Griffin, N., & Wyatt, T. (2021). Speciesism and CITES: What uplistings and downlistings say about views of wildlife. *International Journal of Crime, Law and Social Democracy, 11*(2), 191–209. https://doi.org/10.5204/ijcjsd.1945.
Ibanez Alonso, A., & South, N. (2024). Incorporating a one health approach into the study of environmental crimes and harms: Towards a 'one health green criminology. *British Journal of Criminology, 65*(2), 328–343. https://doi.org/10.1093/bjc/azae047.
Intergovernmental Science-Policy Platform on Biodiversity and Ecosystem Services (IPBES). (2019). *IPBES Global Assessment Summary for Policymakers*. Available at: https://files.ipbes.net/ipbes-web-prod-public-files/inline/files/ipbes_global_assessment_report_summary_for_policymakers.pdf
Iordăchescu, G., Lappe-Osthege, T., Dickinson, H., Duffy, R., & Burns, C. (2023). Political ecologies of green-collar crime: Understanding illegal trades in European wildlife. *Environmental Politics, 32*(5), 923–930.

Iordăchescu, G., & Vasile, M. (2023). Forests of fear: Illegal logging, criminalization, and violence in the carpathian mountains. *Annals of the American Association of Geographers, 113*(9), 2108–2125.
Kurland, J., Pires, S. F., McFann, S. C., & Moreto, W. D. (2017). Wildlife crime: A conceptual integration, literature review, and methodological critique. *Crime Science, 6*, 1–15.
Lampkin, J. A., & Wyatt, T. (2020). Utilising principles of earth jurisprudence to prevent environmental harm: Applying a case study of unconventional hydraulic fracturing for shale gas in the United Kingdom. *Critical Criminology, 28*(3), 501–516.
Lunstrum, E. (2014). Green militarization: Anti-poaching efforts and the spatial contours of Kruger National Park. *Annals of the Association of American Geographers, 104*(4), 816–832.
Lunstrum, E., Givá, N., Massé, F., Mate, F., & Jose, P. L. (2023). The rhino horn trade and radical inequality as environmental conflict. *The Journal of Peasant Studies, 50*(3), 1085–1105.
Lunstrum, E., Massé, F., Ashaba, I. M., Dutta, A., Marijnen, E., Mushonga, T., & Matose, F. (2025). Demilitarizing conservation. *Annals of the American Association of Geographers*, 1–13. https://doi.org/10.1080/24694452.2025.2534550.
Lynch, M. J., & Stretsky, P. B. (2003). The meaning of green: Contrasting criminological perspectives. *Theoretical Criminology, 7*(2): 217–238.
Margulies, J. D., Massé, F., & VandeBerg, B. (2023). Illicit geographies and contested environments: An introduction. *Political Geography 101*: 102818.
Margulies, J. D., Bullough, L-A., Hinsley, A., Ingram, D. J., Cowell, C., Goettsch, B., Klitgård, B. B., Lavorgna, A., Sinovas, P., & Phelps, J. (2019). Illegal wildlife trade and the persistence of "plant blindness". *Plants, People, Planet, 1*(3), 209–220. https://doi.org/10.1002/ppp3.10053
Massarella, K., Krauss, J., Kiwango, W., & Fletcher, R. (2022). Exploring convivial conservation in theory and practice: Possibilities and challenges for a transformative approach to biodiversity conservation. *Conservation and Society, 20*(2), 59–68.
Massé, F., Gardiner, A., Lubilo, R., & Themba, M. N. (2017). Inclusive anti-poaching? Exploring the potential and challenges of community-based anti-poaching. *South African Crime Quarterly, 60*, 19–27.
Massé, F., Dickinson, H., Margulies, J., Joanny, L., Lappe-Osthege, T., & Duffy, R. (2020). Conservation and crime convergence? Situating the 2018 London illegal wildlife trade conference. *Journal of Political Ecology, 27*(1), 23–42.
Massé, F., Givá, N., & Lunstrum, E. (2021). A feminist political ecology of wildlife crime: The gendered dimensions of a poaching economy and its impacts in Southern Africa. *Geoforum, 126*, 205–214.
Massé, F., & Margulies, J. D. (2020). The geopolitical ecology of conservation: The emergence of illegal wildlife trade as national security interest and the re-shaping of US foreign conservation assistance. *World Development, 132*, 104958.
McSweeney, K. (2023). Prohibition geographies: Afterward to special issue on "Illicit geographies and contested environments". *Political Geography, 101*, 102835.
Michalowski, R. J. (1996). Critical criminology and the critique of domination: The story of an intellectual movement. *Critical Criminology, 7*(1), 9–16.
Moreto, W. D. (Ed.). (2018). *Wildlife Crime: From Theory to Practice*. Philadelphia: Temple University Press.
Moreto, W. D., & Pires, S. F. (2018). *Wildlife Crime: An Environmental Criminology and Crime Science Perspective*. Durham, NC: Carolina Academic Press.
Nurse, A., & Wyatt, T. (2020). *Wildlife Criminology*. Bristol: Bristol University Press.
Pearce, F. (1976). *Crimes of the Powerful: Marxism, Crime and Deviance*. London: Pluto Press.
Ramutsindela, M., Matose, F., & Mushonga, T. (2022). Conservation and violence in Africa. In Ramutsindela, M., Matose, F., & Mushonga, T. (eds). *The Violence of Conservation in Africa*. Cheltenham: Edward Elgar Publishing: 2–21.
Robbins, P. (2019). *Political Ecology: A Critical Introduction*. Oxford: John Wiley & Sons.
Roe, D., & Booker, F. (2019). Engaging local communities in tackling illegal wildlife trade: A synthesis of approaches and lessons for best practice. *Conservation Science and Practice, 1*(5), e26.
Ruggiero, V., & South, N. (2013). Green criminology and crimes of the economy: Theory, research and praxis. *Critical Criminology, 21*(3), 359–373.
Schwendinger, H., & Schwendinger, J. (1975). Defenders of order or guardians of human rights? In Taylor, I., Walton, P. and Young, J. (eds). *Critical Criminology*. London: Routledge: 123–158

Sollund, R. (2016). The animal other: Legal and illegal theriocide. In *Greening Criminology in the 21st Century:* 93–113. London: Routledge.
Sollund, R. (2019). *The Crime of Wildlife Trafficking: Issues of Justice, Legality, and Morality*. London: Routledge.
Sollund, R. (2020). Wildlife management, species injustice and ecocide in the anthropocene. *Critical Criminology, 28*(3), 351–369.
Sollund, R. (2022). Wildlife trade and law enforcement: A proposal for a remodeling of CITES incorporating species justice, ecojustice and environmental justice. *International Journal of Offender Therapy and Comparative Criminology, 66*(9), 1017–1035.
South, N., & Brisman, A. (Eds.). (2020). *Routledge International Handbook of Green Criminology*. New York: Routledge.
Travers, M. (2019). The idea of a Southern criminology. *International Journal of Comparative and Applied Criminal Justice, 43*(1), 1–12.
Tushnet, M. (2008). *Critical legal studies: A political history*. Cambridge: Cambridge University Press.
Van Uhm, D., & Moreto, D. (2018). Corruption within the illegal wildlife trade: A symbiotic and antithetical enterprise. *British Journal of Criminology, 58*(4), 864–85.
Vasile, M., & Iordăchescu, G. (2022). Forest crisis narratives: Illegal logging, datafication and the conservation frontier in the Romanian Carpathian Mountains. *Political Geography, 96*, 102600.
Weldemichel, T. G. (2020). Othering pastoralists, state violence, and the remaking of boundaries in Tanzania's militarised wildlife conservation sector. *Antipode, 52*(5), 1496–1518.
White, R. (2011). *Transnational Environmental Crime: Toward an Eco-global Criminology*. London: Routledge.
White, R., & Heckenberg, D. (2014). *Green Criminology: An Introduction to the Study of Environmental Harm*. London: Routledge.
Witter, R. (2021). Why militarized conservation may be counter-productive: Illegal wildlife hunting as defiance. *Journal of Political Ecology, 28*(1), 175–192.
Witter, R., & Satterfield, T. (2019). Rhino poaching and the "slow violence" of conservation-related resettlement in Mozambique's Limpopo National Park. *Geoforum, 101*, 275–284.
Wyatt, T., van Uhm, D., & Nurse, A. (2020). Differentiating criminal networks in the illegal wildlife trade: Organised, corporate and disorganised crime. *Trends in Organised Crime, 23*(4), 350–66.
Wyatt, T. (2021). *Wildlife Trafficking: A Deconstruction of the Crime, Victims and Offenders*. Switzerland: Springer Nature.
Wyatt, T., Friedman, K., & Hutchinson, A. (2021). Are Fish Wild? *Liverpool Law Review, 42*, 485–492.
Wyatt, T., & Hutchinson, A. (2023). Plant blindness and the law on international trade in wildlife. *Erasmus Law Review. 1*, 14–26.
Zabyelina, Y., & van Uhm, D. (Eds.). (2020). *Illegal Mining: Organised Crime, Corruption, and Ecocide in a Resource-Scarce World*. London: Palgrave Macmillan.

PART I

Wildlife crime's interfaces and interconnections

2

INTERFACES OF IL/LEGALITY IN EUROPEAN WILDLIFE TRADE

Hannah Dickinson, George Iordăchescu, and Teresa Lappe-Osthege

Introduction: spotlighting illegal wildlife trade in Europe

Illegal wildlife trade (IWT) is one of the most significant forms of wildlife crime, a major cause of biodiversity losses worldwide, and a growing threat to public health (Halbwax, 2020). IWT involves illicit harvesting, poaching, smuggling, trade, collection, or consumption of protected wildlife – including plants – and their derivatives (Sollund, 2013; Mozer & Prost, 2023). While live animals can be traded for the thriving global exotic pet trade (Collard, 2020), more often, wildlife are illegally killed and transformed into commodities with high market value and appeal. Wildlife trafficking has been catapulted to the forefront of international political discussions due to its global dimensions and purported convergence with other forms of serious and organised crime (Massé et al., 2020). The European Union is prominent in worldwide efforts to stymie wildlife trafficking. In November 2022, the European Commission adopted a revised five-year EU Action Plan against Wildlife Trafficking (2022–2027). The plan recognises the European Union's role as a global hub for wildlife trafficking and sets out four key priorities: preventing wildlife trafficking and addressing its root causes; strengthening the legal and policy framework against wildlife trafficking; enforcing regulations and policies to fight wildlife trafficking effectively; and strengthening the global partnership of source, consumer, and transit countries against wildlife trafficking (European Commission, 2022).

While the revised action plan signals a renewed commitment to targeting IWT in Europe and beyond, for too long European countries, species, businesses, and consumers have been overlooked in scholarly debates and policy discussions about wildlife trafficking (Iordăchescu et al., 2023). Europe has predominantly been viewed as a major transit point and destination for wildlife trafficked from Africa and Asia (Europol, n.d.; IFAW, n.d.). Whereas poaching, illegal trade, and consumption of European species have failed to garner similar levels of policy and enforcement attention. An exception is the trafficking of European eels to Asian markets (Alonso & van Uhm, 2023a), which has seen enhanced transnational enforcement operations since 2016, including annual 'Operation LAKE' coordinated by Europol. Beyond European eels, other commonly trafficked European species include sturgeon and their caviar (Dickinson, 2022; van Uhm & Siegel, 2016; Alonso & van Uhm,

DOI: 10.4324/9781003333906-3

2023b; Ludwig et al., 2023), songbirds (Lappe-Osthege, 2024; Ferns et al., 2022), timber (Iordăchescu & Vasile, 2023; Vasile & Iordăchescu 2022; Neag, 2022), and brown bears (Shepherd et al., 2020; Gomez et al., 2023; Iordăchescu, forthcoming).

This chapter redresses the tendency to overlook Europe as a source of illegally traded wildlife by focusing specifically on illicit trade in protected European species and following trafficking routes occurring within the geographical confines of the European continent. This chapter outlines the mechanisms that sustain wildlife trafficking within and between EU Member States and neighbouring non-EU European countries. Following Alonso and van Uhm's (2023a, p. 294) assessment that European eel trade exhibits 'licit, semi-licit, and overtly illicit economies that are constantly developing points of contact, common interests, and strategies', we argue that IWT in European species more broadly is characterised by similar interfaces of il/legality. The boundaries between legal and illegal activities, actors, and economies have become increasingly blurred across European wildlife trades.

Paying attention to the licit–illicit intersections of European wildlife economies ultimately problematises dominant policy discourses that neatly demarcate between legal and illegal spheres of wildlife trade. Such discourses are often mobilised to justify securitised enforcement agendas that overwhelmingly criminalise marginalised populations at the South-Eastern borders of the EU (Iordăchescu et al., 2023; Iordăchescu & Vasile, 2023; Dickinson, 2022) and in neighbouring non-EU states, while obscuring the role of wealthy Western European consumers and businesses in driving these trade dynamics. 'Enforcement-first' approaches rooted in security logics (Duffy & Massé, 2021; Duffy & Brockington, 2022) are largely predicated upon 'treating the symptoms rather than the root causes' (Lappe-Osthege, 2024, p. 8) of IWT in Europe, ultimately doing little to disrupt the ongoing intertwining of legal and illegal activities in European wildlife trade. Furthermore, IWT intersects with myriad other practices – many of which are legal or at the border of illegality – which lead to harms against wildlife (Gutierrez and Duffy, 2023). Failing to account for the licit–illicit intersections in aquaculture, sports-hunting, ecotourism, and transportation could leave less overt wildlife crimes and harms unaddressed (Duffy et al., 2023).

Our focus on interfaces of il/legality contributes to an emerging body of literature detailing 'grey' wildlife markets (Hübschle, 2017; Dickinson, 2022; Jabin et al., 2019; Basu et al., 2020) and discussions in green criminology on 'green-collar' (Iordăchescu et al., 2023; van Uhm, 2016) and 'corporate' wildlife crimes (Wyatt et al., 2020). A grey market exists when illegal commodities enter a marketplace with the appearance of legal goods. This intertwining of legal and illegal markets is made possible by loopholes in laws and illicit arrangements between criminal 'underworld' and corrupt 'upperworld' actors (Passas, 2002; Mackenzie & Yates, 2016; van Uhm, 2018). Examples of global wildlife markets where legal and illegal commodities have been found to co-exist include cacti (Margulies, 2022), bear gallbladders (Jabin et al., 2019), rhino horn (Hübschle, 2017), abalone (de Greef & Haysom, 2022), lion bone (Williams & t'Sas-Rolfes, 2019), rosewood (UNODC, 2020), and fur (South & Wyatt, 2011).

We outline how grey markets for endangered European wildlife are becoming increasingly commonplace, and thus contribute to nascent critical conservation discussions that spotlight overlooked wildlife trafficking dynamics within Europe (Pons-Hernández, forthcoming). Overall, we argue that the intertwining of legal and illegal wildlife economies in Europe stems from a global conservation apparatus that is overwhelmingly focused on criminalisation and securitisation (Duffy & Brockington, 2022); and a politico-juridical system that is marked by stark regional inconsistencies in the implementation and enforcement of wildlife crime legislation (Schlingemann et al., 2017). Across European countries, conflicting and inconsistent environmental legislations maintain an unclear distinction between administrative and criminal offences, thereby limiting intra-European collaboration between control and enforcement authorities (Engel, 2023). These political and legal factors are further complicated by cultural dynamics which shape levels of compliance with wildlife trade regulations and broader support for anti-trafficking campaigns (Lappe-Osthege, 2024).

In what follows, we characterise the interfaces of il/legality in European wildlife trade according to three interlinked areas. First, we interrogate the routes and actors that facilitate the transformation of illegal wildlife commodities into ostensibly legal goods. Next, we detail how regulations and certification systems are manipulated to obscure wildlife trafficking, while highlighting structural deficiencies in European enforcement and legal systems which allow IWT to become 'invisible' (Engel, 2023). Finally, we describe how moralistic arguments are deployed by various supply-chain actors to justify their involvement in wildlife trafficking and to push back on top-down definitions of il/legality. We conclude by emphasising the necessity of engaging with these overlooked interfaces of il/legality in European wildlife trafficking and also consider the wider implications that a focus on legal–illegal interfaces has for matters of environmental justice, biodiversity conservation, and ecological inequalities in Europe.

Interfaces of il/legality 1: routes and actors

Trafficking routes

Routes used to transport illegal wildlife from Africa and Asia into the EU have been well documented. For example, van Uhm (2016) details how live tortoises, reptiles, and macaques are smuggled from Africa into the EU via sea routes and how international airports (e.g. Schiphol or Heathrow) and harbours (e.g. Antwerp or Rotterdam) act as both destination and transit hubs for trafficked wildlife including ivory, rhino horn, and traditional Chinese medicines.

Less well documented are the routes used for trafficking European species into the EU from neighbouring non-EU countries, as well as smuggling routes used within and between EU Member States. This gap in detailed knowledge likely stems from the way the European Economic Area operates as a free-trade agreement zone, allowing for unhindered movements of goods between the 27 Member States plus Iceland, Liechtenstein, and Norway, thereby facilitating the concealment of cross-border shipments of illicit wildlife (Sina et al., 2016; van Uhm, 2016). Furthermore, differing levels of enforcement and compliance with regulations across European countries also make IWT routes within Europe difficult to ascertain, as criminal actors dynamically shift their trafficking routes (Mozer & Prost, 2023; Musing et al., 2018) to take advantage of legislative-enforcement weaknesses, which can obscure their illegal activities.

What is known, however, is that intra-European wildlife traffickers typically avoid major transport hubs by using lesser-policed Southern and Eastern European land borders or crossing the Black and Mediterranean Seas (Sina et al., 2016). For example, illegal caviar enters the EU via multiple routes. One route is overland through Greece and Bulgaria, following transit via Turkey from the Caspian Sea (van Uhm, 2016); trucks also smuggle caviar from Russia via Belarus and into Poland (van Uhm & Siegel, 2016); and caviar is also trafficked by boat from Ukraine into Romania via the Black Sea and Danube River (Dickinson, 2022).

There is clear evidence that songbirds caught in South-eastern European countries, including Bosnia and Herzegovina, Romania, Croatia, and Serbia, are trafficked to Italy, France, and Malta for consumption in high-end restaurants or pet trade (Sina et al., 2016; Lappe-Osthege, 2024; Engel, 2023). In Serbia alone, between 104,000 and 163,000 birds are illegally killed and transported into the EU each year (Schlingemann et al., 2017). Italian investigators have also established that designated bird couriers use modified cars and ferry crossings to illegally move European greenfinches between the port of Pozzallo, Sicily and Malta's capital, Valletta (Civillini, 2019).

The likelihood of wildlife trafficking going undetected within Europe increasingly hinges on the geographies of the trafficking routes. Indeed, a law enforcement agent interviewed by Alonso and van Uhm (2023a, p. 303) maintains that 'trade routes are the most important thing in eel trafficking; if you have a good route, you can slip tonnes through it'. Oftentimes, a 'good route' for trafficking wildlife means passing through multiple jurisdictions and transboundary spaces – like inland seas or international rivers – to conceal the illegal origins of the shipment. By passing through multiple transnational shipment stages, illegal wildlife can undergo a 'cleaning' of sorts (Mackenzie & Yates, 2016), moving from illegal to 'legal' by obtaining fraudulent documentation and export certificates en route. This 'cleaning' can facilitate the 'legal' export of trafficked European wildlife out of the EU, as in the case of European eels (Alonso & van Uhm, 2023a). Or in reverse, it can enable the 'legal' entry of illegal European wildlife from neighbouring European countries and into the EU's free-trade area. Even where fraudulent documentation is not acquired and IWT takes less organised, more opportunistic forms, the transportation routes nevertheless play a fundamental role in shaping the licit–illicit contours of intra-European wildlife trafficking. The routes reveal well-established criminal links between non-EU European countries and EU Member States, while highlighting where inadequate compliance and poor enforcement of EU wildlife trade regulations occur.

Where enforcement against particular modes of IWT *is* effective, this can cause a phenomenon of 'crime displacement' whereby wildlife trafficking routes and practices shift to different geographies (Iordăchescu et al., 2023; Lappe-Osthege, 2024). Within Europe, this dynamic has seen wildlife crimes relocate from EU Member States to neighbouring European countries with differing legal frameworks. For example, the EU Birds Directive places strict limitations on bird hunting practices within the EU, with criminal penalties and enforceable punishments for infringements. In neighbouring non-EU Western Balkan countries (Albania, Bosnia and Herzegovina, Montenegro, Kosovo, Macedonia, and Serbia), bird hunting is regulated less punitively. EU tourists, therefore, flock to Western Balkan states to engage in bird hunting and trade. Despite regulations being in place, Serbia is a hotspot for illegal wild bird killing and trade (Schlingemann et al., 2017), and legal and illegal hunting activities in the country often overlap. For instance, Lappe-Osthege (2024) describes how legally shot game is illegally trafficked from Serbia to the hunters' home nation, e.g. Italy, and that rules on hunting seasons, illegal equipment, hunting methods, and trade are often not strictly enforced in Serbia.

Illicit actors

Illegal trade in European species is facilitated by a diverse network of actors situated along the 'legal–illegal continuum' (Alonso & van Uhm, 2023a). In other words, European wildlife trafficking is not solely perpetrated by lone criminal individuals or organised criminal groups.

On the contrary, 'legal' actors play an indispensable role in European IWT (Iordăchescu et al., 2023). Yet, this fact is often overshadowed by global policy-enforcement narratives that position wildlife trafficking as a serious organised crime and security issue (Massé et al., 2020; Duffy & Brockington, 2022).

The ongoing lack of consensus on what serious organised crime means in the context of IWT (Wyatt et al., 2020) only serves, however, to divert attention away from the role of legal enterprises and individuals in wildlife trafficking. The focus is typically directed towards spectacularised examples of violent poachers and armed criminal groups, such as Russian caviar traffickers responsible for bombing an apartment block in Dagestan in the North Caucasus in 1996 (Zabyelina, 2014; van Uhm & Siegel, 2016), and a Romanian timber/forest mafia apparently responsible for assaulting and killing foresters in the country (Iordăchescu & Vasile, 2023; Gauriat, 2020; Vasile, 2019). These criminal groups are often said to be linked to other forms of serious crime (Massé et al., 2020; Wyatt et al., 2020), thus catalysing public and policy appetite for 'enforcement-first' action against such offenders (Duffy & Massé, 2021).

However, the reality is far more complex and murkier. The over-reliance on spectacularised depictions of organised 'criminals' diverts attention from pre-existing socioeconomic contexts (such as inequalities and informality, clientelism, patronage) that play significant roles in shaping both participation in, and policing of, IWT on the ground (Neag, 2021). Illegal trade in European species can involve legitimate actors working alongside or in complicity with 'underworld' criminals at each stage of the supply chain (Alonso & van Uhm, 2023a). These arrangements can be symbiotic or antithetical (van Uhm & Moreto, 2018) and take highly organised or entirely disorganised and opportunistic forms (Wyatt et al., 2020). 'Upperworld' actors involved in IWT in Europe include legally registered fishers and hunters, corrupt enforcement officials, and corporate businesses such as zoos, aquaculture enterprises, and tourism operations (Iordăchescu et al., 2023).

For example, registered eel fishers in EU Member States may be involved in illegal eel trade at the harvesting stage of the supply chain (Alonso & van Uhm, 2023a). In 2010, a wholesale ban was placed on the import and export of European eels to and from the EU (UNODC, 2020). However, licensed fishers in Spain, France, and the United Kingdom are currently able to harvest eels for the domestic EU market according to strict, though differing conditions, set by each state (UNODC, 2020; Hanel et al., 2019). This has created opportunities for registered fishers to legally catch eels according to their quotas and channel some of these unreported live animals into the international illegal eel trade. Complete bans on eel fishing in Portugal and regional moratoria in Andalusia have also seen legal eel fishers transformed into illegal poachers overnight, creating shifting boundaries between legality and illegality. Furthermore, licensed eel fishers may sometimes work outside of their legal remit to covertly collaborate with organised eel poaching units (Alonso & van Uhm, 2023a).

The corrupt involvement of enforcement and conservation officials is another common legal–illegal interface of European wildlife trafficking. Corruption is considered endemic to the European illegal caviar trade, with coast guards, police, airline employees and customs officials all reported to have a stake in caviar trafficking (Musing et al., 2019). van Uhm and Siegel (2016) note how Russian poachers bribe police to avoid boat and equipment inspections, and that Polish border guards are bribed to allow safe passage of illegal caviar into the EU from the Caspian Sea region. These actions are tantamount to turning a blind eye in exchange for illicit monetary benefits. More flagrant examples of corrupt enforcement officials engaging in illicit activity include the falsification of export permits and documentation by CITES officers to enable transhipment of wild caviar via legal trade channels (Jeanneret et al., 2023).

Monitoring and reporting on the status of strictly protected species, such as brown bears, is a legal obligation under the EU's Habitats Directive. Nevertheless, scientific uncertainty about population dynamics hinders effective conservation and management in several Eastern European countries with bears, wolves, and lynxes (Iordăchescu, 2023). Across the region, game managers have allegedly exploited imprecise population data to over-report bear numbers. In Romania, where trophy hunting of brown bears has been banned since 2016, the last reported official estimates numbered around 6,000 bears in 2016. However, hunting associations consistently report that bear populations have grown to over 10,000, while environmental groups place populations below 2,000 (Pavalasc, 2021). The over-reporting of bear numbers has fuelled concerns about potential increases in bear attacks (Neagu et al., 2022), prompting successive governments to grant derogations from the hunting ban to prevent depredations and further human–bear conflicts. By elevating population figures, game managers operate in a grey area of legality, effectively facilitating continued access to bears as a resource for trophy hunters, under the guise of population management. This comes with significant profits as hunting operators charge tourists upwards of 15,000 euros per bear trophy (Iordăchescu et al., 2023; Popescu et al., 2016; Pavalasc, 2021).

Furthermore, the imprecise assessment of bear populations allows for other wildlife crimes to go unnoticed in Europe. From Slovakia to Bulgaria, and beyond the EU in other neighbouring countries, bears are caught and held captive to be exploited as tourist attractions in petting or mini-zoos, guest houses, ski resorts, or circuses. Shepherd et al. (2020) also note that live bears are coveted for the exotic pet trade, to stock bear bile extraction facilities (misleadingly known as bear farms) and for illegal bear baiting: a bloodsport that is still thought to occur in Ukraine despite being banned since 2015. Some bears are rescued and rehomed in sanctuaries located in the Carpathian or Rhodope Mountains (see Figure 2.1), but many remain captive within 'legal' private menageries across Europe, to be exploited for illicit activities.

This leads us to another key mechanism by which legal actors engage in IWT in Europe: so-called 'corporate crimes'. Wyatt et al. (2020, p. 355) describe corporate crime as 'one or more corporations using their legitimate formal organization and acting in conjunction with other individuals or outside groups to commit illegal activities to benefit the corporation'. In the context of European IWT, the involvement of legitimate corporate actors and businesses is also referred to as 'green-collar crimes' (Iordăchescu et al., 2023) and typically involves aquaculture facilities, zoos, sport-hunting enterprises, and captive-breeders who engage in forms of 'customary illegality' by tolerating or actively practising illicit economic transactions (Gregson & Crang, 2017).

For example, the involvement of legally registered caviar aquaculture enterprises in illicit caviar laundering has been widely reported (Dickinson, 2022; Musing et al., 2019; van Uhm & Siegel, 2016). Caviar whitewashing involves the laundering of poached live sturgeon or wild caviar into the 'legal' stocks of aquaculture facilities, where the caviar becomes transformed into a 'licit' commodity and sold on the open marketplace. A reverse blackwashing phenomenon has also been reported, whereby aquaculture operators fraudulently sell their legally produced farmed caviar on the black market as wild caviar to obtain higher profits. These corporate actors and organisations that straddle both worlds of legal wildlife trade and illegal wildlife trade are 'Janus figures' who have one face looking into illicit networks

Figure 2.1 Rescued bears show signs of trauma and abuse after years of captivity, Bulgaria 2023.
Source: Author (George Iordachescu, 2023).

and another looking towards the public-legal domain (Mackenzie & Davis, 2014, p. 723; Dickinson, 2022).

The scale of corporate wildlife crime in Europe is likely much higher than expected. Between 2001 and 2010, more than 30% of seizures of illegally traded wildlife in the EU were perpetrated by offenders who had previously legally traded wildlife in their role as zoo and wildlife park operators, breeders, or other wildlife professionals (van Uhm, 2016). Paying sustained attention to corporate crime reveals how 'different legal and illegal actors, motivated by different interests, constantly interact' (Alonso & van Uhm, 2023a, p. 302) in the illegal trade of European species. The licit–illicit interfaces produced by corporate crime offenders are commonly overlooked in discussions surrounding the perpetrators of European IWT, which instead tend to focus on poachers, organised criminal groups, and corrupt officials (Neag, 2022). Nevertheless, green-collar crimes are a fundamental aspect of the il/legal interfaces that sustain wildlife trafficking in Europe and deserve further interrogation.

Interfaces of il/legality 2: law enforcement and certification

The second major area, which facilitates the conjoining of legal and illegal wildlife economies in Europe, is via deficiencies in existing law enforcement and certification systems. Indeed, many of the challenges related to policing and prosecuting against the various wildlife crimes mentioned so far stem from shortcomings in current EU environmental and biodiversity legislation, which can inadvertently provide cover for IWT.

For example, the European Crime Directive, adopted in 2008 as the most important instrument to protect the environment through criminal law, uses imprecise and vague definitions such as 'substantial damage' or 'negligible quantities' while also providing non-binding guidance for the interpretation of ambiguous elements in national legislation (European Commission, 2008). A flagship 2023 report on the prosecution of wildlife crimes in 11 European countries has concluded that enforcement is practically hampered, and cross-border cooperation slowed down, by the imprecise nature of the European Crime Directive (Engel, 2023). In its recently revised version, the Directive recognises the need for enhanced specialisation and training of competent authorities, particularly by improving strategies for data collection on wildlife crimes affecting European species (European Council, 2023).

Similarly, the EU Action Plan against Wildlife Trafficking has long been criticised for not dedicating appropriate financial and human resources to its implementation, thus hindering cross-border collaboration between enforcement and adjudicating authorities (Lemître & Herve-Fornereau, 2020). The newly revised Action Plan emphasises capacity-building along the entire enforcement chain, including by improving data-sharing and strengthening the specialisation of police, prosecutors, and judges (European Commission, 2022). However, the issue of sourcing adequate resources for capacity-building still remains. Thus, as it stands, offenders can take advantage of vague criminal definitions and divergent understandings between EU Member States on what constitutes wildlife crimes to engage in criminal activities that may not be recognised or treated as such.

Besides poor collaboration between Member States, insufficient funding, and lack of specialised personnel, the enforcement of environmental legislation is also hampered by controversies of a scientific nature – which prove to be an important challenge for those tasked with adjudicating against wildlife crimes at both national and European levels. Uncertain scientific knowledge about the ecology and population dynamics of many European species, including European eels, songbirds, and brown bears (Iordăchescu et al., 2023), not only stymies species conservation and management of human–wildlife conflicts (Iordăchescu, 2023) but also prevents judges from offering certainty in the assessment of legal cases involving harms against wildlife (Iordăchescu & Krajewski, 2025). Moreover, insufficient knowledge about European species dynamics combined with a general lack of appreciation among the European judiciary about the broader harms caused by wildlife trafficking often results in notoriously lenient punishments for offenders (Fajardo del Castillo, 2016). This makes wildlife trafficking in Europe a low-risk, high-reward activity, which therefore encourages the engagement of non-traditional criminals, including those 'legitimate' actors described previously.

Other key pieces of legislation, such as the Habitats and the Birds Directives, bring more coherence to the conservation and management of European species and habitats, by imposing mandatory obligations for monitoring and reporting and creating a harmonised protection system across Member States. Nevertheless, endangered species that are strictly protected across the EU can be subject to interventionist management strategies that see national governments (temporarily) remove them from the harmonised protection directives. It is not uncommon for states to issue derogations from EU legislation to allow the translocation, deterrence, and even culling of protected species. We previously outlined how these derogations can function to cover up trophy hunting of brown bears in Slovakia or Romania (Popescu et al., 2016). Similarly, national derogations in Italy and Cyprus have been used to allow for harvesting increased quotas of songbird species which are in high demand as luxury delicacies across multiple southern European countries (Lappe-Osthege, 2024). Finally, in Romania, legislative exemptions are sometimes granted for sturgeon fishing on the basis of 'scientific research'. However, these 'research' derogations have allegedly been exploited for sturgeon laundering and illegal caviar trade (Jahrl, 2013; Dickinson 2022).

To prevent environmental crimes associated with the illicit trade and transport of wildlife species, European states have adopted a range of certification mechanisms, voluntary standards, and/or mandatory requirements for traceability and due diligence. Most obviously, the CITES convention uses a system of permits and certificates to authorise the importation and exportation of wild species of endangered flora and fauna, to ensure their sustainable management and use. In the case of critically endangered species like the European eel, which is widely eaten across Europe and beyond (see Figure 2.2), producing companies have adopted voluntary certification standards to reassure consumers of the licit origins in response to increased demand and reports of widespread illegal trade. Nevertheless, multiple studies have started to question the overall legality and ethics of eel supply (Goymer et al., 2023) claiming that even if voluntary certification mechanisms are put in place, all

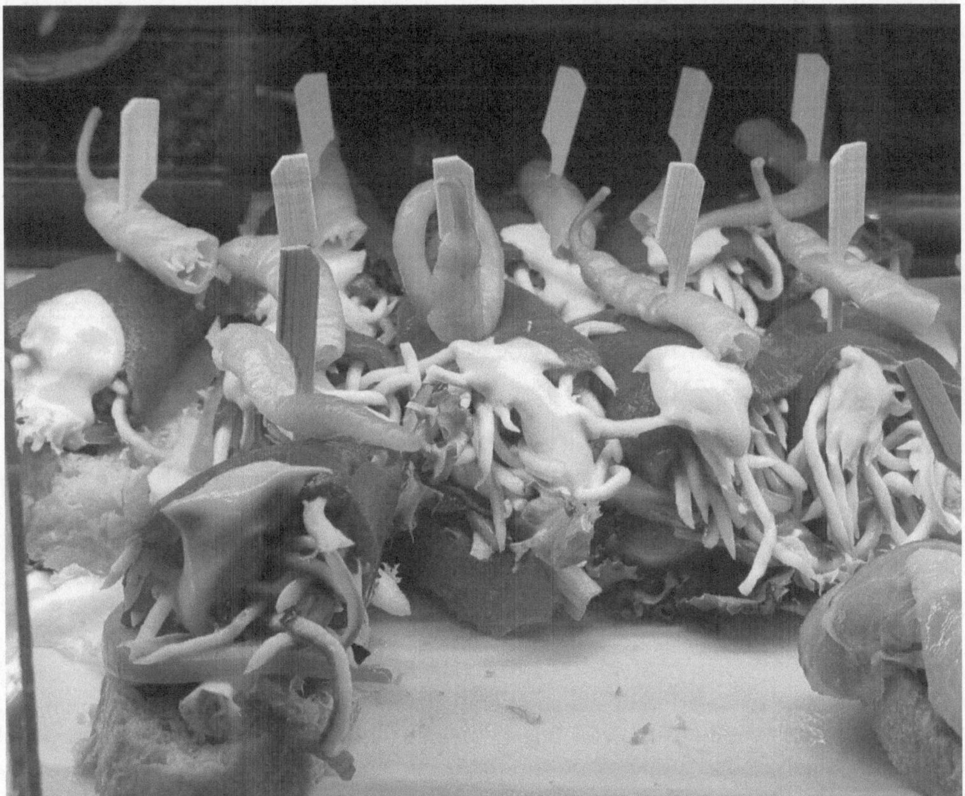

Figure 2.2 Eels tapas in a restaurant in Malaga, Spain 2022.
Source: Author (George Iordachescu, 2022).

European eels are nonetheless sourced from the wild at some point in their life cycle, which inevitably threatens species survival (GoodFish, n.d.).

In contrast, another edible wildlife product – caviar – has been subject to a compulsory EU-wide labelling system since 2006. This system mandates that all caviar containers imported into, exported out of, and sold domestically within the EU must have a non-reusable traceability label. For a long time, policy officials maintained that the caviar labelling system was effective in reducing illegal caviar trade in Europe. However, enforcement checks on caviar are typically limited to checking the presence and placement of labels and the completeness of the CITES codes (Jahrl et al., 2021), with actual verification of the contents of caviar shipments unlikely to take place due to high forensic testing costs (Dickinson, 2022). The faith placed in the caviar labelling system, therefore, provides opportunities for caviar laundering, as offenders are aware that shipments are unlikely to be tested if they appear to be labelled following the legal framework. This has been confirmed by recent research, which revealed that half of caviar products in Europe are illegal, thus casting severe doubt on the efficacy of the labelling system (Ludwig et al., 2023).

To sum up, weaknesses in existing environmental legislation in Europe add to the unsteadiness of policy distinctions between legal and illegal wildlife trade. Vague legal definitions, uncoordinated enforcement efforts, and flawed certification systems create a legal-enforcement landscape that is ripe with opportunities for environmental crime offenders to exploit and manipulate for wildlife trafficking. Rather than reducing incidences of wildlife crime in Europe, a proliferation of certification mechanisms and voluntary standards could end up further blurring the boundaries between legal and illegal wildlife economies, thus reinforcing the need to better tackle the root causes rather than the symptoms of European IWT (Lappe-Osthege, 2024).

Interfaces of il/legality 3: moral economies

The final intersection of il/legality that we explore is the moral economies underpinning European IWT. The term 'moral economy' refers to the social and moral motivations influencing the economic behaviours of individuals or communities (Waters, 2022). Typically, the construction of a moral economy results in the 'widespread acceptability within some societies of economic activities deemed illegal elsewhere' (Gregson & Crang, 2017, p. 207). In their discussion of grey economic markets, Mackenzie and Yates (2016) suggest that perpetrators of criminal market activities commonly use neutralising arguments to justify their illicit actions and to push illegal acts into a morally grey area. This typically involves portraying their behaviours as favourable or heroic, and/or maintaining that the law is intrinsically wrong and so their actions should not be deemed illegal. Mackenzie and Yates call this the 'greying of the moral psychological processes of engagement' (2016, p. 83).

The moral economies of those engaged in IWT globally often display such 'greying' of the processes of engagement (Waters, 2022). In the context of IWT in Europe, various moral economies are at play; asserted by actors operating across the legal–illegal continuum and at different nodes in wildlife supply chains. Perpetrators of IWT in Europe have constructed moral economies around cultural norms, heritage, and traditions. This is particularly prevalent among diverse hunting communities across Europe. According to the European Federation for Hunting and Conservation (FACE), hunting 'is deeply cultural in Europe" (FACE, n.d.) with strong historical roots and centuries-old traditions spread across the continent (Iordăchescu, 2026). Hunting traditions have an important role in European national identity-building processes (Lappe-Osthege, 2024). However, as part of the global agenda on ending wildlife crime, recent years have seen concerted efforts to restrict and curtail traditional hunting methods across Europe, including hunting with dogs, animal baiting, the use of specialised equipment, harvesting of vulnerable or endangered species including songbirds and large carnivores, and imposing limited hunting seasons. Through 'legal definitions that implement dominant social norms and values by characterising certain hunting activities as socially unacceptable' (Nurse, 2017, p. 384), hunting has generally come to be 'perceived as the so-called blood sport not an element of ecology' or biodiversity conservation (Dajczak et al., 2021, p. 803).

Organisations such as FACE are vocal in defending sustainable hunting activities when practised in accordance with the law. FACE maintains that hunting activities, including trapping of mammals and birds, constitute a 'legitimate and indispensable activity for regulating wildlife populations' (FACE, n.d.), while promoting and safeguarding European hunting cultures. Hunting associations and individual hunters have thus built strong moral economies to justify their hunting and trapping activities as both a conservation method and a mode of retaining traditional ecological knowledge and generational solidarities that are at risk of dying out (Dajczak et al., 2021). Overall, a 'widespread opposition to anti-hunting legislation across Europe suggests a shared resistance to legislation and public policy detrimental to their 'sport' among hunters' (Nurse, 2017, p. 383). Furthermore, Dajczak et al. (2021) argue that hunters in Poland are subject to a 'media infodemic', hyperbolic anti-hunting campaigns, and social media vilification that incorrectly portrays the nature of their hunting activities.

With similar dynamics occurring across Europe, hunters can feel vindicated to respond to such anti-hunting rhetoric by engaging in illicit activities and wildlife poaching. At the same time, they mobilise to further the narratives espoused by legitimate organisations such as FACE to effectively 'grey' their illegal behaviours. For example, the protection of wolves has been strengthened by the Spanish government over recent years in various regions across the country. This prompted a backlash from hunters' groups, including demonstrations and threats to kill the excess wolf population (Pettersson et al., 2023). The number of illegal killings has indeed skyrocketed and was generally described as an act of resistance towards top-down and external conservation policies that are at odds with traditional livelihoods based on animal husbandry (Pettersson et al., 2021). European hunters thus construct a moral economy of IWT based on their perceived traditional 'right to hunt' and 'right' to resist what they view as culturally insensitive legal frameworks and campaigns. Nevertheless, despite their moral greying on cultural heritage grounds, 'those engaged in illegal hunting, even with the tacit support of their communities' are perpetrators of wildlife crime (Nurse, 2017, p. 384).

A further dimension to the moral greying of IWT in Europe relates to offenders, policymakers, and enforcement officials relocating responsibility for wildlife trafficking offences to other geographies and actors (both within and beyond Europe). For example, Gutierrez and Duffy (2023, n.p.) describe the 'emergence of racially loaded narratives that cast European participation in the trafficking of European eel as *less criminal* than the participation of Asian actors in the trade' (emphasis added). The authors cite scientists, EU and government officials who suggest that European fishers and businesses are 'not intentionally breaking the law', but unwittingly facilitating illegal eel trade as 'a way to make a living' (2023, n.p.). In contrast, participants in the Asian side of the supply chain are unequivocally labelled as 'cartel people' and 'criminals'. This disavowal of European criminality indicates a problematic and pervasive moral economy that is deeply rooted in race-based 'Othering', whereby criminal responsibility is re-allocated to stereotyped Asian criminals and 'super-consumers' (Margulies et al., 2019). Interestingly, rather than the perpetrators themselves making such justifications for their illicit actions, this example sees official actors construct simplistic, racialised tropes in an attempt to attenuate or obfuscate the role of European citizens in IWT.

Even within Europe, Western European actors go to great lengths to relocate responsibility for IWT to Eastern European communities by exploiting entrenched narratives that frame Eastern Europe as a hotbed of criminality (Šelih & Završnik, 2012). For instance, Dickinson (2022) notes how EU policymakers overlook the European-wide il/legal interfaces of caviar trafficking, to instead portray illegal caviar trade as a criminal problem confined to Eastern European states, including Romania, Bulgaria, Ukraine, and Serbia. In doing so, 'Western European states and actors are absolved of responsibility for sustaining 'grey flows'' of caviar through trafficking networks and into Western European consumer markets (Dickinson, 2022, p. 9). Although not outwardly racialised, the relocation of any culpability from Western Europe to the Eastern European periphery operates according to a similar moral economy of 'Othering', serving to highlight differences in culture and socioeconomic development as the basis for criminality. In other words, the consumption of illegal caviar by elite Western European actors is effectively 'greyed' or neutralised by narratives which spotlight the apparently morally reprehensible behaviours of impoverished poachers and Eastern European organised crime groups.

In sum, 'perceptions of illegality also matter' (Lappe-Osthege, 2024, p. 9) in shaping the blurred boundaries between legal and illegal wildlife trade in Europe. Ranging from hunting tourists who disengage from their usual moral code when on holiday and use various tropes to rationalise their illegal conduct (Tickle & von Essen, 2020); to French chefs serving illegal ortolan bunting dishes to their clientele in the name of preserving centuries-old traditions (Jiguet et al., 2019; Wyatt & Nurse, 2020); the reasons why European actors across the legal–illegal spectrum seek to grey their involvement in IWT are manifold and impossible to cover in their entirety in this chapter. Although the diverse moralistic perceptions and justifications of wildlife crime offenders are often overlooked in discourses about IWT in Europe, they are nevertheless essential in driving and sustaining wildlife trafficking from source to end-user markets. Understanding the varied moral economies of actors engaging in IWT across the legal–illegal continuum is therefore 'crucial to designing prevention and disruption strategies to help declining wildlife populations recover' (Wyatt & Nurse, 2020, p. 44). Rather than tightening regulations further and prioritising enforcement-first approaches, greater emphasis on building nuanced measures against IWT that 'recognise the historical and cultural context in which they are embedded … can reduce EU demand to the point where the grey area in which legal and illegal activity flourishes become insignificant' (Lappe-Osthege, 2024, p. 12).

Conclusion

To understand IWT in Europe, it is first essential to acknowledge that European countries are themselves a source of illegally trafficked wildlife and not just transit or end-user markets for endangered species from Asia, Africa, and Latin America. Furthermore, it is imperative that we examine the multiple interfaces of il/legality that are at work in European wildlife trade and do not perceive the illegal and illicit 'as an aberration, to be bracketed off from the regular economy', but 'think the illegal and illicit as a normal part' (Gregson & Crang, 2017,

p. 207) of how the European wildlife economy functions. Indeed, to achieve the priorities laid out in the EU Action Plan against Wildlife Trafficking, policy-enforcement approaches need to take seriously the multiple dimensions of 'greyness' that conjoin licit and illicit wildlife trades in Europe. To entirely separate the legal and illegal wildlife economies risks ignoring the pivotal interconnections between legal and illegal actors, their use of legitimate trade routes and channels, and the shared moral economies that both 'upperworld' and 'underworld' actors construct and engage with to justify wildlife trafficking operations. Analysing these diverse actors, routes and normative arguments serves to blur the boundaries between il/legality in European wildlife trade in important and far-reaching ways.

Furthermore, the preoccupation with defining IWT as a serious and organised crime has resulted in narrow 'enforcement-first' legal and policing strategies (Massé, 2022) that target poachers and criminal networks, but often miss those businesses and individuals engaged in 'green-collar crimes', as well as wealthy consumers who drive demand for protected European species (Iordăchescu et al., 2023). Despite the persistent policy-enforcement rhetoric on criminalising and securitising wildlife trafficking offences in Europe, practically there are significant shortcomings and inconsistencies in both the design and implementation of legal frameworks and the enforcement and prosecution of IWT across the continent. This effectively means that the networked, 'grey' activities of wildlife traffickers regularly slip the policing net, as actors find innovative ways to exploit the legal-enforcement landscape to conceal IWT or even ostensibly transform their activities into 'legal' transactions.

Thus, drawing attention to the legal–illegal intersections of European wildlife trade is important for pointing to deficiencies in the prevailing European approach to combatting wildlife trafficking and IWT. Aside from bridging policy-enforcement gaps between states and strengthening weaknesses in EU environmental legislation, more efforts need to be targeted at understanding and addressing the root causes, rather than just treating the symptoms of wildlife trafficking in Europe (Lappe-Osthege, 2024). On the one hand, poverty leads to economically marginalised European populations engaging in IWT and poaching as an income source; and on the other, excess wealth sees elite individuals procure luxury illegal wildlife products, including songbirds, caviar, and bear trophies, as a display of their affluence. Efforts to reduce this gaping economic inequality across the European continent and to alter elite consumption practices more concertedly could dramatically curb IWT and the pervasive entangling of licit and illicit wildlife economies. Unfortunately, however, efforts to engage with the cultural and socioeconomic drivers of European IWT have been decidedly lacking thus far, with enforcement against serious organised crime being actively prioritised instead.

However, working collaboratively with, rather than solely against, communities involved in IWT – be it hunting associations, registered fishermen, or individuals engaged in subsistence poaching – has important implications for biodiversity conservation outcomes and environmental justice across Europe. Working with these actors to locate nuanced solutions that support traditional livelihoods, cultural heritage, and biodiversity conservation could lead to lower poaching rates and better outcomes for European species. But as it stands, current modes of criminalisation tend to hone in on the easiest and often most vulnerable targets, such as poachers in South-Eastern Europe. Yet for established wildlife trafficking networks, individual poachers are dispensable and easily replaced parts in their expansive well-oiled machines. As such, enforcement practices can have devastating impacts on impoverished and vulnerable individuals, but barely leave a scratch on the overall wildlife trafficking enterprise. Thus, coupled with on-the-ground cooperation with individuals sourcing wildlife, European and national resources need to be channelled into better disentangling the legal–illegal intersections materialising further along illicit wildlife supply chains. Understanding and actively exploring these il/legal interfaces is ultimately key to effectively targeting IWT in Europe.

References

Alonso, A.I., & van Uhm, D.P. (2023a). The illegal trade in European eels: Outsourcing, funding, and complex symbiotic-antithetical relationships. *Trends in Organized Crime, 26*, 293–307 https://doi.org/10.1007/s12117-023-09490-5

Alonso, A.I., & van Uhm, D.P. (2023b). Blanqueo y 'black-washing' de vida silvestre: Interacciones entre el comercio legal e ilegal de anguila europea y caviar negro desde la perspectiva de la Criminología verde. *Revista Española De Investigación Criminológica, 21*(2), e837. https://doi.org/10.46381/reic.v21i2.837

Basu, S., Jabin, G., Ghosh, A., Singh, S.K., Mitra, A., Chandra, K., & Thakur, M. (2020). Ascertaining suspected wildlife trade from detained parcels under international shipment. *Proceedings of the Zoological Society*, *73*(3), 320–323. https://doi.org/10.1007/s12595-019-00312-7

Civillini, M. (2019). Italy and Malta are hotbeds of songbird poaching. Meet the people fighting to stop it. *National Geographic*. Available at: https://www.nationalgeographic.com/animals/article/songbird-smugglers-in-southern-europe

Collard, R. (2020). *Animal Traffic: Lively Capital in the Global Exotic Pet Trade*. Durham: Duke University Press.

Dajczak, W., Gwiazdowicz, D.J., Matulewska, A. et al. (2021). Should hunting as a cultural heritage be protected? *International Journal for the Semiotics of Law, 34*, 803–838. https://doi.org/10.1007/s11196-020-09763-0

de Greef, K., & Haysom, S. (2022). Disrupting abalone harms. *Global Initiative*. Available at: https://globalinitiative.net/analysis/abalone-south-africa-east-asia/

Dickinson, H. (2022). Caviar Matters: The material politics of the European caviar grey market. *Political Geography, 99*, 102737. https://doi.org/10.1016/j.polgeo.2022.102737

Duffy, R.V., & Brockington, D. (2022). Political ecology of security: Tackling the illegal wildlife trade. *Journal of Political Ecology, 29*(1), 21–35.

Duffy, R.V., & Massé, F. (2021). The integration of conservation and security: Political ecologies of violence and the illegal wildlife trade. In Sowers, J., VanDeveer, S.D & Weinthal, D. (eds.), *The Oxford Handbook of Comparative Environmental Politics*. Oxford: Oxford University Press.

Duffy, R., Iordăchescu, G., Lappe-Osthege, T., & Hutchinson, A. (2023). *Examining Illegal Wildlife Trade in Europe. Final Report,* Beastly Business Project. Sheffield: University of Sheffield.

Engel, K. (2023). *Uncovering the Invisible: Successes and Challenges for Wildlife Crime Prosecution in Europe: European Summary Report*. Successful Wildlife Crime Prosecution in Europe, Bucharest: WWF.

European Commission. (2008). Directive 2008/99/EC on the protection of the environment through the criminal law.

European Commission. (2022). *Revision of the EU Action Plan Against Wildlife Trafficking, COM (2022) 581 Final, 9 November 2022*. Available at: https://eur-lex.europa.eu/legal-content/EN/TXT/PDF/?uri=CELEX:52022DC0581&from=EN

European Council. (2023). Environmental crime: Council and European Parliament reach provisional agreement on new EU law. European Council press releases, Available at: https://www.consilium.europa.eu/en/press/press-releases/2023/11/16/environmental-crime-council-and-european-parliament-reach-provisional-agreement-on-new-eu-law/

EUROPOL (n.d.). *Illicit trafficking in endangered animal species*. Available at: https://www.europol.europa.eu/crime-areas/environmental-crime/illicit-trafficking-in-endangered-animal-species#:~:text=The%20EU%20is%20a%20major,reported%20by%2015%20EU%20countries.

FACE (n.d.). *Trapping*. Available at: https://www.face.eu/hunting-methods-culture/trapping/

FACE (n.d.). *Hunting methods and culture*. Available at: https://www.face.eu/hunting-methods-culture/

Fajardo del Castillo, T. (2016). The European Union's approach in the fight against wildlife trafficking: Challenges ahead. *Journal of International Wildlife Law & Policy, 19*(1), 1–21, https://doi.org/10.1080/13880292.2016.1131516

Ferns, B., Campbell, B., & Veríssimo, D. (2022). Emerging contradictions in the enforcement of bird hunting regulations in Malta. *Conservation Science and Practice, 4*, e12655. https://doi.org/10.1111/csp2.12655

Gauriat, V. (2020). *Romania's virgin forests ravaged by 'wood mafia'. Euronews.* Available at: https://www.euronews.com/my-europe/2020/03/13/romania-s-virgin-forests-ravaged-by-wood-mafia

Gomez, L., Kala, B., & Shepherd, C.R. (2023). Bear trade in Poland: An analysis of legal and illegal international trade from 2000 to 2021. *European Journal of Wildlife Research, 69,* 106. https://doi.org/10.1007/s10344-023-01737-4

Goymer, A. et al. (2023). For R-eel?! Investigating international sales of critically endangered species in freshwater eel products with DNA barcoding. *Food Control, 150,* 109752.

Gregson, N., & Crang, M. (2017). Illicit economies: Customary illegality, moral economies and circulation. *Transactions of the Institute of British Geographers, 42,* 206–219. https://doi.org/10.1111/tran.12158

Gutierrez, L., & Duffy, R. (2023). Harms and the illegal wildlife trade: Political ecology, green criminology and the European eel. *Critical Criminology, 32,* 61–76 https://doi.org/10.1007/s10612-023-09734-4

Halbwax, M. (2020). Addressing the illegal wildlife trade in the European Union as a public health issue to draw decision makers attention. *Biological Conservation, 251,* 108798. https://doi.org/10.1016/j.biocon.2020.108798.

Hanel, R., et al. (2019). *Research for PECH committee - Environmental, social and economic sustainability of European eel management.* Brussels: European Parliament, Policy Department for Structural and Cohesion Policies. Available at: https://www.europarl.europa.eu/RegData/etudes/STUD/2019/629189/IPOL_STU(2019)629189_EN.pdf

Hübschle, A. (2017). Fluid interfaces between flows of rhino horn. *Global Crime, 18*(3), 198–217. https://doi.org/10.1080/17440572.2017.1345680

IFAW (n.d.). *The European Union is widely considered to be the third largest destination for illegal wildlife.* Available at: https://www.ifaw.org/uk/projects/wildlife-crime-prevention-europe

Iordăchescu, G., & Krajewski, M. (2025). Scientific uncertainty before the EU courts, boards of appeal and national courts in environmental matters. In Eliantonion, M & Lees, E. (eds.), *The Legitimacy of EU Environmental Governance,* Oxford: Oxford University Press.

Iordăchescu, G. (2026). Brown bear diplomacy during cold war Romania. In de Bont, R. (ed.), *Globalizing Wildlife*. Chapel Hill: University of North Carolina Press.

Iordăchescu, G., Lappe-Osthege, T., Dickinson, H., Duffy, R., & Burns, C. (2023). Political ecologies of green-collar crime: Understanding illegal trades in European wildlife. *Environmental Politics, 32*(5), 923–930, https://doi.org/10.1080/09644016.2022.2156173

Iordăchescu, G., & Vasile, M. (2023). Forests of fear: Illegal logging, criminalization, and violence in the Carpathian Mountains. *Annals of the American Association of Geographers, 113*(9), 2108–2125, https://doi.org/10.1080/24694452.2023.2209631

Iordăchescu, G. (2023). *Uncertain Scientific Knowledge and European Illegal Wildlife Trade*. Beastly Business Policy Brief. Sheffield: University of Sheffield.

Jabin, G., Singh, S.K., Ghosh, A., Basu, S., Chandra, K., & Thakur, M. (2019). Illegal trade of obscured bear parts: A case study of identifying the suspected bear gall bladders. *Forensic Science International: Report,* 1(March), Article 100001. https://doi.org/10.1016/j.fsir.2019.100001

Jahrl, J., Boner, M., Striebel, B., Ludwig, A., Mihov, S., Stoeva, R., & Caracas, G. (2021). *Evidence for Trafficking of Critically Endangered Sturgeon in the Lower Danube Region of Critically*. Vienna: WWF Austria & WWF CEE.

Jahrl, J. (2013). *Illegal Caviar Trade in Bulgaria and Romania: Results of a Market Survey on Trade in Caviar from Sturgeons (Acipenseridae)*. Vienna: WWF Austria and TRAFFIC.

Jeanneret, A., Hashimoto, Y.Z., & Caneppele, S. (2023). Exploring the factors shaping the illicit caviar trade: A case study of Switzerland's evolution from 1998 to 2021. *Journal of Economic Criminology, 2,* 100019. https://doi.org/10.1016/j.jeconc.2023.100019

Jiguet, F. et al. (2019). Unravelling migration connectivity reveals unsustainable hunting of the declining ortolan bunting. *Science Advances, 5,* eaau2642. https://doi.org/10.1126/sciadv.aau2642

Lappe-Osthege, T. (2024). The ripple effects of compliance: Reconfiguring EU policy effectiveness in transboundary environmental governance. *Journal of Common Market Studies, 62*(3), 653–670. https://doi.org/10.1111/jcms.13519

Lemître, S., & Herve-Fournereau, N. (2020). Fighting wildlife trafficking: An overview of the EU's implementation of its action plan against wildlife trafficking. *Journal of International Wildlife Law, 23*(1), 62–81.

Ludwig, A., Jahrl, J., Congiu, L., Gessner, J., Friedrich, T., Lieckfeldt, D., Peng, Z., & Boner, M. (2023). Poaching and illegal trade of Danube sturgeons. *Current Biology, 33*(22), R1184–R1185, https://doi.org/10.1016/j.cub.2023.09.067

Mackenzie, S., & Davis, T., (2014). Temple looting in Cambodia: Anatomy of a statue trafficking network. *The British Journal of Criminology, 54*(5), 722–740, https://doi.org/10.1093/bjc/azu038

Mackenzie, S. & Yates, D. (2016). Collectors on illicit collecting: Higher loyalties and other techniques of neutralization in the unlawful collecting of rare and precious orchids and antiquities. *Theoretical criminology, 20*(3), 340–357. https://doi.org/10.1177/1362480615607625

Margulies, J.D. (2022). A political ecology of desire: Between extinction, anxiety, and flourishing. *Environmental Humanities, 14*(2), 241–264. https://doi.org/10.1215/22011919-9712357

Margulies, J.D., Wong, R., & Duffy, R. (2019). The imaginary 'asian super consumer': A critique of demand reduction campaigns for the illegal wildlife trade. *Geoforum, 107*, 216–219 https://doi.org/10.1016/j.geoforum.2019.10.005

Massé, F. (2022). Police power in green: Furthering political ecologies of the state. *Political Geography, 99*, 102627. https://doi.org/10.1016/j.polgeo.2022.102627

Massé, F., Dickinson, H., Margulies, J., Joanny, L., Lappe-Osthege, T., & Duffy, R. (2020). Conservation and crime convergence? Situating the 2018 london illegal wildlife trade conference. *Journal of Political Ecology, 27*(1), 23–42, https://doi.org/10.2458/v27i1.23543

Mozer, A., & Prost, S. (2023). An introduction to illegal wildlife trade and its effects on biodiversity and society. *Forensic Science International: Animals and Environments, 3*, 100064. https://doi.org/10.1016/j.fsiae.2023.100064

Musing, L., Harris, L., Williams, A., Parry-Jones, R., van Uhm, D., & Wyatt, T. (2019). *Corruption and wildlife crime: A focus on caviar trade*. A TRAFFIC, WWF, U4 ACRC, Utrecht University, and Northumbria University report. Available at: https://www.traffic.org/site/assets/files/11818/corruption-and-caviar-final.pdf

Musing, L., Shiraishi, H., Crook, V., Gollock, M., & Kecse-Nagy, K. (2018). *Implementation of CITES Appendix II listing of European eel Anguilla anguilla*. CITES AC30 18, 1. Annex 1. 82pp., 2018.

Neag, A. (2021). *The socio-legal construction of organised crime in Romania*. Unpublished PhD Thesis, University of Exeter.

Neagu, A.C., Manolache, S., & Rozylowicz, L. (2022). The drums of war are beating louder. Media coverage of brown bears in Romania. *Nature Conservation, 50*, 65–84.

Nurse, A. (2017). Criminalising the right to hunt: European law perspectives on anti-hunting legislation. *Crime, Law and Social Change, 67*, 383–399. https://doi.org/10.1007/s10611-016-9669-8

Passas, N. (2002). Cross-border crime and the interface between legal and illegal actors. *Security Journal, 16*(1), 19–37.

Pavalasc, M. (2021). If Hunting Bears Is Illegal In Romania, Why Are Hundreds Killed Each Year? *Radio Free Europe Radio Liberty*. Available at: https://www.rferl.org/a/romania-hunting-bears-hundreds-killed-illegal-trophy-environmentalists/31099549.html

Pettersson, H., Holmes, G., Quinn, C., Sait, S., & Carlos Blanco, J. (2023). Who must adapt to whom? Contested discourses on human-wolf coexistence and their impact on policy in Spain. *People and Nature*, Online First 15 October 2023.

Pettersson, H., Quinn, C.H., Holmes, G., Sait, S., & Lopez-Bao, J.V. (2021). Welcoming wolves? Governing the return of large carnivores in traditional pastoral landscapes. *Frontiers in Conservation Science, 2*–2021.Pons-Hernández, M. (forthcoming). Tráfico de Especies Silvestres en España: una exploración desde la criminología verde crítica no-especista. *Revista Española de Investigación Criminológica*. https://doi.org/10.46381/reic.v21i2.815

Popescu, V., et al. (2016). Assessing the biological realism of wildlife population estimates in data-poor systems. *Journal of Applied Ecology, 53*(4), 1248–1259. https://doi.org/10.1111/1365-2664.12660

Schlingeman, L., de Bortoli, I., Favilli, F., Egerer, H., Musco, E., & Lucius, I. (2017). *Combating wildlife and forest crime in the danube-carpathian region*. A UN Environment. Eurac Research and WWF Report. Available at: http://www.carpathianconvention.org/tl_files/carpathiancon/Downloads/02%20Activities/Biodiveristy/UN%20Environment-WWF-Eurac%20Booklet%20on%20wildlife%20and%20forest%20crime%20in%20the%20Danube%20Carpathian%20Region.pdf

Šelih, A., & Završnik, A. (2012). *Crime and Transition in Central and Eastern Europe*. New York: Springer. https://doi.org/10.1007/978-1-4614-3517-4

Shepherd, C.R., Kufnerová, J., Cajthaml, T. et al. (2020). Bear trade in the Czech Republic: An analysis of legal and illegal international trade from 2005 to 2020. *European Journal of Wildlife Research, 66*, 92 https://doi.org/10.1007/s10344-020-01425-7

Sina, S. et al. (2016). Wildlife Crime. Study for the ENVI Committee, IP/A/ENVI/2015-10, European Parliament, Brussels. Available at: https://www.europarl.europa.eu/RegData/etudes/STUD/2016/570008/IPOL_STU(2016)570008_EN.pdf

Sollund, R. (2013). Animal trafficking and trade: Abuse and species injustice. In Walters, R., Westerhuis, D.S., Wyatt, T. (eds.), *Emerging Issues in Green Criminology. Critical Criminological Perspectives*. London: Palgrave Macmillan. https://doi.org/10.1057/9781137273994_5

South, N., & Wyatt, T. (2011). Comparing illicit trades in wildlife and drugs: An exploratory study. *Deviant Behaviour, 32*(6), 538–561. https://doi.org/10.1080/01639625.2010.483162

Tickle, L., & von Essen, E. (2020). The seven sins of hunting tourism. *Annals of Tourism Research, 84*, 102996. https://doi.org/10.1016/j.annals.2020.102996

UNODC. (2020). *World Wildlife Crime Report 2020: Trafficking in Protected Species*. Available at: https://www.unodc.org/documents/data-and-analysis/wildlife/2020/World_Wildlife_Report_2020_9July.pdf

van Uhm, D.P. (2016). *The Illegal Wildlife Trade: Inside the World of Poachers, Smugglers and Traders*. Cham: Springer.

van Uhm, D.P., & Moreto, W.D. (2018). Corruption within the illegal wildlife trade: A symbiotic and antithetical enterprise. *British Journal of Criminology, 58(4)*, 864–885. https://doi.org/10.1093/bjc/azx032

van Uhm, D., & Siegel, D. (2016). The illegal trade in black caviar. *Trends in Organized Crime, 19*(1), 67–87. https://doi.org/10.1007/s12117-016-9264-5

Vasile, M. (2019). Fiefdom forests: Authoritarianism, labor vulnerability and the limits of resistance in the Carpathian Mountains. *Geoforum, 106*, 155–66. https://doi.org/10.1016/j.geoforum.2019.08.001.

Vasile, M. & Iordachescu, G. (2022). Forest crisis narratives: Illegal logging, datafication and the conservation frontier in the Romanian Carpathian Mountains. *Political Geography, 96*, 102600. https://doi.org/10.1016/j.polgeo.2022.102600

Waters, H. (2022). Building merit: The moral economy of the illegal wildlife trade in rural, post-socialist Eastern Mongolia. *Comparative Studies in Society and History, 64*(2), 422–445. https://doi.org/10.1017/S0010417522000081

Williams, V.L., & t' Sas-Rolfes, M.J. (2019). Born captive: A survey of the lion breeding, keeping and hunting industries in South Africa. *PLoS One, 14*(5). https://doi.org/10.1371/journal.pone.0217409

Wyatt, T., van Uhm, D., & Nurse, A. (2020). Differentiating criminal networks in the illegal wildlife trade: Organized, corporate and disorganized crime. *Trends in Organized Crime, 23*(4), 350–366. https://doi.org/10.1007/s12117-020-09385-9

Wyatt, T., & Nurse, A. (2020). *Wildlife Criminology*. Bristol: Bristol University Press. https://doi.org/10.2307/j.ctv10kmc42.7

Zabyelina, Y.G. (2014). The "fishy" business: A qualitative analysis of the illicit market in black caviar. *Trends in Organized Crime, 17*(3), 181–198. https://doi.org/10.1007/s12117-014-9214-z

3

WILDLIFE CRIMES, HARMS, AND THE ROLE OF MASCULINITY IN LEGAL AND ILLEGAL HUNTING

Ragnhild Sollund

Introduction

This chapter[1] addresses gender perspectives connected to the legal and illegal hunting of animals, whether they are meant to be kept alive or die through the hunting act. Some conceptual clarifications are necessary: since I take a harm perspective and harms against freeborn nonhuman animals are as painful and destructive for individuals and the species they belong to as crimes, this is how I will refer to them, even if the act in question is not a breach of law. This is in line with a green criminology approach, which acknowledges that harms against the natural environment and those who populate it are as important as those harms which, for the moment, are defined as criminal acts (e.g. South, 2008; Sollund, 2015; White, 2013; Sollund 2024b). One reason for this is that hunting can be both legal and illegal. At times, the methods used to kill (or attempts to kill) the animal are illegal, the sex of the victim is wrong, the species in question is not "huntable", or the number of individuals killed or the time of the year the killing (or attempted killing) takes place is surpassed.

There are many rules for hunting and trapping, such as relating to season and methods. These vary in-between geographical locations, in relation to what species are "huntable" species, and depending on whether there are international nature conventions that interplay with national legislation when it comes to what species may be traded or killed. Examples of such are CITES (The Convention for International Trade in Endangered Species of Wild Fauna and Flora, 1973) and the Bern Convention (The Council of Europe's Convention on the Conservation of European Wildlife and Natural Habitats, 1979).

Hunting is a male-dominated activity, although countries aiming to achieve gender equality, such as Norway and Sweden, see an increasing number of women taking to weapons (Bye, 2003; Kheel, 2007; Sollund, 2019, 2021). Trophy hunting is also mainly a male activity with colonial roots, although increasingly female hunters are also portrayed posing with, for example, dead giraffes.[2] Before I delve further into my topic, being a gender approach to hunting, some further linguistic and philosophical clarifications are required.

The term "animal" is a word that serves to juxtapose all other animal species against the human animal, which is extremely imprecise and anthropocentric. For example, the human animal, a mammal, has far more in common with a pig than a pig has in common with a

 DOI: 10.4324/9781003333906-4

crab or fish, yet we group them together and position ourselves as something different. We are different, but so are all other species from each other, although some categories have similarities. For example, mammals living in the sea have similarities, species with gills have similarities, amphibians have similarities, birds have similarities, and mammals have similarities, but we are also mostly the same when it comes to our capacities for suffering and joy (e.g. Regan, 2004; Bekoff, 2010). In addition, we all have different individual features and personalities (e.g. Trillmich & Hudson, 2011). These observations have several implications that are important for this chapter: hunting entails individual suffering, species loss and extinction, and contributes to the nature crisis. Despite this, people, and men in particular, frequently conduct such acts of harm and destruction; their motivations for which I will discuss in this chapter.

Scholars have in various ways tried to deal with the speciesist implications of words like "animal", e.g. by using the term nonhuman animal, but this does not solve the problem since the human remains the yardstick. Terms like "animals other than the human animal" AOTHAS (e.g. Gregory, 2023) and "compan" (Sollund, 2011) for companion animal other than a human companion are constructs that remain somewhat heavy or awkward, and the "animal" problem remains. In this chapter, I use the name of the species or the category of species in an endeavour to be more precise, but regretfully also use the "animal" term for simplicity.

"Wildlife" is also an anthropocentric term that often covers both animals and plants, and as a mass term, it disguises that "wildlife" is composed of individuals. The word "wild" also has associations with savageness and unpredictability, even lack of reason, which is highly unjustified for cognitively well-equipped "animals" such as polar bears, wolves, and whales, but also for mice and birds, who generally have very advanced cognitive abilities (Ackerman, 2017; Martinho-Truswell, 2022). Life is a neutral term, in the sense that it is not speciesist per se. When I use the term "wildlife" in the following, it refers to animals other than humans who are born free or bred in captivity without having achieved features connected with humans' "domestication". Parrots, monkeys, and reptiles that are bred to be companions for humans, for example, are often still "wild" in the sense that they are not adapted to a life in captivity, neither physiologically nor psychologically. Finally, when I write about humans, I refer to the species that has self-congratulatory named itself *Homo sapiens*, even though this is the species on Earth that has caused the greatest destruction, putting all species at peril. I will describe three categories/cases of hunting, and then discuss these in relation to gender ideals, particularly those connected to masculinity.

First case: killing "huntable" species in Norway

In Norway, hunting traditions are strong (Sollund, 2021). This is reflected in the hunting statistics that account for how many animals are killed during the hunting seasons. According to Statistics Norway, in 2022–2023, 27,487 elk, 49,317 deer (*Cervus elaphus*), and 32,900 roedeers (*Capreolus capreolus*) were killed in hunts (SSB, 2023). In addition, many other animal victims were hunted for "sport", like 109,700 lily grouse, 47,800 mountain grouse, 32,110 doves (*Columba palumbus*), 6,860 ducks, 14,310 hares, and 19,610 red foxes. Thousands of jays, crows, magpies, ravens, squirrels, badgers, and beavers (to mention some) are trapped and killed only for entertainment. So are red foxes, who are exempt from the rule of the wildlife law in Norway that determines that animals cannot be killed during breeding season. Whether these thousands of animals are shot directly or after having suffered in traps, i.e. beavers will usually drown rather than being killed by the trap, they often experience a lot of stress and injuries before they eventually die. Animals that are shot at distance will often suffer injuries without being killed, such as when the hunter is drunk and misses his target. According to a Danish Study (see NOAH, 2012), 36% of short-billed geese, 32% of greylag geese, 34% of eider, 25% of foxes, 8% of hares, and 5% of roedeers had shells from shotguns in their bodies.

Species that are regularly hunted in Norway are also listed by the International Union for Conservation of Nature (IUCN), like hare and eider, which are listed as near threatened and vulnerable, respectively (Artsdatabanken n.d.). Still, hunters are allowed to kill them in thousands. Animals are killed to become meat, but this is often only an excuse for getting out in the woods to enjoy masculine comradeship with fellow hunters (Borgen & Skogen, 2021; Presser & Williams, 2011). Such company may include alcohol – for instance, the Norwegian police have distributed a folder among hunters where they warn against hunting in combination with alcohol (NTB, 2020). While the species listed above generally are regarded as "huntable" species, the large carnivores are protected by law because they are endangered species, but are hunted nonetheless, a point to which I return. In relation to gender, the hunter statistics in Norway is quite revealing in that the large majority of Norwegian hunters are men, and hunting is a way of doing masculinity responding to masculinity ideals (Kheel, 2007; Sollund, 2020b).

Second case: killing large carnivores and trophy hunting

The Norwegian hunter and fisher organisation Norsk jeger og fiskerforbund (NJFF) has trophy judges who establish the trophy value of dead animals killed by hunters (NJFF, 2023). Trophies that are judged include the antlers of deer, teeth of wild boar, cranium of red fox, badger, raccoon dog (an alien species in Norway), lynx, wolverine, wolf, and brown bear. Trophy hunting motivates many hunters' participation in the large carnivore hunts in Norway, as exemplified in social media platforms and news media where men pose with dead lynx, wolves, bears and wolverines in combination with their weapons (e.g. Sviggum, 2022).

Large numbers of hunters line up to shoot the wolves in highly abusive ways, using loose dogs and fencing in the wolf groups through kilometre-long lines that the wolves dare not cross. The state hunters themselves recur to helicopter hunting and automatic shotguns, disrespecting the prohibition of the Bern Convention to hunt by means of aircraft and automatic weapons (Appendix IV, of the Bern conventions). For example, in 2015, a wolf was killed after having killed 40 sheep in Valdres by State Nature Surveillance (SNO) hunters who used a helicopter and automatic shotguns (TV2, 2015). The economic drivers of this speciesist politics and practice are the interests of the farmers in producing meat and those of the forest owners who sell hunting rights, while populist politicians use the mythological wolf hatred to gain votes (Sollund & Goyes, 2021). Forest owners mean that if there are wolves on their land, the value of the hunting licences is reduced, since the wolves also prey on elks. This entails there will be fewer elk to hunt, therefore decreasing their willingness to pay for the hunting permissions. Hunters also find that wolves constitute a threat to their hunting dogs; consequently, there are several reasons for the conflict with wolves (Sollund, 2017; Sollund & Goyes, 2021; Skogen, Krange & Figari, 2017).

Generally, and as shown, how the SNO executes Norwegian "nature (rifle) management" is no less cruel than how ordinary hunters kill because the SNO enjoys exceptions to the rules that apply to hunting under the Wildlife Act. On the other hand, after a wolf hunt in 2023, when a female wolf was shot by licensed hunters, three hunters were fined between 12,000 and 15,000 NOK for having hunted in breach of the Animal Welfare Act (AWA), shooting her at too long a distance (Østre innlandet tingrett, 2024). It took several shots to kill her, and this, according to the police, breached § 3 of the AWA, stating that animals shall be protected from unnecessary suffering, and the Wildlife Act, which determines that hunting shall be "humane" (Sandberg, 2023). The hunters also lost their "right" to participate in licence hunts on large carnivores for one year. This is, however, an exception to the rule. When animals are hunted by hunters with dogs, this will unavoidably entail suffering. During this hunt, the wolves were chased by 60 hunters. Given the cowardliness of 60 hunters chasing one wolf with weapons from a long distance, it would have been interesting to know how the hunters themselves regard this act in view of their ideals of masculinity and toughness. Perhaps, the empowerment that the weapons offer them is what they regard as male prowess (Gonçalves, 2020).

As exemplified by the evaluation of trophies by the Norwegian hunters' organisation (NJFF), Norwegian hunters value trophies, and they are for many a motivation for hunting (Sollund & Runhovde, 2020; Sollund, 2019). Trophy hunting is, as mentioned, a typically masculine activity through which hunters aim to enhance their status through the admiration they may receive from fellow hunters and other people, within and outside their social networks. For example, a Norwegian trophy hunter who travelled to Africa, where he participated in a hunt of a lion, later boasted about how the entire village carried him around in admiration (Sollund, 2019, p. 69). Norwegian trophy hunters, who also travel to South Africa, Canada, and Alaska to shoot trophies from endangered species, including lions in canned hunts, and polar bears and wolves, save a lot of money doing the trophy hunting on their own turf. Hunters do not receive bounties for killing animals of these species anymore – the practice of paying bounties for these species was abandoned in the 1970s but is upheld in many municipalities for other species, however, recently a Norwegian municipality, Lyngen, decided they would even pay 10,000 NOK (870 Euro) for the killing of lynx, which is listed as endangered in Norway (NRK, 2024, Statistics Norway, 2023a). So rather than paying for killing their trophies in Norway, they may be rewarded. This is, of course, preferable for them; the price for killing an elephant in Mozambique in a hunt lasting for 18 days is US$44,000 (Bookyourhunt n.d.). The website advertises the prices for killing elephants in different countries and says, "leopard and buffalo combos are available". The website advertises a 21-day licence safari, with the following price list: "SUCCESS FEES: Leopard: $5000, Lion: $12000 and Elephant: $16000 – Success fees are to be paid by the hunter only when an animal is successfully harvested not wounded" (Bookyourhunt, 2023). Trophy hunters, whether they kill domestic endangered carnivores or travel to Alaska, Canada, and South Africa, hunt for the thrill of the kill (Gonçalves, 2020). Every month, trophies from these countries arrive to Norway, legally and illegally (Sollund 2025). Hunters killing wolves in Norway do not have to pay anything other than for the annual hunter registration, and to further encourage trophy hunting, the state allows them to keep the pelts of the wolves.

In 2022/2023, 1,570 hunters participated in wolverine hunts, 880 in wolf hunts, and 500 in bear hunts in Norway (Statistics Norway, 2024). What the state-licensed hunters who shoot large, endangered carnivores and the regular hunters in Norway have in common, is their gender. For instance, in 2015, the state issued licences for the killing of 16 wolves, for which 11,571 hunters registered (Løchen, 2015) for the state-organised theriocide (Beirne,

2014; Sollund, 2017; Sollund & Goyes, 2021). The same year, 10,930 hunters registered for a licence to kill brown bear and 10,820 registered for a licence to kill wolverine (Løchen, 2015). Of all these hunters, only about 500 were women. The large carnivore hunters are almost always men, middle-aged, live in rural areas, are hunters also of other species, have only a lower education, and are farmers, fishers, foresters, or work with other manual labour. They are resentful to the authorities, embrace outdated masculinity ideals as being "tough", protective of women, insensitive, strong, and brave (Skjerden, 2024).

When it comes to the gender balance of registered hunters in Norway more generally, as mentioned, men dominate. In the hunting year 2022–2023, in total, there were 537,375 hunters in Norway, of whom 85,822 (16%) were women (SSB, 2023c). Hunters may be in the register but choose not to go hunting, but of those who paid the hunting fee the same season, 173,688 (90.1%) were men and 19,100 (9.9%) were women. The number of registered hunters increased, but the number of active hunters in Norway decreased with 4.9% from 2018–2019 to 2022–2023 (SSB, 2023c). Of the overall population older than 16 years, 0.9% are registered women hunters, while 7.8% of the total population (5,514,042 inhabitants) are registered men hunters (SSB, 2023c).

Overall, a question is why this relatively limited portion of the Norwegian population is allowed to set the agenda regarding the management policy relating to large, critically endangered carnivores, and to hunt even vulnerable species on an annual, regular basis despite the risk that this will drive them to extinction. This is likely due to the strong impact of the hunters on public conservation policy in Norway, and the strong influence the hunters' association has on Norwegian leisure activities in nature more generally (Sollund & Goyes, 2021). For instance, the NJFF, the Norwegian Hunters and Fishers' Association, directly advised the government to introduce economic compensation to people who teach children and teenagers to trap and hunt (Sollund, 2021).

To fully comprehend the harm done by men who hunt with a licence from the state, it is necessary to show the conservation status of the large carnivores in Norway. Like men were behind the hunt to extinction in earlier times, extinction is still a goal for many men. When they go wolf hunting, they demonstrate their attitudes with bumper stickers on their cars with the text "zero vision" to show they want zero wolves, and a picture of a wolf head in the crossline of a telescopic sight, as well as stickers saying, "Real men hunt wolves". While the Norwegian state has committed itself to preserving the large carnivores, the hunters' zero vision is encouraged by the state and consequently so is the trophy hunting (Sollund & Goyes, 2021). In order to illuminate the absurdity in allowing trophy hunters to nearly eradicate the large carnivores in Norway, some details about their conservation status are in order.

The conservation status of the large carnivores in Norway

The critically endangered large mammal carnivores that live in Norway, brown bear, grey wolf, Eurasian lynx, and wolverine, give Norway a particular responsibility to protect them according to the Bern Convention (1979 n.d.), which Norway joined in 1986. Fifty European and four African states and the European Union are signatories to the Convention and committed to promoting national conservation policies, considering the impact of planning and development on the natural environment, promoting education and information on conservation, and coordinating research (Bern Convention n.d.). The Bern Convention aims to preserve the natural environment and wild animals and plants for future generations and acknowledges in its preamble that animals have intrinsic value. The Bern Convention operates with an appendix system where species are listed depending on their threat of extinction. In Norway, wolves are listed in Appendix II of the Convention as a species that should have special protection. However, according to a report written for the Bern Convention that assesses the situation for the European wolves.

> The management policies (low population goals and high hunting/control quotas) in Norway (and recently proposed in Sweden) dramatically affect the future conservation prospects of the overall population, and there is no common management plan at the population level.
>
> *(Boitani et al., 2022)*

While the wolf in Europe has reached the status as "least concern", in Norway it is listed as critically endangered (Artsdatabanken, 2021). The Norwegian monitor of large carnivores, *Rovdata*, established in the winter of 2022–2023, that there had been a significant reduction in the Scandinavian wolf population, with only 43–44 wolves who only lived in Norway, while 46–48 wolves live in border territory (with Sweden). The reason for this crisis in the Norwegian wolf population is the result of a political aim and will to keep the wolf population low to satisfy wolf opponents such as hunters and sheep farmers (Sollund, 2017; Sollund & Goyes, 2021). The mythological character of the wolf seems to stir a particular conflict, which is impacted by superstitions and folklore, from which populist politicians' profit at the cost of species rights. In Norway, every year the authorities invite hunters who, as shown, are mainly men to shoot many wolves (seen in relation to the total number), and in the fall of 2023, the local predator boards, who are delegated the responsibility to make decisions concerning large predator populations, decided that 32 wolves should be shot in 2023–2024. This would leave only about ten wolves, which, of course, is unsustainable, particularly in view of the inbreeding of the Norwegian wolves (Boitani et al., 2022).

The brown bear in Norway is listed by IUCN as endangered. In 2022, there were approximately 175 bears in the country, of whom 79 were females. In the summer of 2023, despite the bears' protection under Appendix II as a strictly protected species, the Norwegian Ministry of Climate and the Natural environment decided that a female bear and her two cubs should he killed, because she had taken sheep (Hatlevik, 2023). The bear family live partly in a predator zone where bears should enjoy particular protection (Sollund & Goyes, 2021), and partly on grazing land for sheep. The quota for licensed killing of brown bear in 2023–2024 is three (Oen, 2023). In writing moment, October 10, 2023, there has been no evidence of her taking sheep during the summer, but the then minister of Climate and Environment, Espen Barth Eide, still upheld the killing order.

The lynx population counted in 2023, before the yearly quota hunt and before breeding, was between 351 and 491 individuals from 71.5 family groups. Despite the status as endangered, the state yearly organises "quota hunts" attracting trophy hunters who, in 2023, were allowed to kill 51 lynx. The Norwegian parliament has established political (not scientifically based) goals for how many large carnivores shall be allowed to live in Norway. The goal for lynx in 2023 was breeding of 65 sibling groups, for wolf, 4–6 sibling groups, for brown bears, 13 sibling groups, for wolverine, 39 sibling groups (Rovdata n.d.). Lynx is a species which is very attractive to trophy hunters, and men regularly pose with dead lynx after the quota hunts on Facebook groups promoting the hunting of endangered species.

The last endangered species in Norway, which is subject to licensed hunting, is the wolverine. There were about 350 wolverines in Norway in 2023. The goal is to have 39 sibling groups born every year. The permitted number for 2023–2024 is 37. Since the state licences the killing of large carnivores every year (Oen, 2023), it is to be expected that the stated goals of large carnivore populations are often not met. The state, through the SNO, has hunters who also execute the killing, such as in wolf hunts and bear hunts in so called "skadefelling" ("damage killing"), executed if a predator has taken sheep or is considered to be a risk for sheep and other livestock) and through "den withdrawals" (Sollund, 2020a). During "den withdrawals", the hunters of the SNO will follow the wolverine female's tracks in the snow to the den, and when she is there with her pups, they will take out the pups and kill them in front of her and shoot her next. The role of the state in encouraging hunting and in orchestrating and executing the killing of endangered carnivores in Norway cannot be underestimated; even national parks are hunting grounds. In line with the strong Norwegian hunting traditions, with "harvesting" from nature through killing animals as a favourable activity, the Norwegian state proceeds with encouraging hunting for recreation, even of critically endangered species. This particular part of the hunting culture in Norway is, as mentioned, strongly tinted with archaic masculinity ideals through which the men "conquer" nature and protect their livestock. In reality, the animals are given no chance of escape, the hunt is extremely uneven in terms of the power of the numerous hunters, their dogs, and the use of cars and flag lines to fence wolves in. In the next section, I turn to another form of hunting and wildlife crimes, the legal and illegal hunting of night monkeys for medicinal experimentation.

Third case: wildlife trafficking in the Amazon for the pharmaceutical industry

In 2023, I visited the Amazon and had the opportunity to talk with persons who had previously been involved in the trafficking of night monkeys for them to be exploited in research experiments in the laboratory of Dr Manuel Elkin Patarroyo (Maldonado et al., 2009; Goyes, 2015; Sollund, 2019). The primatologist Dr Angela Maldonado, in charge of the NGO Entropika, has employed the former animal traffickers in her organisation to provide them with an alternative source of income that protects the natural environment and their inhabitants, rather than exploiting and destroying them. For several decades the late Patorroyo tried to develop a vaccine against malaria, which was only partly successful. During more than four decades, 50,000 monkeys were used in the three-border area of the Amazon (Charity & Ferreira, 2020). Monkeys of the species *Aotus nancymaae, Aotus nigriceps,* and *Aotus vociferans* were taken either from Brazil and Peru illegally or from Colombia with and without permits (Maldonado et al., 2009). The men who used to catch, abduct, and sell the night monkeys in Vista Alegre, Peru, acting on the indirect initiative of Patarroyo, now show tourists around and are engaged in their conservation.

At our visit to Vista Alegre, I asked the men about their motivations for their involvement in the trafficking, and it appeared they were purely economic; it gave "dinero en el bolsillo" (money in the pocket). One of them, Alfredo (anonymised), explained how he moved out of his parents' house and left school at 14. He married as a teenager and had to earn money to provide for his wife and children, but the job options were scarce. Patarroyo offered this opportunity through this initiative. Nevertheless, it was hard and dangerous work, the former traffickers told me. They would go out to catch the monkeys at night, and even though the men knew where the monkeys had their nests, they would be hard to catch. Concerning the methods used, Maldonado and Peck (2014, p. 7) detail:

> Owl monkeys are captured by teams of 5–7 people. The capture of 1–3 owl monkeys from their sleeping tree (nest) requires clearance of a 15–30 m radius around the tree. The collectors leave a "tree-bridge" forcing the primates to the ground where a 50 m double nylon fishing-net (3 cm mesh size) prevents animals from escaping once they have descended across the bridge to the floor. One or two collectors climb the tree, making noise to scare the animals out. Once on the ground the primates are captured in sacks before being transferred to individual wooden cages. Extraction methods have also had environmental impacts on the composition of the forest with the loss of some estimated 65,000 trees.

The monkeys would then be collected by a woman who paid the hunters, and the animals were transported to the laboratory in Leticia, in Colombia, by boat. Patarroyo used so many monkeys from Colombia that he exhorted the source, which was why his suppliers would take monkeys, not only from Colombia but also from Peru (Maldonado & Peck, 2014; Charity & Ferreira, 2020). This was a breach of CITES, but in this border zone between the three countries, there is little, if any, border control. Anybody can cross borders freely, and therefore, this part of the Amazon is a free zone for crime of many kinds, including cocaine, weapons, and wildlife trafficking (personal information, Entropika).

The methods of capturing described by Maldonado and Peck, and which were confirmed in the narratives of the former traffickers, were evidently very harmful for the monkeys, the forest, and extremely challenging and risky for the traffickers themselves. Aware of the harm done to the monkeys, I asked my informants whether they ever felt sorry for them, to which all three responded no. The financial gains of abducting the monkeys outweighed the compassion for the monkey victims, who after the experiments were released without a spleen, vulnerable themselves, and constituted a danger to other animals to whom they could transfer diseases. Yet, the men now expressed contentment for the fact that the monkeys were more numerous now that they had stopped abducting them. One said with a smile: "I can hear them on the roof at night". In addition, trees were gradually growing back. They took us to a spot in the woods near the village, where they, with pride, showed us a family of night monkeys in a tree, like they do to tourists who come. The overall reason why the men were engaged in the trafficking of night monkeys for Patarroyo was that they as men had the role as provider in the family. When other income was scarce, catching the monkeys was a way to ensure their families had bread on the table. Rather than regarding this part of nature with care and concern, they were regarded in economic terms, each monkey constituting money that could support the family.

Wildlife crimes and harms are not gender neutral

As discussed elsewhere (Sollund, 2015, 2019, 2020b; Nurse, 2020), the perpetrators/offenders/actors behind wildlife crimes, whether wildlife trafficking or the illegal killing of endangered species, are generally men. The Norwegian hunting statistics presented earlier confirm that there are far more men hunters than women, a fact supported by research in other countries (Birdsong et al., 2022), since men by far outnumber women when it comes to the legal killing of animals of many species. There are several reasons for this related to the socialisation into hunting cultures and masculinity ideals and the responsibilities that are laid on humans depending on their gender. Male hunters express that they may hunt to get access to the other factors that surround the hunt, such as being in nature, the excitement of the hunt, the hunt for trophies and the social happening around the campfire (Presser & Williams, 2011; Gigliotti & Metcalf, 2016). Female hunters are more inclined to hunt because they want to kill their own meat and find it more ethical than eating industrially produced meat (Gigliotti & Metcalf, 2016; Birdsong et al., 2022).

Norwegian studies support that hunting is an activity to which one is socialised that includes masculinity ideals, which often exclude women (Bye, 2003; Borgen & Skogen, 2021). In hunting subcultures, they enhance the relational aspects of hegemonic masculinity (Sollund, 2020b). Whether the hunt is legal or illegal, they increase their social identity as hunters and their social status among their significant others in constructing the normative climate for masculinity that permits them to kill nonhuman animals pertaining to endangered species (Sollund, 2020b, p. 30). Hunting, while being a way to source meat for humans (and extinct human species) for millions of years (Sollund, 2024a) does, however not come that naturally to modern people living in urban areas, which is why education and socialisation to become a hunter is crucial (e.g. Birdsong et al., 2022).

While generally men dominate the hunting arenas, women are gradually making their way into the field. This is likely connected to gender equality ideals. Norway is quite successful in its attempts to achieve gender equality. This is exemplified by women's participation in politics, e.g. there are ten women ministers in the current Norwegian (Oct, 2023) government, while ten are men, including the prime minister. In higher education, 65% of those who defend a PhD in medicine and health disciplines are women (Kifinfo, 2023). Men participate largely in child upbringing (Sollund, 2010).

Hunting is by many regarded as "sport", even trophy hunters who leave the animals that are bred to be killed in canned hunts, regard this killing as awardable "sport" (Gonçalves, 2020). As gender equality is improved in many other areas of society, hunting has also become an attraction to many women, e.g. they can demonstrate through hunting that they can also perform in this area of life. In Norway, most female hunters work in male-dominated professions (Rundtom & Steinset, 2016). The traditional view of males as hunters and females as gatherers has also been challenged through archaeological research that shows that in prehistoric times, women partook widely in hunting (Anderson et al., 2023). However, for decades in modern times, women were confined to the home, and female hunters were regarded as an anomaly. As gender equality ideals developed (or reappeared in the Western world) women have slowly increased their participation also in this domain, and it is a wanted development (Gigliotti & Metcalf, 2016; Birdsong et al., 2022).

The former traffickers in Vista Alegre engaged in the trafficking to earn money and provide for their families, and in this way fulfil masculinity expectations towards them (Asencio, 2012). Similar expectations to men as providers were documented in relation to the illegal killing of rhinos in South Africa, indicating that such expectations of men's roles in the family may be a motivating factor for the involvement in wildlife trafficking (Masse et al., 2021). This research also underlines the importance the involvement in wildlife trafficking has for family subsistence and that poachers must be offered an alternative income to make them abstain from the crime. This does not mean women are not involved, for example large number of animals are abducted from the forests in Latin America to become women's "pets" (Sollund, 2019). Women who are taking up hunting because they want to shoot their own meat may also do this because they want to provide their family with more sustainable and healthier meat than the one they can purchase in the supermarket (Gigliotti & Metcalf, 2016). Nevertheless, they may also take up hunting in an attempt to gain social status and become more gender equal to men.

As noted (Sollund, 2019; 2025), trophy hunting usually requires a lot of economic wealth; to kill large animals of endangered species in South Africa is a very expensive "hobby". As with the Visa Alegre case, the norms around masculinity and economic motivations are clear. These men respond to ideals relating to masculinity, which can be referred to as being "big spenders". In certain social circles, trophy hunting will also give status among peers, through which the hunter receives positive attention and elevated status for displaying how "strong" s/he is when s/he is "combatting" and "conquering" nature and an animal and that s/he possesses the will and capacity to do so, expressing male prowess (Kheel, 2007; Krier & Swart, 2016; Sollund, 2019; Batavia et al., 2019; Gonçalves, 2020). The fact that many trophies are endangered adds to the status rather than decreases it, unfortunately. To get a rare animal to the collection is a motivating factor per se, whether dead or alive (Courchamp et al., 2006; Gonçalves, 2020), rather than what it should be: a dissuading factor.

The trophy hunters who are annually invited to kill endangered carnivores in Norway enjoy the encouragement of the state to proceed with the hunting practices of the past in the current nature crisis. These state-organised theriocides mock scientific data concerning the seriousness of the situation and that most of the Norwegian population wants to have a country where large carnivores exist (Krange et al., 2017). In the last survey from 2023, 73% state they like or like a lot that there are lynxes in Norway, while for bear, wolverine, and wolf, corresponding percentages are 67, 61, or 60 (Krange & Figari, 2024).

A reason why antiquated attitudes to nature and wildlife still govern at the political level is that, despite Norway being a quite gender equal country in some respect and a modern country in others, its relationship to nature is characterised by perceiving it through its use value, or with animosity, something which can be destroyed to accommodate humans. Such attitudes to nature merit the term ideological inertia (Sollund & Goyes, 2021). Despite being a country basing its fortune on oil and gas, it is still a country where the values of previous farmers, fishers, and hunters prevail. When these are combined with econocentric values, nature conventions such as the Bern Convention (Goyes, 2021), make little difference, and the senseless hunting of freeborn animals proceeds to the satisfaction of a limited number of the population. Similarly, in Vista Alegre, in view of the norms of patriarchal structures of society in Latin America generally with the cultural values associated with machismo, it is likely that even though the particular abductors of night monkeys mentioned in this research have abandoned their practices and have turned to be carers and protectors of the monkeys due to education, on a larger scale men will probably proceed as the most prevalent offenders behind such crimes. This is likely to be the case if the underlying economic challenges faced by men and families persist, and wildlife hunting and trafficking present an opportunity for men to ease these pressures. Such attitudes to nature and to those who compose the ecosystems on which we all depend, and their dire consequences in the form of animal suffering and ecosystem destruction, are shaped by and prolong andro-anthropocentric values that support hunting, not least trophy hunting. Such values are rooted in patriarchal structures of society.

This chapter has provided a gender analysis of aspects related to hunting, specifically concerning trophy hunting. I have argued that antiquated masculinity ideals are drivers for both the legal and illegal trophy hunting and hunting more broadly. Unless the underlying structural conditions of societies change, as those implying andro-anthropocentrism, and econocentric values that are drivers for wildlife crimes, the direction for the conservation status of the world's wildlife continues to be a path towards extinction. It is therefore crucial that educational steps are taken that will teach children and consumers to conserve and protect, rather than to consume and destroy, and that such measures are gender sensitive. For instance, in Norway there is a massive influence by the state and the Norwegian Hunters and Fishers' Association (NJFF) to teach *children* to hunt and trap. This is exemplified through hunters bringing the dead carcasses of bears and lynx to kindergartens, to early confront children with hunting, allegedly for educational purposes (e.g. Holstad, 2023). Instead, children and youths should be taught to care for wildlife, like in Colombia, where they have campaigns for children to teach them that wildlife is better off in nature than in their home.

Notes

1 I am grateful to the editors of this volume for comments on previous versions of the chapter.
2 See, e.g., https://www.cbc.ca/news/trending/sabrina-corgatelli-giraffe-facebook-photos-1.3178579.

Bibliography

Ackerman, J. (2017). *The Genius of Birds*. London: Penguin.

Anderson, A., Chilczuk, S., Nelson, K., Ruther, R., & Wall-Scheffler, C. (2023). The Myth of Man the Hunter: Women's contribution to the hunt across ethnographic contexts. *PLoS One, 18*(6), e0287101.

Artsdatabanken (n.d.). *Hare Lepus Timidus. Linneaus, 1758*. Accessed on 7 October 2023 from https://artsdatabanken.no/taxon/Lepus%20timidus/47963

Artsdatabanken (2021). *Vurdering av ulv, canus lupus*. Accessed on 25 September 2023 from https://artsdatabanken.no/lister/rodlisteforarter/2021/2251

Asencio, M. (2012). Machismo/Macho. *Journal of Orthopsychiatry, 72*, 163–181.

Batavia, C., Nelson, M. P., Darimont, C. T., Paquet, P. C., Ripple, W. J., & Wallach, A. D. (2019). The elephant (head) in the room: A critical look at trophy hunting. *Conservation Letters, 12*(1), e12565.

Bekoff, M. (2010). *The Emotional Lives of Animals: A Leading Scientist Explores Animal Joy, Sorrow, and Empathy—and Why They Matter*. Novato, California: New World Library.

Beirne, P. (2014). Theriocide: Naming animal killing. *International Journal for Crime, Justice and Social Democracy, 3*(2), 50–67.

Bern Convention (n.d.). *About the Convention*. Accessed on 25 September 2023 from https://www.coe.int/en/web/bern-convention/presentation

Birdsong, M., Morse, W., Steury, T., & Smith, M. (2022). Socialization and motivational pathways among different groups of non-traditional hunters in Alabama reveal unique recruitment and retention opportunities. *Human Dimensions of Wildlife, 27*(5), 407–421.

Boitani, L. et al (2022). *Assessment of the conservation status of the Wolf (Canis lupus) in Europe. Document prepared by Large Carnivore Initiative for Europe, a Specialist Group of the IUCN Species Survival Commission with assistance of the Instituto Ecologia Applicata, Roma, for the Standing Committee 42nd meeting 28 November - 2 December 2022*. Accessed on 25 September 2023 from https://rm.coe.int/inf45e-2022-wolf-assessment-bern-convention-2791-5979-4182-1-2/1680a7fa47

Bookyourhunt (n.d.). *Where to hunt elephants. Price distribution*. Accessed on 7 October 2023 from https://www.bookyourhunt.com/en/elephant-hunting

Bookyourhunt (2023). *About this hunt*. Accessed on 7 October from https://www.bookyourhunt.com/en/Tour/24088?species=Elephant

Borgen, O. Å., & Skogen, K. (2021). Gutta på jakt. In G. Ødegård & W. Pedersen (eds.), *UNGDOMMEN* Oslo: Cappelen Damm Akademisk (173–203). https://doi.org/10.23865/noasp.142.ch8

Bye, L. M. (2003). Masculinity and rurality at play in stories about hunting. *Norsk Geografisk Tidsskrift - Norwegian Journal of Geography, 57*(3), 145–153, DOI: 10.1080/00291950310002125

Charity, S., & Ferreiara, J. M. (2020). *Wildlife trafficking in Brazil. TRAFFIC*. Accessed on 13.october 2023 from https://www.traffic.org/site/assets/files/13031/brazil_wildlife_trafficking_assessment.pdf

Courchamp, F., Angulo, E., Rivalan, P., Hall, R. J., Signoret, L., Bull, L., & Meinard, Y. (2006). Rarity value and species extinction: The anthropogenic Allee effect. *PLoS Biology, 4*(12), e415.

Gigliotti, L. M., & Metcalf, E. C. (2016). Motivations of female Black Hills deer hunters. *Human Dimensions of Wildlife, 21*(4), 371–378.

Gonçalves, E. (2020). *Trophy Leaks: Trophy Hunters and Industry Secrets Revealed.* London: Future books.

Goyes, D. R. (2015). Denying the harms of animal abductions for biomedical research. In Sollund, R. (ed.), *Green Harms and Crimes: Critical Criminology in a Changing World* (170–188). London: Palgrave Macmillan.UK.

Gregory, R. K. (2023). Animals other than human animals and their claim to equal consideration in coronavirus and criminological study: Examining harm to domesticated dogs during COVID-19 in the UK. *Social Sciences, 12*(4), 195.

Hatlevik, S. E. (2023). *SNO leter etter bjørnebinne med unger i Trøndelag.* Rovdyr.org. Accessed on 2 September 2023 from https://www.rovdyr.org/aktuelt/sno-leter-etter-bjornebinne-med-unger-i-trondelag/

Holstad, G. (2023). *Reaksjonene haglet da barnehagen fikk se død gaupe.* Addresseavisen. Accessed on 13, October 2023 from https://www.adressa.no/nyheter/i/15V17K/reaksjonene-haglet-da-barnehagen-fikk-se-doed-gaupe

Kheel, M. (2007). *Nature Ethics: An Ecofeminist Perspective.* Lanham Maryland: Rowman & Littlefield Publishers.

Kifinfo. (2023). *Statistikk om kjønnsbalanse og mangfold.* Accessed on 13. October 2023 from https://kifinfo.no/nb/content/statistikk-0

Krange, O., & Figari, H. (2024). *Nordmenns holdninger til store rovdyr: 2010, 2017 og 2023. NINA Rapport 2407.* Norsk institutt for naturforskning.

Krange, O., Skogen, K., & Helland, H. (2017). *Nordmenns holdninger til store rovdyr-endringer fra 2010 til 2017.*

Krier, D., & Swart, W. J. (2016). Trophies of surplus enjoyment. *Critical Sociology, 42,* 371–392. https://doi.org/10.1177/0892930514528819

Løchen, P. (NRK 2015). *11.571 jegere vil skyte ulv.* Accessed on 7 October 2023 from https://www.nrk.no/osloogviken/11.571-jegere-vil-skyte-ulv-1.12671226

Lynch, M. J., & Genco, L. J. (2023). Deer slayers: Examining the scope of and arguments for and against legal deer theriocide in the US. *Sustainability, 15*(7), 5987.

Maldonado, A. M., Nijman, V., & Bearder, S. K. (2009). Trade in night monkeys Aotus spp. in the Brazil–Colombia–Peru tri-border area: International wildlife trade regulations are ineffectively enforced. *Endangered Species Research, 9*(2), 143–149.

Maldonado, A. M., & Peck, M. R. (2014). Research and in situ conservation of owl monkeys enhances environmental law enforcement at the Colombian-Peruvian border. *American Journal of Primatology, 76*(7), 658–669.

Martinho-Truswell, A. (2022). *The Parrot in the Mirror. How Evolving to be Like Birds Made Us Human.* Oxford: Oxford University Press.

Masse, F., Givá, N., & Lunstrum, E. (2021). A feminist political ecology of wildlife crime: The gendered dimensions of a poaching economy and its impacts in Southern Africa. *Geoforum, 126,* 205–214. ISSN 0016-7185.

NOAH (2012). Jakt – lidelse for dyrene. NOAH for dyrs rettigheter. Accessed on 13 October 2025 from Jakt - lidelse for dyrene - NOAH - for dyrs rettigheter

Norsk jeger og fiskerforbund [NJFF] (n.d.). *Troféer.* Accessed on 7 October 2023 from https://www.njff.no/jakt/trofeer

Norwegian Environment Agency (n.d.). *Tilskudd til tiltak for å gi barn og unge økt forståelse for bruk av utmarksressurser.* Accessed on 13, October 2023 from https://soknadssenter.miljodirektoratet.no/TilskuddBarnOgUnge%C3%98ktForst%C3%A5elseBrukUtmarksressurserSkjema/Startside/Index?s%C3%B8knadstypeId=62

NRK (2024). https://www.nrk.no/tromsogfinnmark/innforer-skuddpremie-pa-gaupe-1.16765948

NTB (2020). *Politiet ber jegere ligge unna alkoholen.* Accessed on 7 October 2023 from https://jaktogfiske.njff.no/jus-smavilt-storvilt/politiet-ber-jegere-ligge-unna-alkoholen/146043

Nurse, A. (2020). Masculinities and animal harm. Men and masculinities, 23(5), 908–926.

Oen, I. A. (2023). *Lisensjakt på jerv og bjørn i 2023/2024.* Troms Bondelag. Accessed on 25 September 2023 from https://www.bondelaget.no/troms_3/lisensjakt-pa-jerv-og-bjorn-i-2023-2024

Presser, L., & Taylor, W. V. (2011). An autoethnography of hunting. *Crime, Law and Social Change, 55,* 483–494.

Regan, T. (2004). The case for animal rights. Berkeley: University of California Press.

Rodríguez Goyes, D. (2021). Contending philosophical foundations in international wildlife law: A discourse analysis of CITES and the Bern convention. *Revista Catalana de Dret Ambiental, 12*(1), 1–35.
Rovdata (n.d.). *Nasjonal leverandør av bestandstall.* Accessed on 27 September 2023 from https://rovdata.no/
Rundtom, T. O., & Seinset, T. A. (2016). Hvem er jeger? *Samfunnsspeilet, 3*(2016), 3–7. Accessed on 17.Oct. 2023 from https://www.ssb.no/jord-skog-jakt-og-fiskeri/artikler-og-publikasjoner/_attachment/280788?_ts=157be115730
Sandberg, T. (2023). *Politiet: Ulvejakt i strid med loven.* Dagsavisen 5.6.2023. Accessed on 7 October from https://www.dagsavisen.no/nyheter/2023/06/05/politiet-ulvejakt-i-strid-med-loven/
Skogen, K., Krange, O., & Figari, H. (2017). *Wolf Conflicts: A Sociological Study* (Vol. 1). Oxford New York: Berghahn Books.
Sollund, R. (2010). Regarding au pairs in the Norwegian welfare state. *European Journal of Women's Studies, 17*(2), 143–160.
Sollund, R. (2011). Expressions of speciesism: The effects of keeping companion animals on animal abuse, animal trafficking and species decline. *Crime, Law and Social Change, 55,* 437–451.
Sollund, R. (2015). Green harms and crimes. *Critical criminology in a changing world.*London: Palgrave McMillan
Sollund, R. (2017). Perceptions and law enforcement of illegal and legal wolf killing in Norway: Organized crime or folk crime?*Palgrave. Communications, 3*(1), 1–9.
Sollund, R. A. (2019). *The Crimes of Wildlife Trafficking: Issues of Justice, Legality and Morality.* Routledge.
Sollund, R. (2020a). Wildlife management, species injustice and ecocide in the Anthropocene. *Critical Criminology, 28*(3), 351–369.
Sollund, R. (2020b). Wildlife crime: A crime of hegemonic masculinity? *Social Sciences, 9*(6), 93.
Sollund, R. (2021). *Bare et dyr? [Only an animal?].* Oslo: Solum Bokvennen.
Sollund, R. (2024a). Eating ET: Carnism and speciesism. *International Journal for Crime, Justice and Social Democracy. 13*(2), 78–87
Sollund, R. (2024b). Anthropocentrism, speciesism and speciecide. In Ronco, A. D & Carrabine, E. (eds.), *Criminological Connections, Directions, Horizons: Essays in Honour of Nigel South.* London: Routledge, 116–132.
Sollund, R. (2025). *Wildlife Trade and Animal Victimization. Parallel Harms and Crimes.* London: Routledge.
Sollund, R. A., & Runhovde, S. R. (2020). Responses to wildlife crime in post-colonial times. Who fares best? *The British Journal of Criminology, 60*(4), 1014–1033.
Sollund, R., & Goyes, D. R. (2021). State-organized crime and the killing of wolves in Norway. *Trends in Organized Crime, 24*(4), 467–484.
South, N. (2008). Nature, Differende, and the Rejection of Harm: Expanding the Agenda for Green Criminology. In Sollund, R. (ed), *Global Harms. Ecological Crime and Speciesism.* New York: Nova sc Publishers,
SSB Statistics Norway (2023a). *Jakt i din commune [Hunting in your municipality].* Accessed on 7 October 2023 from https://www.ssb.no/jakt-i-din-kommune/kommune
SSB Statistics Norway (2023b). *07514: Felte småvilt, etter region, småvilt, statistikkvariabel og intervall (år).* Statistikkbanken (ssb.no) Accessed on 7 October 2023 from https://www.ssb.no/statbank/table/07514/tableViewLayout1/?loadedQueryId=10072844&timeType=top&timeValue=1
SSB Statistics Norway (2023c). *Registrerte jegere.* Accessed on 7 October 2023 from https://www.ssb.no/jord-skog-jakt-og-fiskeri/jakt/statistikk/registrerte-jegere
Statistics Norway (2024). 09497: Lisensjakt, etter dyreart, statistikkvariabel og intervall (år) Accessed on February 28, 2024 from https://www.ssb.no/statbank/table/09497/tableViewLayout1/
Sviggum, S. K. (2022). *Bjørn skutt i Snåsa: – Følelsene kommer i etterkant.* VG. Accessed on 7 October 2023 from https://www.vg.no/nyheter/innenriks/i/kEW8Oa/bjoern-skutt-i-snaasa-foelelsene-kommer-i-etterkant
Trillmich, F., & Hudson, R. (2011). The emergence of personality in animals: The need for a developmental approach. *Developmental Psychobiology, 53*(6), 505–509.
TV 2 (2015). *Jeger med pumpehagle skyter ulv fra helikopter.* Accessed on 7 October 2023 from https://www.tv2.no/video/nyhetene/jeger-med-pumpehagle-skyter-ulv-fra-helikopter/927364/
Østre innlandet tingrett (2024). 23-115868MED-TOIN/TELV.
White, R. (2013). Environmental harm: An eco-justice perspective. Bristol: Bristol University Press.

4
UNDERSTANDING CULTURAL USE AND SPIRITUALITY
Lion body part trade in southern Africa

Peter Coals, Vincent Naude, Nolwazi S. Mbongwa, and Vivienne L Williams

Introduction

Contested illegality (Hübschle, 2017) may arise when the need, or desire, to use natural resources conflicts with the regulation of such resources. As opposed to clearly delineated and widely accepted laws (e.g. homicide), such contested illegality can occur where societal practices or traditions do not align with statutory legislation, and where regulations are not perceived as fair or equitable by impacted communities (e.g. possession/use of some narcotics) (Sollund, 2019; Dore et al., 2022). Continued and widespread disregard for such statutory regulations and alternative or parallel forms of authority (e.g. cultural, religious, or even corporate governance) may serve to delegitimise the established law (Hübschle, 2017) and/or could lead to the potentially unjust criminalisation of marginalised and vulnerable peoples (Duffy, 2010).

Natural resource use and its formal (i.e. statutory) or informal (i.e. non-governmental and social organisation) regulation may often fall into such contested space. Potential 'grey' areas of legitimacy can arise in the face of historical practice, disenfranchisement, and displaced resentment where there is a lack, or failure, of appropriate community and stakeholder engagement (Fabricius et al., 2004; Tanner et al., 2010). Cultural and religious practices may face heightened susceptibility to conflicts with regulations when these regulations are insufficiently understood or disregarded. In addition, the deep-seated nature of beliefs and complex systems of tradition and spirituality that inform and drive the use of certain natural resources make them prone to conflicts, misunderstandings, and misinterpretations (Willow, 2018; Coals et al., 2024). In the case of longstanding practices, more recent regulation may construct criminality as it alters access to and trade in natural resources – this may be purposeful, or not (Calavita, 2007; Naude et al., 2020; Arias et al., 2024). The use of lion (*Panthera leo*) derivatives for traditional purposes in southern Africa, for example, presents an insightful case of such contested, and constructed, illegality.

Wild lions are undergoing widespread population declines in all but four African countries (Botswana, Namibia, South Africa, and Zimbabwe), having been eradicated from over 90% of their historic range (Bauer et al., 2015; Nicholson et al., 2023). Major threats to remaining wild populations include prey depletion, livestock encroachment, illegal killing

DOI: 10.4324/9781003333906-5

due to a real or perceived conflict over livestock depredation, poisoning, and habitat transformation (Gore et al., 2020; Dunford, 2022; Becker et al., 2022; Nicholson et al., 2023). In addition, concerns have been raised that consumptive use and trade in lion body parts pose a threat to some wild lion populations (Williams et al., 2017a; Everatt et al., 2019; Coals et al., 2022). Wild lions in South Africa (ca. 3,000 mature individuals; African Lion Database, 2024, unpublished data) encompass large, free-ranging populations in the Kruger and Kgalagadi National Parks, as well as smaller satellite populations maintained in a series of fenced reserves and protected areas that are managed as a collective metapopulation (Miller et al., 2015, 2023). In addition, South Africa has a sizeable captive-bred lion population that has been commercially farmed, primarily for use in trophy hunting (sometimes called 'canned' hunting) and the associated intercontinental bone trade to Southeast Asia (Williams et al., 2017b; Williams & 't Sas-Rolfes, 2019; Coals et al., 2019). The complexity of balancing threats to, and commercial exploitation of, lions with their entrenched cultural significance in southern African traditions underscores the multifaceted challenges to their conservation and the attendant, often contested, legal issues.

Ethnographic records suggest that lions have occupied a longstanding role in southern African cultural traditions (Morton & Hitchcock, 2014; Val et al., 2020). Cultural uses of lion include traditional spiritual medicine (*umuthi*), cultural regalia and talismans, ornamentation, food, and other ritual practices (which seek to use the symbolic or totemic meaning of lions through their derivatives) (Williams & Whiting, 2016; Williams et al., 2025). Within southern African spiritual practices, the selection of animals for specific healing, ritual and religious practices often aligns with traits imbued through the 'Doctrine of Signatures' philosophy. This philosophy holds that the perceived attributes and traits of the living animal are considered transferable from the focal species used to the human recipient (Naude et al., 2020; Simelane & Kerley, 1998; Thornton, 2009; Williams & Whiting, 2016). While these practices are common throughout the continent (Williams et al., 2017a; Williams et al., 2025) and increasingly so in southern Africa, commercially and traditionally driven trade in lion derivatives in South Africa is a well-documented and overtly contested space from a socio-political, legal and transnational wildlife trade perspective (Williams & 't Sas-Rolfes, 2019; Williams et al., 2021).

Here, we use the lion body part trade in southern Africa as a case study to illustrate considerations pertaining to the cultural use of wildlife products against a variable backdrop of legislative policy, governance, and enforcement, with a particular focus on the spiritually motivated use of and trade in lion derivatives. We describe the cultural uses of lion parts and explore interactions between the practice of cultural traditions in southern Africa, commercial farming of lions, and wild lion conservation (Coals et al., 2019; Hiller & MacMillan, 2021; Coals et al., 2022). We aim to elucidate barriers to organised legitimacy and legality faced by traditional users, expose the potential for constructed criminality in the cultural use of lion parts through ongoing changes in accessibility of natural derivatives, and discuss the potentially (un)fair treatment of traditional spirituality and its practitioners in southern Africa.

Study system

Traditional cultural trade

Much of this chapter is based on the findings of a survey we conducted between September and November 2019 of traditional healers (n = 20) and informal commercial *muthi* traders

(n = 10) in two major traditional wildlife markets in South Africa where lion products were known to be traded (Johannesburg, Gauteng Province and Nongoma, KwaZulu-Natal Province) (Coals et al., 2022). Traders and healers were asked about the symbolism, uses, and sourcing of lion parts in relation to traditional cultural–spiritual practices. These respondents are the primary users of lion products, whose traditions of practice also require these parts to be passed down through generations (i.e. 'transgenerational cultural equity') and contribute to livelihoods through trade. However, law enforcement raids on traditional markets have directly impacted and further criminalised this sector.

Informal traders (i.e. individuals with permanent or recurring stalls in traditional markets) are typically engaged in the commercial and subsistence sale of bioresources used in botanical and zootherapeutic cultural healing practices (Cook, 2009; Thornton, 2009; Williams & Whiting, 2016; Coals et al., 2022). Most of the species traded in these markets are linked to traditional medicine and other cultural uses, such as traditional attire. While informal traders are not typically trained to use lion body parts, they are usually involved in the system due to socioeconomic pressures and, through experience, have learnt the important cultural uses of lion body parts. These traders thus serve various consumer needs such as supplying *muthi* shops, other markets, traditional healers, traditional leaders, and walk-in customers (Coals et al., 2022).

Traditional healers, for example, diviners and herbalists, perform sacred rituals integral to their cultural–spiritual medicinal practices, and revere their ancestors as divine messengers (Gumede, 1990; Nigosan, 1994; van Dyk, 2001; Mlisa, 2009). Lions, known as *Ndau* by some ethnic groups in southern Africa, are predominantly associated with the ancestral spirit *umndawu*, a 'go-between' spirit that transcends ethnic boundaries (Muzondi, 2014). Diviners are chosen by ancestors and undergo training and initiation rituals, enabling direct communication with their ancestors for diagnosis and treatment prescriptions (Berg, 2003; Mokgobi, 2014; Thobane, 2015). Herbalists, however, acquire expertise with wildlife products through familial elders, other traditional healers, or guidance from visions and dreams (Veenstra, 2006; Mlisa, 2009; Washington, 2010).

Legislation and enforcement

In order to explore the system of contested and created illegality, it is important to set out the primary pieces of legislation, and decision-making and enforcement entities, involved. A major piece of legislation relating to the domestic trade in lion in South Africa is the National Environmental Management: Biodiversity Act of 2004 (NEMBA No. 10, 2004), which requires Threatened or Protected Species (ToPS) permits and formal registration to carry out restricted activities with ToPS-listed species – such as the legal possession of, and trade in, live lions and lion body parts (Williams et al., 2015; Miller et al., 2023). International trade in lion is covered by the Convention on International Trade in Endangered Species of Wild Fauna and Flora (CITES), to which South Africa is a signatory. Lions are listed on CITES Appendix II – allowing international trade at a level that is not detrimental to the survival of wild populations (Bauer et al., 2018). South Africa was a legal exporter of lion skeletons internationally; however, since 2018, there has been no legal commercial export of lion parts (see Section: *Captive-bred lions and the legal lion skeleton trade*).

Determination of legislation and policy relating to wildlife trade, which is enacted by the South African Parliament, falls under the purview of the Department of Forestry, Fisheries and the Environment (DFFE), which is headed by a cabinet minister. Local regulatory responsibilities are held at the province level by provincial conservation departments

(Williams et al., 2015). Law enforcement personnel dealing with wildlife trade crimes and compliance issues are primarily from the DFFE and the South African Police Service.

Enforcing bodies have long been cognisant of the potential for conflict within traditional resource use systems in South Africa. Nevertheless, while there have been attempts to ensure the sharing of information concerning what constitutes legal and illegal practices for traditional users and traders of lion parts, there still exists significant potential for contested and created illegality, which we discuss under *Criminality and barriers to legality*.

The landscape of South African lion body part trade

Captive-bred lions and the legal lion skeleton trade

Exact numbers of captive-bred lions in South Africa have always been uncertain, and estimates vary over time and with changes in commercial demand and legislation – both domestic and international. It is likely that by the end of 2016, South Africa had over 8,000 captive-bred lions in over 300 commercial breeding and rearing facilities, representing the high point of a boom in captive-bred lion numbers (Williams & 't Sas-Rolfes, 2019). Since the 1990s, in South Africa, lions were chiefly commercially bred for hunting or tourism purposes (e.g. game viewing or cub interactions) (Hunter et al., 2013; Williams & 't Sas-Rolfes, 2019; Hiller et al., 2022).

From 2008 to 2018, South Africa permitted the commercial export of lion skeletons to Southeast Asia (Williams & 't Sas-Rolfes, 2019). The majority of skeletons originated from captive-bred lions that were either hunted – with the skeletons being a by-product of the hunt for which additional value could be realised – or euthanised directly for the bone trade. Numbers of legally exported skeletons originating from South Africa rose from approximately 50 individuals in 2008 to a maximum of 1,771 skeletons in 2016 (Williams & 't Sas-Rolfes, 2019). Latterly, the numbers of exported skeletons were capped and controlled by a quota, the imposition of which was agreed at the CITES Conference of the Parties in 2016 (CoP17). The South African government set the first export quota at 800 lion skeletons in 2017 and maintained it at 800 in 2018 (Williams et al., 2021). In August 2019, based on legal action taken by the National Council of The Society for Prevention of Cruelty to Animals, the South African High Court determined that lion skeleton export quotas from South Africa in 2017 and 2018 were unlawful and unconstitutional due to inadequate consideration for the welfare of captive-bred lions in the process of quota determination (The Republic of South Africa, 2019). Since that ruling, to-date, no further export quotas have been confirmed (Miller et al., 2023), and, at the time of writing, the South African government is seeking a permanent end to lion farming (Miller et al., 2023).

There remains considerable uncertainty, however, as to the fate of an unknown number of captive-bred lions and an unknown number of stockpiled skeletons privately held on current and former breeding and keeping facilities. Concerns were raised that an end to legal bone trade and lion farming could incentivise illegal trade of lion bones as owners of stockpiles sought to recoup costs or to capitalise on unmet demand (Williams & 't Sas-Rolfes, 2019; Williams et al., 2021) – some of which could come from domestic cultural uses.

Muthi trade – *cultural uses and significance of lion*

In African spirituality, lions are attributed important symbolic meanings. The most common associations of lions in *umuthi* (southern African traditional cultural–spiritual medicinal practice) include power, protection, strength, protection from witchcraft, and patience. These interpretations align with the Doctrine of Signatures' interpretation of the symbolism of animal products. Lion is also a symbol of royalty and forms part of ceremonial tribal regalia associated with status and ancestral linkages (Williams et al., 2015; Coals et al., 2022; Williams et al., 2025).

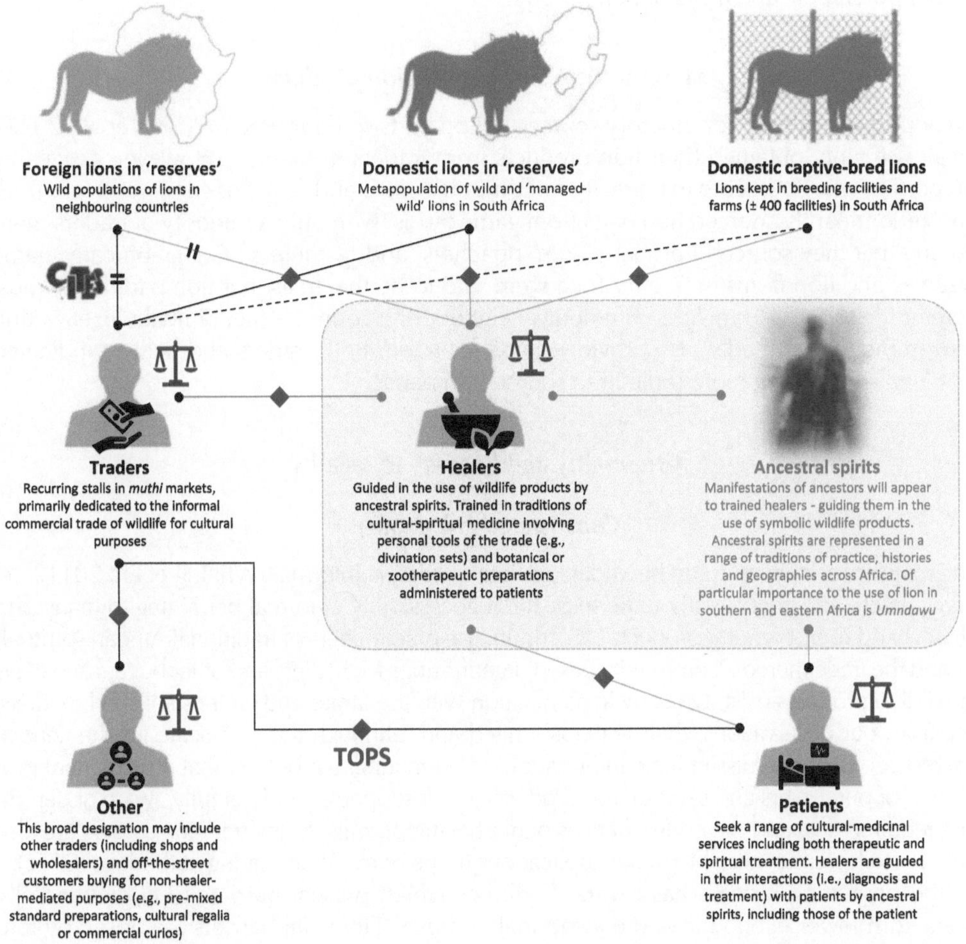

Figure 4.1 Diagram of trade and use linkages for lion body parts in traditional medicine and spiritual practice (*muthi*). Linkages demonstrate the position of ancestral spirits (primarily, but not limited to, *umndawu*) as adjunct stakeholders and 'gatekeepers' mediating lion part use and trade. Solid lines indicate relationships revealed by responses to 2019 research; dotted lines represent relationships which have been previously recorded in the literature; light grey lines show interactions mediated and gate-kept by ancestral spirits; trade interactions requiring CITES permits are shown with parallel bars – N.B. In this system, all trade is contrary to CITES; diamonds indicate trade and possession requiring a ToPS permit – N.B. Not all trade in this system is necessarily in contravention of ToPS permitting. Scales of Justice highlight areas where illegality may be contested. (Figure based on Figure 4.1 of Coals et al., 2022)

Lion is often, but not exclusively, associated with the *umndawu* ancestral spirit (Coals et al., 2022). Based on exploration of the role of ancestral spirits in the decision-making of healers as to which wildlife products to use, we found that ancestral spirits may be conceived of as active stakeholders in the *umuthi* system (see Figure 4.1) (Coals et al., 2022). Ancestral spirits appear to function as mediators for the use of lions by healers through their role in diagnosis and suggestions of materials for treatment. A deeper understanding of the motivation for the use of lion in *umuthi* may be gained by acknowledging the spiritual direction of purchasers and users of lion products.

Muthi trade – *sourcing of lion products*

We found a variety of contributory sources of lion parts sold for *umuthi* (Coals et al., 2022). Healers mainly obtained their lion products from traders in markets. However, a sizeable proportion also sourced parts directly from lion breeders and farms or game reserves. Traders predominantly sourced lion parts from game farms, with only a minority of traders mentioning that they sourced lion parts from 'poachers' and 'gatherers'. Geographically, game reserves and lion farms in South Africa were said to be the areas that lion parts were most commonly sourced from. Less commonly, neighbouring countries (i.e. Zimbabwe, eSwatini, Mozambique) reportedly also provided parts for trade. Both traders and healers indicated that it was becoming more difficult to source lion parts.

Criminality and barriers to legality

Constructed criminality

The majority of cultural trade in wildlife in South Africa is informal (Whiting et al., 2011), i.e. unregulated, and occurs without permits. The inaccessibility of formal permitting to many purchasers and users of *muthi* products can criminalise practitioners of traditional African spirituality and the trade thereof. Community based organisations for healers and other buyers and users of wildlife products exist, typically in association with the larger and well-established markets; yet, from our observations, their reach is limited and, although they advocate for the general interests of users and disseminate information on permitting, we believe that a substantial proportion of purchasers and users of lion products are unsupported and not fully aware of permitting requirements and the consequences of not having permits. Many traditional users are also unlikely to have the means to make applications to the relevant permitting authorities (DFFE).

The permitting process has several attributes which present barriers to compliance by users from lower economic and educational contexts. Financial barriers are posed by fees payable for permits (between ZAR 50 and 1,000; to be paid via bank transfer). Other barriers include language barriers (English language application form), no specific category for traditional traders on the application for standing permits, and a requirement to detail the scientific (binomial) name of species for which permission is sought (DFFE n.d.). Furthermore, a person carrying out a restricted activity, such as transporting lion parts between provinces, needs a ToPS permit for every province where the restricted activity will take place. The regulation on restricted activities poses additional challenges for healers, as they typically treat patients from various parts of the country; it is impractical for them to obtain permits for each province, especially considering the distances they must travel to permit offices. This is particularly challenging for healers based deep in the rural areas with limited access to transportation.

Penalties for non-compliance may be fines of up to ZAR 5 million and/or up to five years imprisonment, rising to ZAR 10 million and/or ten years imprisonment for second offences (DFFE n.d.). A lack of accessibility to legal permitting for traditional trade and use in cultural–spiritual practices thus risks the criminalisation of a considerable number of people, often among the most impoverished and vulnerable in the country and contributes to feelings of marginalisation and disenfranchisement from wildlife conservation. These feelings have not been aided by poorly timed enforcement raids that preceded the dissemination of permitting and legality information via DFFE and provincial conservation department-run awareness programmes (pers. obs.).

Accessibility, wild lion conservation, and lion farming

Permitting procedures thus limit access of cultural users to legal means of trading and possessing lion products. Criminality may also be generated through accessibility to, and provenance of, lion parts. An uncertain proportion of lion parts in *muthi* are sourced from the wild and across borders in contravention of CITES. The most in-demand lion body parts in the *muthi* trade show similarities to parts most frequently removed from wild lions, notably: paws, claws, and head (Everatt et al., 2019; Coals et al., 2020, 2022). Cultural uses may therefore be a driving factor in the removal of body parts from deceased lions (Arias et al., 2024). However, it is considerably harder to draw inferences about what motivates the initial killing and poaching of lions. It is likely that trade in parts from wild lions is, for the most part, opportunistic (Coals et al., 2020), with mortalities due to human–lion conflict having the potential to contribute substantial amounts of material (Arias et al., 2024). Nevertheless, the lion trade in South Africa demonstrates that supply chains linked to cultural use criminally cross international boundaries (i.e. in contravention of CITES). Hence, cultural use is an as-yet largely unquantified risk factor for regional lion populations (Williams et al., 2025).

The impact of wildlife farming on species conservation is a contested topic in conservation discourse, with substantial debate and research focusing on whether legalised, farmed sources of wildlife can contribute to easing pressure on wild populations (e.g. Phelps et al., 2014; Damania & Bulte, 2007; Rizzolo, 2021). A common economic argument is that a farmed source of lion parts would reduce incentives for the poaching of wild lions (see Coals et al., 2019 for analysis of competing arguments). Such arguments are contingent, among other factors and nuances, upon the wild and farmed product being perceived as being sufficiently similar by purchasers and users (Phelps et al., 2014). Our research has found that the use of farmed lion parts is acceptable to many traders and users. However, a minority prefer wild lion, mainly because captivity is believed to dilute or alter its symbolic properties. Nevertheless, many users of lion products (especially traditional healers) appear to source them from lion keeping and breeding facilities. Captive lion breeders have been recorded to knowingly sell lion body parts (such as fat, internal organs, paws, and bones) for use in traditional medicine (Williams & 't Sas-Rolfes, 2019). Indeed, anecdotally, we have observed that some facilities have maintained relationships with users and valued their sales or donations of parts as contributions to the maintenance of cultural practices in South Africa.

The closure of the captive-bred lion industry thus raises considerable concerns for some traditional users as to how they will continue to source material for use in their cultural practices – some of which are immutable (i.e. require lion and cannot be substituted for another item). Specific consideration is being given to how the needs of traditional users might be catered for in South Africa's cessation of the captive lion breeding industry. At the time of writing, there have been no formal assurances given to users and practitioners as to whether there will be legal mechanisms whereby they can sustainably and legally obtain lion parts for cultural practice. Suggestions include instigating mechanisms whereby mortalities from managed-wild populations of lions might be used to meet demand. Further quantitative work is required to establish whether such sources could sustainably meet demand. Any mechanism would need to ensure that it is sufficiently accessible by practitioners and users across geographic regions and social strata, and adequately addresses the barriers to legality established by current ToPS permitting.

Spiritual compulsions and hard choices

Inaccessible permitting and the closure of a major source of products (i.e. the lion farming industry) thus present traditional users with potentially difficult choices that place their desire to practice their ancestral spirituality in conflict with the legality of sourcing and possessing lion products. Such circumstances are not unique to lions and are writ large across histories of indigenous resource use. Traditional healers may be presented with compelling reasons to use lions, especially if it comes as part of the request or guidance of a powerful ancestral spirit. Determining the role of spirituality in directing the use of wildlife products allows a better understanding of the quandaries experienced by practitioners when faced with legal prohibitions and/or reduced supplies of the product. Whether lion is used or not may not be a choice within the tangible (mortal) part of the system (see Figure 4.1).

Decision-makers in governance and law enforcement are thus faced with how to best balance deeply held spiritual beliefs with potential risks posed to wild lion populations by trade and use. The fair use of biological resources is a long-running theme in conservation discourse. Just resolutions may be difficult but are not necessarily intractable (Redpath et al., 2013; Vucetich et al., 2018; Coals et al., 2019; Coals et al., 2024). Indeed, such ethically challenging questions are not as polarising as may be initially supposed, as individuals may hold multiple (potentially competing) priorities (Vucetich et al., 2021a) and a variety of participatory mechanisms exist for arriving at fair, just decisions that balance the requirements of humans and wildlife (Redpath et al., 2013; Vucetich et al., 2021b).

The deep-set cultural importance of lions to traditional southern African spirituality can be acknowledged alongside the necessity to prevent unsustainable exploitation of wild lion populations and to be sensitive to the welfare of individual lions (both wild and captive). Achieving such outcomes requires an understanding of the cultural importance of lions in order to fairly represent the value of use to traditional practitioners and users, alongside understanding the often-considerable challenges facing wild lion populations.

The investigation of spiritually motivated uses of lion parts opens new avenues for understanding people's relationship with the wildlife resources that they use. In particular, consulting with traditional users highlighted the role of ancestral spirits in directing the use of particular wildlife products. Although we advocate avoidance of cultural relativism and the blind acceptance of cultural motivations as reasons not to regulate the use of at-risk wildlife, we are cognisant that the position of spiritual guidance in the use of lion parts may necessitate a deeper understanding of notions of culpability among those who are instructed to use lion. Culpability is concerned with the apportioning of responsibility for actions based on an actor's choice to risk others' legally protected interests (Alexander et al., 2009). Culpability is linked closely to the determination of punishment if an actor is found guilty of carrying out the illegal act. Cultural defences hold that the following of certain illegal cultural practices should lead to the accused being considered less culpable due to the cultural compulsion they feel to carry out the illegal act and hence be considered less deserving of punishment (Tunick, 2004).

The influence of cultural practices and traditions on culpability is a contested topic that continues to be debated and is beset with contextual complexities and normative (moral) interpretations based on participants' world views (Tunick, 2004). What is clear, as exemplified by consideration of the defence of marital rape in South Africa, is that care should be taken to ensure that cultural defences do not excuse or facilitate damaging practices that serve the interests of criminal actors to the detriment of victims; however, cultural beliefs may be necessary to establish the context of an offence and aiding understanding of motivations (Mwambene & Kruuse, 2018). Readers will doubtless find examples which they feel more comfortable with, and others to which they have a greater aversion – depending on their ethical stance. Here, we do not intend to resolve the debate around the use of cultural defence for lion body part trade in South Africa. Rather, we highlight it as an ongoing and unresolved consideration when investigating the nexus of law enforcement, traditional customs, and resource use.

Conclusion

The use and trade of lion products for traditional medicine and cultural–spiritual purposes in South Africa demonstrates the often contentious relationship between regulation and the needs and desires of people. In particular, spiritual compulsions in the use of wildlife have the potential to exert conflicting pressures on users as they seek to obtain and use wildlife products legally. There is a need to understand the scope, complexity, and importance of cultural uses and their influence on wildlife resource use and ensuing impacts upon criminality. In particular, accessibility and understandability of permitting processes are important factors in the construction of criminality, and large-scale changes to supply chains (in this case, national policy regarding the farming of lions) may present additional challenges in the legal accessibility of material. In addition, exploration of spiritual compulsions may serve to augment dialogue around culpability and the understanding of the cultural context in which wildlife crime occurs.

Fair regulation of the use of wildlife resources is complex and prone to generating conflict between regulatory bodies and (often vulnerable) people (Redpath et al., 2013). As such, the management of conservation conflicts frequently depends on cross-disciplinary approaches to provide more accurate and nuanced characterisations of complex systems (Macdonald, 2019). A fuller understanding of the traditional–spiritual complex of wildlife use provides a sounder base from which to explore fair and just outcomes for people and wildlife; this approach emphasises dialogue, co-design of regulatory systems, and effective adaptive management (Redpath et al., 2013; Vucetich et al., 2018; Coals et al., 2024).

References

Alexander, L., Ferzan, K.K., & Morse, S.J. (2009). *Crime and Culpability: A Theory of Criminal Law*. Cambridge University Press, Cambridge.

Arias, M., Coals, P., Ardiantiono, Elves-Powell. J., Rizzolo, J.B., Ghoddousi, A., Boron, V., da Silva, M., Naude, V., Williams, V., Poudel, S., Loveridge, A., Payán, E., Suryawanshi, K., Dickman, A. (2024). Reflecting on the role of human-felid conflict and local use in big cat trade. *Conservation Science and Practice*, 6(1), e13030.

Bauer, H., Chapron, G., Nowell, K., Henschel, P., Funston, P., Hunter, L.T.B., Macdonald, D.W., & Packer, C. (2015). *Lion (Panthera leo) populations are declining rapidly across Africa, except in intensively managed areas*. Proceedings of the National Acadamy of Sciences, USA, 112, 14894–14899.

Bauer, H., Nowell, K., Sillero-Zubiri, C., & Macdonald, D.W. (2018). Lions in the modern arena of CITES. *Conservation Letters, 11*(5), e12444.

Becker, M.S., Almeida, J., Begg, C., Bertola, L., Breitenmoser, C., Breitenmoser, U., Coals, P., Funston, P., Gaylard, A., Groom, R., et al. (2022). Guidelines for evaluating the conservation value of African lion (Panthera leo) translocations. *Frontiers in Conservation Science, 3*, 963961.

Berg, A. (2003). Ancestor reverence and mental health in South Africa. *Transcultural Psychiatry, 40*, 194–207.

Calavita, K. (2007). Immigration, law, race, and identity. *Annual Review of the Law Society, 3*, 1–20.

CITES. Committee 1. Conservation of and the trade in the African Lion. (2016). *CoP17 Com. 1*. 29. https://cites.org/sites/default/files/eng/cop/17/Com_I/E-CoP17-Com-I-29.pdf

Coals, P., Burnham, D., Loveridge, A., Macdonald, D.W., 't Sas-Rolfes, M., Williams, V.L., & Vucetich, J.A. (2019). The ethics of human–animal relationships and public discourse: A case study of lions bred for their bones. *Animals, 9*(2), 52.

Coals, P., Dickman, A., Hunt, J., Grau, A., Mandisodza-Chikerema, R., Ikanda, D., Macdonald, D.W., & Loveridge, A. (2020). Commercially driven lion part removal: What is the evidence from mortality records? *Global Ecology and Conservation, 24*, e01327.

Coals, P.G.R., Mbongwa, N.S., Naude, V.N., & Williams, V.L. (2022). Contemporary cultural trade of lion body parts. *Animals, 12*(22), 3169.

Coals, P.G.R., Williams, V.L., Benitez, G., Chassagne, F., & Leonti, M. (2024). Ethnopharmacology, ethnomedicine, and wildlife conservation. *Journal of Ethnopharmacology, 333*: 118399.

Cook, C.T. (2009). Sangomas: Problem or solution for South Africa's health care system. *Journal of the National Medical Association, 101*, 261–265.

Damania, R., & Bulte, E.H. (2007). The economics of wildlife farming and endangered species conservation. *Ecological Economics, 62*(3–4), 461–472.

DFFE (n.d.). *Threatened Or Protected Species (TOPS) permit and registration application process for the carrying out of restricted activities in terms of the National Environmental Management*. Biodiversity Act 2004 (No. 10 of 2004). Retrieved August 28, 2024, from https://www.dffe.gov.za/tops-permit-and-registration-application-process-carrying-out-restricted-activities-terms-nemba-10

Dore, A., Hübschle, A., & Batley, M. (2022). Towards environmental restorative justice in South Africa: How to understand and address wildlife offences. In: Pali, B., Forsyth, M. & Tepper, F. (eds) *The Palgrave Handbook of Environmental Restorative Justice*, pp. 333–359. Springer International Publishing, Cham.

Duffy, R. (2010). *Nature Crime: How We're Getting Conservation Wrong*. Yale University Press, New Haven and London.

Dunford, C.E. (2022). Poisoning of big cats. *Current Biology, 32*(24), R1334–R1335.

Everatt, K.T., Kokes, R., & Lopez Pereira, C. (2019). Evidence of a further emerging threat to lion conservation; targeted poaching for body parts. *Biodiversity and Conservation, 28*, 4099–4114.

Fabricius, C., Koch, E., Magome, H., & Turner, S. (2004). *Rights, Resources and Rural Development*. Community-Based Natural Resource Management in Southern Africa. Earthscan, London.

Gore, M.L., Hübschle, A., Botha, A.J., Coverdale, B.M., Garbett, R., Harrell, R.M., & Bowerman, W.W. (2020). A conservation criminology-based desk assessment of vulture poisoning in the great Limpopo transfrontier conservation area. *Global Ecology and Conservation, 23*, e01076.

Gumede, M.V. (1990). *Traditional Healers: A Medical Doctor's Perspective*. Skotaville Publishers, Cape Town.

Hiller, C., Davies-Mostert, H.T., Friedmann, Y., Taylor, W.A., & Macmillan, D.C. (2022). *Understanding South Africa's captive lion sector and the trade in captive lions (Panthera leo). EWT Research Technical Paper No.* 2. *South Africa*. https://h8l0bb.p3cdn1.secureserver.net/wp-content/uploads/2022/10/WIT-Captive-Lion-Report-hi-res.pdf?time=1674396447
Hiller, C., & MacMillan, D.C. (2021). How worldview and personal values can shape conservation conflict –the case of captive-bred lions. *Biological Conservation, 258,* 109151.
Hübschle, A. (2017). Contested illegality. The architecture of illegal markets: Towards an economic sociology of illegality in the economy, 177–197.
Hunter, L.T., White, P., Henschel, P., Frank, L., Burton, C., Loveridge, A., & Breitenmoser, U. (2013). Walking with lions: Why there is no role for captive-origin lions Panthera leo in species restoration. *Oryx, 47*(1), 19–24.
Macdonald, D.W. (2019). Mammal conservation: Old problems, new perspectives, transdisciplinarity, and the coming of age of conservation geopolitics. *Annual Review of Environment and Resources, 44,* 61–88.
Miller, S.M., Harper, C.K., Bloomer, P., Hofmeyr, J., & Funston, P.J. (2015). Fenced and fragmented: Conservation value of managed metapopulations. *PLoS ONE, 10,* e0144605.
Miller, S.M., Kotze, A., Harper, C.K., Bishop, J.M., Williams, V.L., Rossouw, C., Schoeman, J., de Bruyn, M., Dalton, D.L., & Selier, S.J. (2023). Genetic diversity and origin of captive lion (Panthera leo) in South Africa: An assessment and comparison to wild populations. *Conservation Genetics, 24,* 721–737.
Mlisa, L.S. (2009). *Ukuthwasa, Initiation of Amagqirha: Identity Construction and Training of Xhosa women as Traditional Healers*. PhD Thesis, University of the Free State, Bloemfontein, South Africa
Morton, F., & Hitchcock, R. (2014). Tswana hunting: Continuities and changes in the Transvaal and Kalahari after 1600. *South African Historical Journal,* 66(3), 418–439.
Mokgobi, M.G. (2014). Understanding traditional African Healing. *African Journal for Physical Health Education, Recreation and Dance, 20,* 24–34.
Muzondi, N. (2014). *The Contribution of Traditional Belief Systems in Biodiversity Conservation among the Ndau People in the Nyagadza Community of Chipenge*. Masters Dissertation, Midlands State University, Gweru, Zimbabwe.
Mwambene, L., & Kruuse, H. (2018). Marital rape and the cultural defence in South Africa. *Stellenbosch Law Review, 29*(1), 25–47.
Naude, V.N., Balme, G.A., Rogan, M.S., Needham, M.D., Whittington-Jones, G., Dickerson, T., Mabaso, X., Nattrass, N., Bishop, J.M., & Hunter, L. (2020). Longitudinal assessment of illegal leopard skin use in ceremonial regalia and acceptance of faux alternatives among followers of the Shembe Church, South Africa. *Conservation Science and Practice, 2,* e289.
Nicholson, S., Bauer, H., Strampelli, P., Sogbohossou, E., Ikanda, D., Tumenta, P.F., Venktraman, M., Chapron, G., & Loveridge, A. (2023). *Panthera leo. The IUCN Red List of Threatened Species 2023:* e.T15951A231696234. https://dx.doi.org/10.2305/IUCN.UK.2023-1.RLTS.T15951A231696234.en
Nigosan, S.A. (1994). *World Faiths*. Martin's Press, New York.
Phelps, J., Carrasco, L.R., & Webb, E.L. (2014). A framework for assessing supply-side wildlife conservation. *Conservation Biology, 28*(1), 244–257.
Redpath, S.M., Young, J., Evely, A., Adams, W.M., Sutherland, W.J., Whitehouse, A., ... & Gutierrez, R.J. (2013). Understanding and managing conservation conflicts. *Trends in Ecology & Evolution, 28*(2), 100–109.
Rizzolo, J.B. (2021). Effects of legalization and wildlife farming on conservation. *Global Ecology and Conservation, 25,* e01390.
The Republic of South Africa (2019). *CASE NO: 86515/2017 In the matter between – National Council of The Society for Prevention of Cruelty to Animals v Minister of Environmental Affairs and Others*. Retrieved October 24, 2022 from https://conservationaction.co.za/wp-content/uploads/2019/08/Judgment-Lion-Bone-case-6-August-2019-1.pdf.
Simelane, T.S., & Kerley, G.I.H. (1998). Conservation implications of the use of vertebrates by Xhosa traditional healers in South Africa. *South African Journal of Wildlife Research, 28,* 121–126.
Sollund, R.A. (2019). *The Crimes of Wildlife Trafficking: Issues of Justice, Legality and Morality*. Routledge, London.
Tanner, R., Freimund, W., & van Wyk, E. (2010). Legitimacy and the use of natural resources in Kruger National Park, South Africa. *International Journal of Sociology, 40*(3), 71–85.

Thobane, M.S. (2015). Armed robbers: Creating a perception of invisibility and invincibility through mysticism – are sangomas providing protection? *Acta Criminologica: Southern African Journal of Criminology, 4/2015*, 151–168.

Thornton, R. (2009). The transmission of knowledge in South African traditional healing. *Africa, 79*, 17–34.

Tunick, M. (2004). 'Can culture excuse crime?' Evaluating the inability thesis. *Punishment & Society*, 6(4), 395–409.

Val, A., Porraz, G., Texier, P.-J., Fisher, J.W., & Parkington, J. (2020). Human exploitation of nocturnal felines at Diepkloof Rock shelter provides further evidence for symbolic behaviours during the middle stone age. *Scientific Reports, 10*, 6424.

van Dyk, A.C. (2001). Traditional African beliefs and customs: Implications for AIDS education and prevention in Africa. *South African Journal of Psychology, 31*, 60–66.

Veenstra, S.J. (2006). *Healing Ministry among the Zulu Speaking People Troubled by Evil Spirits*. Master's Dissertation. North West University, Potchefstroom.

Vucetich, J.A., Bruskotter, J.T., & Macdonald, D.W. (2021b). Can deliberative democracy favor a flourishing relationship between humans and carnivores? *Frontiers in Conservation Science, 2*, 680925.

Vucetich, J.A., Bruskotter, J.T., van Eeden, L.M., & Macdonald, E.A. (2021a). How scholars prioritize the competing values of conservation and sustainability. *Biological Conservation, 257*, 109126.

Vucetich, J.A., Burnham, D., Macdonald, E.A., Bruskotter, J.T., Marchini, S., Zimmermann, A., & Macdonald, D.W. (2018). Just conservation: What is it and should we pursue it? *Biological Conservation, 221*, 23–33.

Washington, K. (2010). Zulu traditional healing, Afrikan worldview and the practice of Ubuntu: deep thought for Afrikan/Black psychology. *Journal of Pan African Studies, 3*, 24–39.

Whiting, M.J., Williams, V.L., & Hibbitts, T.J. (2011). Animals traded for traditional medicine at the Faraday market in South Africa: Species diversity and conservation implications. *Journal of Zoology, 284*, 421–473.

Williams, V.L., Coals, P.G.R., de Bruyn, M., Naude, V.N., Dalton, D.L., & Kotzé, A. (2021). Monitoring compliance of CITES lion bone exports from South Africa. *PLoS ONE, 16*(4), e0249306.

Williams, V.L., Drouilly, M., Coals, P.G.R., & Whittington-Jones, G.M. (2025). Pan-African review of cultural uses of carnivores. *PLoS ONE, 20*(3), e0315903.

Williams, V.L., Loveridge, A.J., Newton, D.J., & Macdonald, D.W. (2017a). Questionnaire survey of the pan-African trade in lion body parts. *PLoS ONE, 12*(10), e0187060.

Williams, V.L., Loveridge, A.J., Newton, D.J., & Macdonald, D.W. (2017b). A roaring trade? The legal trade in Panthera leo bones from Africa to East-Southeast Asia. *PLoS ONE, 12*(10), e0185996.

Williams, V.L., Newton, D.J., Loveridge, A.J., & Macdonald, D.W. (2015). *Bones of Contention: an Assessment of the South African Trade in African lion Panthera leo Bones and other Body Parts*. Joint Report TRAFFIC, Cambridge, UK & WildCRU, Oxford. https://www.traffic.org/site/assets/files/2474/bones_of_contention_report.pdf

Williams, V.L., & 't Sas-Rolfes, M.J. (2019). Born captive: A survey of the lion breeding, keeping and hunting industries in South Africa. *PLoS ONE, 14*(5), e0217409.

Williams, V.L., & Whiting, M.J. (2016). A picture of health? Animal use and the Faraday traditional medicine market. *South Africa. Journal of Ethnopharmacology, 179*, 265–273.

Willow, A.J. (2018). *Understanding ExtrACTIVISM. Culture and Power in Natural Resource Disputes*. Routledge, London.

5
CORPORATIONS AND WILDLIFE CRIME
Causes, contexts, and consequences

Angus Nurse

Introduction

Despite the existence of environmental protection legislation, corporate environmental crime remains widespread, with the transnational corporation at the forefront of contemporary environmental harm. Corporations are responsible for various harms to the natural environment including 'illegal emissions to air, water and land; hazardous waste dumping; and illegal manufacturing practices' (Tombs & Whyte, 2015, p.47). Corporate environmental crimes have both simple and complex explanations. In simple terms, corporate environmental crimes take place when corporations choose to break the law in order to maximise profits, taking direct and deliberate action to bypass or ignore environmental laws. Corporate environmental crimes can also occur when corporations are negligent in respect of their obligations under environmental protection legislation. But there is also a more complex explanation that can apply when the very nature of corporate activity and drive for profits places undue pressure on staff who have no choice but to break the law.

This chapter examines the involvement of corporations in wildlife crime (broadly defined) and the adequacy of enforcement and policy approaches to dealing with illegal corporate activity. A range of different corporations are involved in wildlife crime, from small corporations occasionally involved in dealing in wildlife products as a supplement to their legitimate business through to major corporate activity involved in corruption and transnational shipment of wildlife and wildlife products in contravention of both local and international regulations. The binary divide between legal and illegal corporations arguably no longer applies, and in respect of wildlife crime, the reluctance of some governments and failure of legal systems to effectively deal with corporate wrongdoing highlights an inadequacy in enforcement and policy approaches. This chapter assesses different perspectives on dealing with corporate involvement in wildlife crime.

Defining corporate environmental and wildlife crime

For the purposes of this chapter's discussion, wildlife crime is to be taken as a broad range of non-compliant activity situated within environmental and animal law discourse and

DOI: 10.4324/9781003333906-6

incorporating a broad conception of illegal activities aimed at nonhuman animals living in the wild and including non-native species who are present as 'visitors' living in a wild state. This goes beyond a potentially reductive consideration of wildlife trafficking to also consider issues of fraud, failure to pay duties/taxes on wildlife products and deceiving end consumers on the provenance of wildlife products. It also includes a wider notion of wildlife crimes that involve harms that are not directly linked to the illegal trade in wildlife. For example, shooting estates that are involved in the lawful and regulated sporting activity of game shooting are frequently linked with the illegal killing of birds of prey (Nurse, 2013a). Despite being illegal under conservation laws (e.g. the UK's Wildlife and Countryside Act 1981 and pesticides legislation), illegal predator control, such as the use of poisoned baits to kill birds of prey, is considered a widespread method used on shooting estates to kill recognised predators for game birds such as grouse. Many shooting estates operate as corporations, and thus, wildlife crime, such as illegal killing of protected wildlife, has become normalised within the otherwise lawful employment of game management staff. This would also include indirect wildlife crimes such as the killing and harm caused to marine wildlife during the Gulf Oil spill, an event that also involved the 'traditional' crime of manslaughter in the deaths of 11 oil rig workers largely attributed to corporate negligence (Mufson, 2012; Solomon & Janssen, 2010).

While various conceptions of wildlife-related illegality exist, Nurse and Wyatt proposed a holistic definition of 'wildlife crime' that include various harmful practices 'that encompass direct and indirect acts or omissions and that indicate failure to comply with legal obligations or legislative restrictions on wildlife use, irrespective of whether the legislation or its associated sanction is distinctly criminal in nature' (Nurse & Wyatt, 2020, pp.7–8). This classification goes beyond the general definition of environmental crime proposed by Situ and Emmons, which defined environmental crime via a potentially restrictive focus on criminal law classifications, primarily as those acts defined as such by the criminal law and subject to criminal sanctions (Situ & Emmons, 2000). Thus, the act comprising the offence is broadly negative in its wording, relating to the prohibitive acts commonly defined in criminal law, while the nature of the sanction is generally punitive and based on retributive punishment approaches with fines and prison as the primary response (Horder, 2016). The global trade in wildlife occupies a potentially paradoxical position in law, as laws governing wildlife trade are arguably separate both in intent and content from laws that aim to conserve wildlife and/or protect it for the common good by strictly preventing trade from taking place. The more expansive definition of wildlife crime, adopted by this chapter, reflects the reality that legislation relevant to corporate involvement in wildlife crime incorporates a range of legal perspectives. Wildlife trade law, for example, is more regulatory than criminal in nature, something that emerges consistently in the research literature, and particularly in the literature that is critical of the law's relative failure to address wildlife trafficking. Wildlife trade is a regulated activity, with policy arguably concerned with managing the trade along sustainable lines, albeit the criminal activity that is synonymous with wildlife trafficking raises questions about the efficacy of trade and regulatory law to effectively deal with criminal activity. However, in respect of corporate involvement in wildlife crime, it is also relevant to consider the nature of business regulation, private contracts (between buyer and seller of wildlife products), property offences (e.g. in relation to wildlife trophies) and even tort law. Thus, the extent to which much business activity falls outside the remit of mainstream criminal law and enforcement by criminal justice agencies is an important consideration.

Green criminological perspectives are critical of criminology's relative silence on the harms caused to ecosystems and nonhuman nature, and the lack of both scholarly and law enforcement attention to wildlife trafficking (including that of fish, plants and timber) as one of the most serious global crime problems (Lynch & Stretesky, 2014; Sollund, 2015). Green criminology's species justice perspective contends that justice systems need to adapt to consider the needs of nonhuman nature, while law enforcement and policy perspectives also need to develop a more holistic view of wildlife crime offending to effectively address illegal activity that impacts negatively on nonhuman nature (Beirne, 2009; Benton, 1998; Nurse, 2015). This includes ensuring policy and enforcement approaches consider wildlife crime as a multifaceted contemporary problem involving both crime and harm, and not just an environmental concern. This includes actively considering the corporation and the role of consumers as integral components in environmental crime (Nurse, 2015, 2022).

The demands of modern consumerism and the operation of markets result in production practices that inevitably cause environmental harm. For example, 'industrial pollution of the air, water and land, toxic waste sites, deforestation, species extinction, excessive pesticide use and pollution, climate change, the excessive use of fossil fuels, acid rain, a growing reliance on coal and oil, the environmental effects of drilling for oil or mining coal are the problems of the modern world' many of which are linked to corporate activity (Lynch & Stretesky, 2014, p.11). Several of these harms are the result of legal corporate activity, although the legal and the illegal often go side by side, where corporations may, for example, go beyond permitted levels of pollution. Green criminology recognises that

> eco-global crimes are often transnationally organised, with damage being caused by legal transnational corporations, such as British Petroleum in the Gulf of Mexico, and the destruction of the soil and forest in the Ecuadorian Amazon caused by Chevron and Texaco.
>
> *(Sollund, 2012, p.3)*

In addition, corporations are involved in transnational wildlife markets at various levels. As discussed later in this chapter, corporations of various types operate at different levels within wildlife crime activity. A key aspect in this regard is the definition of the corporation as a legal entity whose core purpose is to conduct a lawful, ethical and profitable business (Millon, 1990).

Corporations are generally seen as benign actors with the dominant policy paradigm being one that sees corporations as beneficial to society. Thus, corporate environmental crime is almost by default classified as an aberration or accidental wrongdoing, either the fault of 'rogue' employees that have failed to comply with corporate policies, or as the unintended consequences of governance failures which allow blame to be apportioned to individuals rather than the corporation (Bell et al., 2013). In assessing corporate liability for crime or harm, a significant challenge exists in determining whether the crime is sufficiently representative of the corporate identity to be considered representative of the company. This is both a legal problem, linked to identifying the 'controlling mind' for any unlawful activity, and a practical problem associated with proving that the person committing the unlawful act held or acted with sufficient seniority to be considered indicative of the corporation.

Wildlife trafficking, as an aspect of wildlife crime, is an offence that involves several different offences and actors. First, it is consumer-led with consumers as the end-users of wildlife and wildlife-related products, parts, and derivatives. Consumer needs drive wildlife trafficking to satisfy consumer demand, which arguably drives all activities further down in the trade supply chain and frequently creates a situation in which legal and illegal markets operate within a symbiotic relationship, as the later discussion of wildlife crime networks indicates. Suppliers are primarily driven by economic needs, 'ranging from small-scale subsistence needs for some to major high-profit business for others' (Felbab-Brown, 2017, p.89). While retailers, many of whom operate within a potentially legitimate corporate structure (e.g. pet shops, importers and retailers) may make large profits from the end retail value of wildlife products (such as the high value placed on ivory and rhino horn). However, the hunters and collectors of wildlife (including plants and timber) who occupy the bottom and first stage of the supply chain can involve comparatively poor actors, albeit professional hunters present in wildlife trafficking discourse. Professional hunters and intermediaries are integral where intermediaries are connected to trafficking networks and transnational or global markets. Felbab-Brown notes that 'both middlemen and top traffickers can also establish networks of corruption, such as customs officials, and cultivate political patrons' (2017, p.93). This can include the use of diplomats or payments to officials to organise and facilitate transit and falsified permits, invoking considerations of state-corporate crime where corporations and the state are 'partners in crime' in respect of offending (Tombs & Whyte, 2020).

From a criminological perspective, wildlife trafficking aligns closely with the criminological gaze in respect of the focus on identifiable 'criminal' actors in respect of organised crime groups who have increasingly become involved in wildlife crime and perspectives on corporate crime when otherwise legitimate corporations become involved in wildlife offences. The links with other forms of crime such as drugs, human trafficking, and the illicit trade in weapons allow for comparison of trade routes, methods of operation and the operation of neutralisation techniques through which offenders are able to claim wildlife trafficking as a victimless crime, or may condemn their condemners (Wyatt et al., 2020). This may be particularly effective where enforcers are those outside of mainstream law enforcement activity and can be demonised as NGOs looking to raise funds or pursuing wildlife crime campaigns from an ideological position (Nurse, 2013a, 2013b). White collar crime discourse is also relevant where otherwise law-abiding actors or at least those perceived as legitimate actors are involved in the illegal activities of wildlife crime.

Species-related research provides for understanding of the dimensions of contemporary wildlife crime problems, while other forms of research have clarified the nature of wildlife trafficking's processes and behaviours. Wyatt's (2022) research into wildlife trafficking provided an in-depth deconstruction of how wildlife crime occurs and the various actors and stages involved in wildlife trafficking. This illuminates that a range of different actors, from those involved on the ground who are sometimes themselves marginalised by the trade as procurers and trappers, through to the intermediaries involved in transporting wildlife and the retailers ultimately selling wildlife, are involved in wildlife trafficking. Thus, policy and enforcement approaches need to consider the different motivations and justifications for those involved in wildlife trafficking and tailor the enforcement response accordingly. Research has identified that a range of wildlife offenders exist, not just those who might be considered rational choice offenders dealing with wildlife purely for profit, but also those whose engagement with wildlife might be linked to otherwise law-abiding employment and corporate activity (Nurse, 2013a). In the context of Global South countries, where sensibilities towards wildlife may well differ from Global North perspectives, and food and security imperatives may also be motivations, participation in aspects of wildlife trafficking might be a subsistence necessity, particularly in those contexts in which the benefits of income raised through sport tourism or trophy hunting may fail to reach the local communities most affected by wildlife.

Prior research has examined the involvement of organised crime in wildlife trafficking, with evidence of organised crime groups diversifying into wildlife crime, sometimes using supply routes developed for other forms of illicit commerce to facilitate wildlife trafficking. Crucially, research has identified a range of organised and corporate crime groups involved in wildlife trafficking. In 2003, Zimmerman identified three categories of organised crime groups as being involved in wildlife trafficking, namely:

1 Low level – groups of farmers who sell species illegally in order to supplement their incomes.
2 Mid-level – mafia-style groups that purchase species from peasants to sell at a large profit.
3 High level – major international smuggling rings are also involved in other illegal trades and utilise the same smuggling routes and use violence and intimidation to facilitate their trade (Zimmerman, 2003).

The second and third categories of Zimmerman's classification fall outside of normative definitions of corporate activity where the underlying structure of the groups does not conform to the definition of a 'legal' entity. However, a cooperative of farmers suitably constructed to be a legal entity and complying with corporate norms would meet the definition of a corporation. South and Wyatt's 2011 research drew comparisons between wildlife trafficking and the illegal drugs trade, identifying several key actors as follows:

1 Trading charities – collectors and Traditional Chinese Medicine (TCM) practitioners involved because of an ideological attachment to TCM and enterprises involved in drugs because of quasi-ideological commitments to drugs. Profit is a secondary motive when dealing in drugs or wildlife, the ideological attachment is the primary motivator.
2 Mutual societies – traffickers and traders whose lifestyle revolves around their work and who effectively deal to order in wildlife. This category also includes friendship networks of user dealers who support each other and deal drugs to order and within the framework of their mutually beneficial society.
3 Business sideliners – legal entrepreneurs who view drugs or wildlife as lucrative sidelines to their main often legal business. Thus, legitimate wildlife dealers may deal in illegal specimens such as exporting more wildlife than their permits or the quota system allows, or an importer whose legal business provides cover for an illegal drugs shipment.
4 Criminal diversifiers – traditional criminals who identify profitable opportunities in either drugs or wildlife trafficking and exploit these as an adjunct to their main business.
5 Opportunistic irregulars – street-level operatives who take advantage of opportunities whether in the form of drugs or wildlife, whenever the opportunity arises (South & Wyatt, 2011).

Again, several of these categories can be characterised as corporate actors and identify that overlap between drugs and wildlife trade sometimes occurs outside of the large-scale wildlife trafficking of organised crime and is not confined to transnational crime. While profit is a significant motive in illegal wildlife trade, it sometimes occurs as a secondary business or is the product of opportunity rather than specific intent to deal in illegal wildlife (Nurse,

2013a). Historically, the concept of a 'criminal network' or 'serious organised crime' is amorphous, with the terminology used interchangeably and describing all crime that is structured rather than solely reflecting crime that fits within normative definitions of 'organised' crime. More recently, Wyatt, van Uhm, and Nurse further identified variation in organised crime groups, classifying these as follows: (1) organised crime groups, (2) corporate crime groups, and (3) disorganised criminal networks. This conception further identified the reality of criminal networks as being more diverse, concluding that with globalisation and expanding markets, a range of criminal networks have become involved in different stages of the illegal wildlife trade (Wyatt et al., 2020). Wyatt et al.'s conception of organised crime groups, corporate crime groups, and disorganised crime networks, as well as collaborations between such groups (in addition to unattached individuals that also function in this market), are an endemic part of the global wildlife trade. Each of the criminal networks discussed is distinctive in terms of levels of organisation, the degree to which they act opportunistically, have links to other crimes, and employ corruption. However, research consistently shows corruption as a key facilitator and integral part of laundering processes involved in organised crime. In wildlife trafficking, corruption is manifested in different forms: evidence exists of officials accepting bribes during border crossings, the making of false claims of captive breeding to subvert permit and documentation requirements and the involvement of government officials in the smuggling process. Wyatt et al note that 'organized crime groups are highly organized, sometimes act opportunistically, often if not always have links to other crimes (particularly other illicit markets) and rely on corruption as a tool to perpetrate their illegal activities' (Wyatt et al., 2020). In contrast, corporate crime groups are also highly organised and sometimes act opportunistically, but generally they are less likely to be linked to other crimes and are less reliant on public corruption to facilitate their activities, though they themselves can be considered corrupt. However, linkage with business crimes such as fraud, failure to pay taxes, evasion of important duties, falsification of documentation, are factors common in respect of corporate illegality in wildlife trafficking. The disorganised crime networks identified by Wyatt et al. are the least organised, are the most likely to act opportunistically, can sometimes be linked to other crimes, and have the lowest use of corruption as a tool of perpetration.

Causes of corporate environmental crimes

Research suggests that the causes of corporate environmental crime lie in the operation of neoliberal markets and the drive for profits that lie at the heart of the modern corporation. It has been said that 'the systematic causal chains that underpin much environmental harm are located at the level of the global political economy – within which the transnational corporation stands as the central social force' (White, 2012, p.15). Globalisation is a significant factor in corporate crime because international systems of production, distribution and consumption encourage corporations to see environmental harms as something that can either be considered as an 'externality' something that their system of production does not need to consider, while in fact the nature of production processes continue to generate environmental harms (Lynch et al., 2020). In respect of wildlife crime, for example, wildlife is simply a commodity to be traded to serve the corporate drive for profit. Arguably, the first responsibility of many corporate actors is to the company's shareholders, which means first ensuring the profitability and ongoing sustainability of the corporation. Secondly, managers and directors have a responsibility to provide for sustainable growth and profits that will increase shareholder dividends and increase the company's market share. Adam Smith's concept of 'the invisible hand' holds that markets operate by self-regulation where a healthy market is one in which there is a steady supply of buyers and sellers, and the market regulates itself (Harrison, 2011). But the market is concerned only with the sale and purchase of products and less so with 'hidden' or external costs such as environmental harm. So, the market considers that unprofitable corporations cannot survive and that self-regulation weeds out weak corporations. But the market does not always charge the 'real' cost of goods, for example the costs of extinction of select species from trade, the negative costs of deforestation and illegal logging or the cost to society of not reducing the pollution caused by production. But the corporation may consider some of these costs and seek to reduce them and make choices about the extent to which they are prepared to comply with environmental laws. In this sense, corporate environmental crime is 'a product of motivation and opportunity conditioned by the quality of law enforcement' (Situ and Emmons, 2000, p.67), and there is an inherent conflict between drivers to make money and the costs of regulatory compliance.

Corporate crime as a form of innovation

The conflict between compliance costs and environmental (and other) harms has been recognised by UK and US governments, who employ the principles of risk-based regulation, which seeks a regulatory approach to enforcing corporate compliance with environmental and wildlife laws rather than the traditional criminal justice-based retributive punishment approach. However, the regulatory approach is in part based on the idea that there is a distinction between 'productive corporate innovation' and 'unproductive activities such as organized crime' (Baumol, 1990, p.893). But corporate compliance with environmental regulations operates along a continuum where environmental non-compliance can be a form of business innovation that allows corporations to be creative in when and how they comply with the law. In part, the commission of offences and the willingness to commit offences or ignore wildlife protection law are influenced by the quality of law enforcement, the willingness of enforcers (and state policy) to prosecute, the likelihood of such prosecution, and the relatively low level of penalties that may be imposed.

Many corporations that have Corporate Social Responsibility (CSR) or Corporate Environmental Responsibility (CER) policies in place and that purport to consider the impact of their activities on the environment and nonhuman nature continue to commit environmental crimes. McBarnett (2006) identifies 'creative compliance' as a tool through which corporations circumvent and undermine legal regulation by essentially engaging in symbolic compliance while continuing with harmful and sometimes illegal business practices. Corporations may also engage in constructive compliance, using their sustainability reporting as a tool to appear compliant with regulation while at the same time making it all but impossible for staff to comply with the law and meet their objectives. Where this happens, the 'rogue employee' argument can be used, so that in the event of enforcement action, the company is able to protect itself by pointing to its CSR/CER policies and the employee's failure to comply with these.

Enforcement approaches

Economic instruments, informational devices, voluntary agreements and command and control regulation are some of the techniques modern states use to protect the environment and deal with corporate environmental crimes. The principle of enforcement that is regularly implemented rests on the notion that corporate norms and the natural drive of corporations to behave ethically will automatically provide for self-regulation. As a result, corporate environmental crime is primarily dealt with via regulatory mechanisms that view corporate offending as an aberration rather than the norm. With this perspective in place, fines and administrative sanctions are thought to be more effective than prosecution with the aim of impacting corporate profits while allowing the corporation to continue business. In addition, the principle of 'risk-based regulation' is adopted by Global North governments, with the aim of reducing the regulatory burden on business in line with the principle that corporations should not have to allocate excessive resources to complying with legislation and should be allowed to concentrate on their core business of innovation and providing products (Rothstein et al., 2006; Black & Baldwin, 2010). The risk-based regulation concept is predicated on inspections and other regulatory activity as tools to help or encourage compliance. This enforcement action takes place only when necessary, minimising any burdensome impact on business, commensurate with the idea that regulatory justice is about achieving compliance (Nurse, 2022). Whereas criminal enforcement and prosecution is the dominant paradigm in respect of individual offenders and recognised criminal groups, the corporate enforcement approach is designed to work alongside corporate activity so that use of settlements, administrative fines and allowing corporations to amend their behaviour are favoured approaches with criminal prosecution used as a last resort (Macrory, 2013). The idea is that sanctions imposed on businesses should be proportionate and meaningful and not overly punitive (Baldwin & Black, 2016). This approach to enforcement differs from the retributive justice approach of mainstream criminal enforcement, where prison is arguably the main sanction for serious offending. There are debates concerning the 'benefits' of sending corporate managers to prison. In one sense, this would signal to corporations that their behaviour is considered serious, unacceptable and will be punished, invoking deterrence principles. However, given the lawful activity carried out by corporations, a punishment visited on individual managers is potentially ineffective, as that manager can simply be replaced and the 'rogue employee' argument can be deployed. Thus, the corporate entity is protected and distanced from the offence, and a neutralisation that apportions blame to the individual staff member rather than the corporate entity is deployed. Individualised punishment in the form of prison sentences also fails to address wildlife harms, and many transnational corporations can simply absorb fines as the cost of doing business.

Seizures are a primary enforcement tool and mechanism through which the scale of illegal markets can be assessed, combined with distinct studies of illegal items in trade. While it is beyond the scope of this chapter to provide an exhaustive analysis of research in this area, some specific studies are available that illustrate the nature of certain illegal markets where corporate actors operate. For example, the Environmental Investigation Agency (EIA) and Indonesian NGO Kaoem Telapak (KT) investigated illegal logging of merbau in Papua and West Papua and its trade to Surabaya (East Java) and identified inconsistencies in enforcement action. This included inconsistency in timber trafficking court proceedings and some inadequacies in the treatment of corporate offenders. Failings identified in their analysis included: several instances where no action was taken against companies identified as having been clearly involved in illegal timber; lack of follow-up on court verdict, with some companies continuing to operate despite being ordered to close following convictions; and inconsistency in sentencing, with some companies apparently going unpunished despite their directors being found of trafficking offences (Environmental Investigation Agency & Kaoem Telepak, 2021, pp.19–23).

While much research is targeted at the causes and scope of wildlife crime and the criminality involved, approaches based on addressing the consumer end and potentially targeting corporate profits have been proposed. Schneider's 2008 research advocated a market reduction approach to wildlife crime, recognising that markets might consider the legality of trading in wildlife products as a market externality but market-based analysis might identify 'hot' products, reveal who is involved in their criminal exploitation, provide for subsequent analysis of the reasons for their involvement in the illegal activity and then identify ways to disrupt the market and provide for appropriate intervention (Schneider, 2008). For example, a detailed understanding of the reasons why rhino horn, ivory and wild plants continue to be traded despite campaigns aimed at encouraging consumers not to purchase these products and the likelihood that most traders know they are breaking the law. In addition to actors on the supply side, e.g. local subsistence users, commercial hunters, and forest concessionaires, demand side actors, particularly corporate actors of different sizes, provide services to consumptive end-users in markets and restaurants, and non-consumptive users such as tourists (Nurse, 2013a). In the case of plants, timber, and derivatives and animal-derived consumables such as wildlife derived clothing (e.g. shawls, coats) or food (e.g. shark fin soup, tiger wine) consumers are not traditionally the subject of law enforcement attention, but unless consumers and the nature of the market are also addressed, wildlife trafficking likely persists as a means of satisfying consumer demand. Accordingly, there is a need to consider how best to influence consumer behaviour to eliminate the consumption of illegal wildlife products through effective demand reduction strategies. Policy also needs to consider approaches that might strengthen legal and enforcement deterrent effectiveness by raising awareness of legal protection and penalties for the sale and consumption of illegal wildlife products, where these exist.

Market-based strategies to prevent the illegal trade in wildlife products include a range of potential tactics, as follows:

- Imposing taxes or other levies to raise consumer prices or reduce producer profitability.
- Lowering tax rates on (sustainable) substitute products.
- Increasing the profitability of sustainable-harvested production through subsidies, value adding, certification, and labelling.

(UNODC, 2012, p.346)

For example, market-based approaches have been deployed to address the demand for and supply of timber products by targeting prices and markets for timber products such as sustainably harvested resources. Such mechanisms arguably operate effectively within legal markets for timber products where consumers may be influenced by price and ethical considerations. Thus, certification schemes and price manipulation through taxes or incentives may impact positively on consumer behaviour, and encourage engagement with 'responsible' corporate retailers, thus reducing demand for illegal products. However, market-based instruments rely on clearly established property rights, potentially a contested area in many source countries for wildlife products and the cooperation of governments to apply financial and economic instruments to the wildlife trade. Potentially, certification schemes and robust trade regulation provide for social crime prevention by addressing consumer and retailer behaviour. Arguably, the focus on producers as criminal actors could be shifted or expanded to also encompass the corporate retailer selling illegally sourced timber products and the consumer who purchases and ultimately possesses these products.

Conclusion: addressing corporate wildlife crime

Contemporary green criminology, environmental law discourse and corporate governance scholarship raise questions about the adequacy of the criminal law as the appropriate enforcement tool for corporate crime (Nurse, 2022). First, the reality is that much corporate environmental harm and wildlife crime activity falls outside the remit of the criminal law and would be caught by other justice systems (civil, administrative, and regulatory). Second, within wildlife crime, the symbiotic relationship between legal and illegal trade presents difficulties in respect of applying criminal law principles to legal actors committing wildlife crime. In respect of otherwise legal (corporate) actors, the effectiveness and use of criminal penalties may be limited and inconsistently applied at the discretion of environmental regulators operating within a non-punitive enforcement environment. The adequacy of the criminal law to deal with wildlife crime problems is questionable in respect of corporations. The criminal law generally punishes and applies retributive justice, whereas green criminological perspectives on corporate environmental crime contend that the focus should be on repairing the harm caused by corporate wrongdoing and preventing repeat offending and the ongoing activities of corporations.

The reality is that the individualistic nature of criminal prosecution potentially lacks a coherent deterrent effect, particularly in the case of the transnational corporation. In addition, major corporations with the resources to frustrate prosecution activity can arguably withstand multiple prosecutions, either factoring in the cost of fines as the cost of doing business or having the resources to settle cases. Addressing the specific characteristics of activities like wildlife trafficking arguably requires more than an international law approach based on regulating the trade. Policy needs to consider the various motivations for different corporate actors involved in wildlife crime, as well as the actions of consumers who continue to drive the trade and the issues of corruption. This chapter contends that addressing corporate wildlife crime requires a more holistic approach situated within preventative and restorative principles and that adequately considers the diverse offences involved in corporate wildlife crime.

References

Baldwin, R. and Black, J. (2016). Driving Priorities in Risk-based Regulation: What's the Problem? *Journal of Law and Society*, *43*(2), 565–595.

Baumol, J. (1990). Entrepreneurship: Productive, Unproductive and Destructive. *Journal of Political Economy*, *98*(5), 893–921.

Beirne, P. (2009). *Confronting Animal Abuse: Law, Criminology, and Human-Animal Relationships*. Rowman & Littlefield Publishers.

Bell, S., McGillivray, D., & Pedersen, O. (2013). *Environmental law*. Oxford University Press.

Benton, T. (1998). Rights and Justice on a Shared Planet: More Rights or New Relations? *Theoretical Criminology*, *2*(2), 149–175.

Black, J. and Baldwin, R. (2010). Really Responsive Risk-Based Regulation. *Law & Policy*, *32*(2), 181–213.

Environmental Investigation Agency and Kaoem Telepak. (2021, January). Criminal neglect: Failings in enforcement undermine efforts to stop illegal logging in Indonesia. https://eia-international.org/wp-content/uploads/Criminal-Neglect-2021-SINGLE-PAGES.pdf

Felbab-Brown, V. (2017). *The Extinction Market*. Oxford University Press.
Harrison, P. (2011). Adam Smith and the History of the Invisible Hand. *Journal of the History of Ideas, 72*(1), 29–49.
Horder, J. (2016). *Ashworth's Principles of Criminal Law* (8th ed.). Oxford University Press.
Lynch, M., Stretesky, P. and Long, M. (2020). The Treadmill of Production and the Treadmill of Law: Propositions for Analyzing Law, Ecological Disorganization and Crime. *Capitalism Nature Socialism, 31*(1), 107–122.
Lynch, M. and Stretesky, P. (2014). *Exploring Green Criminology*. Ashgate.
Macrory, R. (2013). Sanctions and Safeguards: The Brave New World of Regulatory Enforcement. *Current Legal Problems, 66*(1), 233–266.
McBarnet, D. (2006). After Enron will 'Whiter than White Collar Crime' Still Wash? *The British Journal of Criminology, 46*(6), 1091–1109, https://doi.org/10.1093/bjc/azl068
Millon, D. (1990). Theories of the Corporation. *Duke Law Journal, 39*(2), 201–262.
Mufson, S. (2012). BP Settles Criminal Charges for $4 Billion in Spill; Supervisors Indicted on Manslaughter. Washington Post. Available at: https://www.washingtonpost.com/business/economy/bp-to-pay-billions-in-gulf-oil-spill-settlement/2012/11/15/ba0b783a-2f2e-11e2-9f50-0308e1e75445_story.html
Nurse, A. (2013a). *Animal Harm*. Ashgate.
Nurse, A. (2013b). Privatising the Green Police: The Role of NGOs in Wildlife law Enforcement. *Crime Law and Social Change, 59,* 305–318.
Nurse, A. (2015). *Policing Wildlife: Perspectives on the Enforcement of Wildlife Legislation*. Palgrave Macmillan.
Nurse, A. (2022). *Cleaning up Greenwash: Corporate Environmental Crime and the Crisis of Capitalism*. Palgrave Macmillan.
Nurse, A. and Wyatt, T. (2020). *Wildlife Criminology*. Bristol University Press.
Rothstein, H., Irving, P., Walden, T. and Yearsley, R. (2006). The Risks of Risk-Based Regulation: Insights from the Environmental Policy Domain. *Environment International, 32*(8), 1056–1065.
Schneider, J. L. (2008). Reducing the Illicit Trade in Endangered Wildlife: The Market Reduction Approach. *Journal of Contemporary Criminal Justice, 24*(3), 274–95.
Situ, Y. and Emmons, D. (2000). *Environmental Crime*. Sage.
Solomon, G. and Janssen, S. (2010). Health Effects of the Gulf Oil Spill. *JAMA, 304*(10), 1118–1119.
Sollund, R. A. (2015). The Illegal Wildlife Trade from a Norwegian Outlook: Tendencies in Practices and Law Enforcement. In R. A. Sollund (ed.) *Green Harms and Crimes: Critical Criminology in a Changing World* (pp 147–169). Palgrave Macmillan.
Sollund, R. (2012). 'Introduction' In R. Ellefsen, R. Sollund and G. Larsen (eds.) *Eco-Global Crimes: Contemporary Problems and Future Challenges* (pp 3–14) Ashgate.
South, N. and Wyatt, T. (2011). Comparing Illicit Trades in Wildlife and Drugs: An Exploratory Study. *Deviant Behavior, 32,* 552–555.
Tombs, S. and Whyte, D. (2015). *The Corporate Criminal*. Routledge.
Tombs, S. and Whyte, D. (2020). The Shifting Imaginaries of Corporate Crime. *Journal of White Collar and Corporate Crime, 1*(1), 16–23.
UNODC. (2012). *Wildlife and Forest Analytic Toolkit*. UNODC.
White, R. (2012a). The foundations of eco-global criminology. In R. Ellefsen, R. Sollund, & G. Larsen (Eds.), *Eco-global crimes: Contemporary problems and future challenges*. Ashgate
Wyatt, T, (2022). *Wildlife Trafficking* (second edition). Palgrave Macmillan
Wyatt, T., van Uhm, D. and Nurse, A. (2020). Differentiating Criminal Networks in the Illegal Wildlife Trade: Organized, Corporate and Disorganized Crime. *Trends in Organised Crime 23,* 350–366.
Zimmerman, M. (2003). The Black Market for Wildlife: Combatting Transnational Organized Crime in the Illegal Wildlife Trade. *Vanderbilt Journal of Transnational Law, 36,* 1668.

6
CAMEL RACES – TRADITION, SPORT, AND POLITICS

Dina Siegel

Introduction

Much of green criminology research is concerned with flora and fauna in general and the relationship between humans and nonhuman animals in particular. Different green criminology authors have made efforts to identify how nonhuman animals are manipulated through laws that perpetuate their exploitation. Their status as 'property' is especially problematic (White & Heckenberg, 2014). Hunting, fishing, and shooting are carried out legally in some situations, but are prohibited in others, or take place as 'underground sports' (Nurse & Wyatt, 2020; Siegel & Uhm, 2021). This article aims to analyse various political and cultural meanings of a specific case in the field of green criminology, namely camel racing. The analysis is based on ethnographic data collected during fieldwork research conducted between 2020 and 2023. Research methods included interviews, participant observation in the Negev and Sinai, media analysis, and ethnographic research on various social media platforms. This chapter also hopes to contribute to the discussion in green and cultural criminology about questions such as the following: Is camel racing a sport or animal abuse? How do we know what is harmful for animals and who decides this? And, how did major differences in views on this phenomenon develop?

Historical continuity

For centuries, camels were divided into two main categories: riding camels and pack camels, depending on their suitability for caravan transportation and military purposes. However, since the second half of the 20th century, this classification has expanded to include several more categories, such as meat camels, dairy camels, dual-purpose (combining meat and dairy) camels, and race camels (Wardeh, 2004, p. 5). The traditional role of camels as *safinat al-sahra* ('ship of the desert') has evolved, and nowadays camels are also used for their milk, meat, and wool. In many Arab countries camel, urine has become a commercial product. In Yemen and Saudi Arabia, it is considered to be a remedy for a low libido, and good for cleaning wounds, strengthening the heart, promoting bone growth in children and preventing hair loss (Wardeh, 2004, p. 201).

 DOI: 10.4324/9781003333906-7

But camels have always had several other functions. Historically, in many countries, camels were used during military operations. Endurance and strength are the camel's most remarkable qualities: the animal can move under the hot desert sun without water for up to 15 days. In the past, this made the camel the most suitable companion for soldiers in the Middle East. The great success of the Assyrian invasion of Egypt in 671 B.C. has been attributed to the army's use of camels. In 484 B.C., the Persian ruler Xerxes had 10,000 camels in his army during his march into Europe (Robinson, 1936, p. 61).

Camel racing has been held in Africa and the Middle East since medieval times and was considered a traditional sport, a respectable form of entertainment, a social event, and an inevitable part of various religious celebrations. Traditionally, camel races were held during Eid al-Fitr, the important Islamic festival that marks the end of the holy month of Ramadan.

Various rituals and ceremonies were performed before and during the competitions to bring good luck and ward off evil spirits, including adorning the camels with colourful blankets and tying bells and ornaments around their necks. Special prayers were said and offerings brought. Camels were trained for months before the competitions. For example, in ancient Egypt, the camels had to wear weighted blankets to increase their strength (Khomeiri & Yam, 2015).

While camel racing has been popular in the Middle East for many centuries, on other continents it is a relatively new phenomenon. Thus, in Australia, it developed only in the 19th century. Today, Australia is the country where the largest camel racing events take place. The first major events in Alice Springs were launched in 1971, and since then, camel races have been organised there several times per year. Other major camel races take place in Dubai (at the Al Marmoom Camel Racing Track), India (during the Bikaner Camel Festival), Kazakhstan, Oman and California (in Palm Springs). The United Arab Emirates and the Gulf countries also organise large-scale camel races, where the phenomenon even 'reaches the level of a national industry' (Khalaf, 1999, p. 87). Camel racing is a recognised and respected sport in these countries, the races are televised, and wealthy sheikhs invest a fortune in professional trainers, jockeys, and special breeding stocks. Annual camel festivals are organised to preserve tradition and, according to some authors, in response to modern developments in the consumerist, materialistic culture (Khalaf, 2000).

Many countries are making efforts to regulate and promote camel races. There is an influential Federation of International Camel Racing with its headquarters in Tanzania. In 2020, camel racing was inscribed on the Representative List of the Intangible Cultural Heritage of Humanity at UNESCO.[1] In other countries, especially in Europe and the United States, camel racing remains controversial; it is not banned but is subject to various animal welfare laws. In some countries, such as Israel, it is illegal and punishable. Banning an activity, however, does not mean that it no longer exists, as illegal markets usually emerge after the criminalisation of a specific product, service or phenomenon (South, 1998; Siegel & Van Uhm, 2021). Organised crime has a rich history of exploiting opportunities, from the Prohibition era in the United States to the proliferation of illegal markets for smuggled and counterfeit cigarettes, often in response to tightening regulations in various EU countries. As a rule, organised crime exploits the deficiencies in legal markets to generate profits by supplying 'forbidden' goods in a parallel economy.

Camel races: feast, sport, gambling

Camel racing is almost always associated with nomadic peoples and in the Middle East specifically with the Bedouins, who wander around the desert in search of water and food sources for themselves and for their livestock, which consists mainly of camels, sheep, and goats. Ten major Bedouin tribes are today divided across borders and spread across Sinai, Gaza, and the Negev. The tribes are different in origin, customs and language, and there is fierce inter-tribal competition over land and water resources. In both Israel and Egypt, the marginalisation and discrimination of Bedouins lead to tensions with the governments. The prevailing image of a Bedouin in both countries is that of a bandit or a terrorist (Abu-Guider, 2018).

Northern Sinai is one of the poorest regions of Egypt, with fishing and agriculture being the dominant economic activities in the region. The Bedouins of Sinai have been neglected for years, not allowed to join any government position or the Egyptian army. Their limited economic opportunities lie mainly in agriculture and tourism, but also in illegal practices such as cannabis cultivation and smuggling drugs and weapons to Gaza and Israel (Tuitel, 2014). Similar to other Arab countries, camel racing is an important social activity in Sinai, during which the government has an opportunity to promote its political messages and romanticise the centuries-long tradition, while the wealthy sheikhs who invest in these events can showcase their power and influence.

Some authors who have written about camel racing draw a distinction between races that are held during social celebrations, such as religious festivals, weddings, or circumcision feasts, which do not involve financial rewards or betting practices, and competition races which have a fixed *shara* (prize) and follow specific rules, including agreements on relatively short racing distances (Khalaf, 1999). Other authors have argued that competition races are also held during social celebrations. For instance, the anthropologist Emanuel Marx noted that during circumcision feasts, young men from both nearby and distant areas in the Negev would gather to participate in camel races. 'Some wished to compete in the races, while others laid bets on their outcome. This inflammable mixture of competitiveness and cupidity caused numerous squabbles among young guests' (Marx, 1973, p. 422). Some men attended social celebrations, particularly in order to participate in races. 'Thus, one young man of the Hazail tribe attended as many feasts as possible and persuaded young camel-owners to compete with his well-trained mount' (Marx, 1973, p. 422).

Criminologist Jay Albanese listed betting on animals as one of the four types of illegal gambling activities of organised crime (Albanese, 2018). The tradition of betting on animals, from dog fighting to camel racing, is an important source of income for both legal and illegal entrepreneurs. In Islam, gambling is forbidden (*haram*), but an exception is made for horse and camel racing and archery (Kizar, 2018). Gambling is common in the Sinai during official events. The Jewish owners of camel farms in the Negev do not participate in camel races, but their Bedouin business partners, employees and their families engage in it with an unyielding argument: 'It may be illegal, but it is a tradition, and no one can beat tradition'.[2] In the Negev, the illegal camel races usually take place between September (the beginning of the Bedouin wedding season) and December. In December, it is more difficult to organise competitions because the weather is usually bad, and it can be too slippery and wet for the camels. The participants are glued to the weather sites to see if rain or a storm is on the way. Organising such an event far in advance is almost impossible, due to its illegal nature. Communications about the day and location are sent to the participants as late as possible, but a race can also be cancelled at the last minute.

Some of the camel racers come from Jordan a day in advance. All participants gather in the evening to eat and drink together, and the race starts very early in the morning to avoid police attention. The meal provides an opportunity to exchange opinions and the latest news. According to the participants, camel racing is a sport that is well suited to the animals because they need a lot of exercise and freedom to remain healthy. They are not only intended for short tourist trips with mothers and toddlers on their backs but are also strong animals that need challenges and can carry up to 300 kilos. According to the participants, it is the government that is progressively reducing their area of freedom and constraining their living space.[3] In contrast to Israel, no strict animal welfare law has been enforced in Egypt. According to PETA (People for the Ethical Treatment of Animals), there is evidence of camels being beaten with sticks in order to force them to carry tourists for entertainment.[4] However, there is no consensus on camel riding among animal rights groups. While PETA considers all forms of camel riding as abusive, other organisations do not view it as a harmful activity.[5]

When the race begins and the camels start running, they are accompanied by jeeps and motorcycles with cheering fans. Especially at the finish, there is a lot of shouting and gunshots are fired into the air. During the entire race, participants and spectators run the risk of injury because the accompanying cars are driven at such high speeds. Falls and camel bites pose additional risks. The high risk of accidents during the events is often used as an argument to justify a ban on camel racing in Israel. The first prize is awarded immediately after the last camel crosses the finish line, usually it is around 15,000 shekels (approximately equivalent to 4,000 euros). But participants in illegal camel races are not motivated solely by financial gain. More important is the desire to uphold the power and prestige of their tribe. Participating in camel races, and especially winning the race, is an opportunity to manifest tribal wealth and power, as well as loyalty and family duty. In contrast to the official camel races in Egypt and other Arab countries, where elaborate prize-awarding ceremonies are held in the presence of wealthy sheikhs and heads of state, the ceremony in the Negev is brief and modest. Israeli police often arrive at the end of the event but appear to turn a blind eye to it.

> These events happen 'under the nose' of police. However, the police are not interested in getting involved. Sometimes they arrest somebody symbolically, because it is illegal, but they release him the same day. The police have no interest in getting involved because the lawyers and activists will later claim that it is tradition and sport.[6]

Criminalisation of camel racing

Several NGOs have reported on the inhumane living conditions at camel farms (*azbas*) in some Arab countries. These reports highlighted concerns about sexual abuse and the labour exploitation of children who are trafficked to work as camel jockeys. 'The ideal camel jockey would be between the ages of five to eight years weighing about 15 kg' (Asgar et al., 2005, p. 39). The lighter the jockey, the faster the camel can run. Hence, there is a demand for young boys to undergo training and participate as camel jockeys in these races. Camel racing has come under scrutiny after the appearance of these reports and media coverage on the human trafficking of young boys from Bangladesh, India, Sudan, and Pakistan. Poverty, coupled with a dearth of educational and employment prospects, is the main reason why parents entrust their sons to 'agents' who facilitate their migration to wealthier Arab countries such as the UAE, Qatar, Bahrain, Kuwait, and Saudi Arabia. They do so in the hope that this will ensure a better economic future for both the boys and the rest of the family.

This has led to the rules becoming stricter and a ban on jockeys under the age of 15. In many countries, human jockeys have been replaced by robots, which are controlled remotely. Whips are used by pushing a button. This has given rise to new ethical questions. Can a robot feel the camel and its needs in the same way as a human sitting on its back? Does this new technique really improve the well-being of these animals?[7] In Israel, arguments for the criminalisation of camel racing primarily revolve around animal protection and public health laws. The significant risk of serious injuries and fatalities resulting from falls off camels running at high speeds, sometimes reaching up to 60 km per hour, is cited as a major concern for both humans and animals. Additionally, since the event takes place illegally in a remote area, there is no prompt access to medical assistance, as is the case with official camel races in other countries.

There are also other arguments for banning camel races. While the camel population has been growing in recent years and gaining popularity in the rest of the world, including European countries (Wardeh, 2004), the situation in Israel has been the opposite. In the Negev, the number of camels has been declining, primarily due to the restriction of free passage between the Negev in Israel and the Sinai in Egypt after the fence was built on the border (Degen et al., 2019, p. 197). Camel herding is not officially recognised by the Ministry of Agriculture as a distinct sector of animal production. Consequently, camel owners are unable to obtain permits to graze their camels on state lands, in contrast to sheep owners who do receive such permits. As a result, many Bedouins resort to grazing their camels illegally. This does not motivate them to continue the business, as it is easier to herd sheep and goats. Young Bedouin therefore prefer 'shoats' (sheep and goats) to camels (Degen et al., 2019, p. 198).

In Israel, camels are mainly a tourist attraction, especially popular in eco-tourism. Camel meat and milk have been commercialised to a much lesser extent. In the Negev, most camel farms are owned by Jewish families, who have run them for generations. There is only one camel farm in the village of Tarabin, known as Bereshit (Genesis), which is jointly owned by a Bedouin from the Tarabin-el-Sana tribe and a Jewish Israeli. Other Jewish camel farm owners employ one or two Bedouins as the 'face of the farm' and as guides on camel treks, to convey authenticity and the 'desert feeling', the traditional link between Bedouins and camels. Some Bedouins who work for these camel farms come from far away and have to travel for hours by bus to their workplace in the early morning and return late in the evening to their families. These families (*hamulas*) are scattered across the Negev and the Sinai. Particularly with the latter, they maintain contact only through WhatsApp, as a wall separates Israel and Egypt, and there is only one official border crossing point, making communication between family members difficult. The traditional and unbreakable bond between Bedouins and camels is showcased to the outside world, but the Bedouins themselves struggle to benefit commercially from this image.

Camel racing as a political symbol

In Egypt, camel races were suspended in 2008 due to violent conflicts between the Egyptian military and various terrorist groups. In 2014, militants in the Sinai pledged allegiance to ISIS leader Abu Bakr al-Baghdadi, claiming responsibility for numerous attacks in North Sinai, resulting in hundreds of casualties among Egyptian police officers and soldiers. However, by 2018, Egypt's military achieved significant success in anti-terror operations, coinciding with the gradual renewal of camel racing. In 2022, one of the largest international events took place in the tourist destination of Sharm El-Sheikh in South Sinai as part of the Arab Heritage Festival, attracting over 900 participants and a large audience of Egyptians and tourists. Additionally, in El-Arish in North Sinai, where camel racing took place twice a year until 2008, authorities emphasised that resuming the sport was a crucial step towards uniting tribes to collectively counteract terrorist threats. The camel races were also intended to encourage millions of Egyptian citizens to relocate to Sinai, where they could live in peace and security and invest in the local economy. The political motivations behind reinstating this tradition were repeatedly emphasised. The resumption of regular camel races was viewed by politicians quoted in the media as a sign of normalisation and a victory for the Egyptian army over terrorist groups in Sinai (El-Balad News, August 13, 2022). In Egypt, camel racing is also seen as a contribution to resolving conflicts between various tribes. Ongoing disputes, particularly between the Tarabin, the Sawarka, and Ramilat, find a form of resolution through camel races, interpreted as a crucial unifying factor.

There is a completely different political significance to camel racing in Israel. In contrast to the unifying effect of the phenomenon in Egypt, camel racing is prohibited by Israeli law, making it a political symbol of what my respondents called 'discrimination against the cultural heritage of Bedouins' by the Israeli authorities. According to them, authorities are attempting to criminalise camel owners as a distraction from the actual issue: the state policy of dispossessing Bedouins and their camels. This involves closing off land and water sources to force them out, even though camels require space, food, and water. Camels are integral to Bedouin identity, and the authorities determine the fate of these animals and the attitude towards them. Because raising and herding camels are not officially recognised as a branch of the Ministry of Agriculture, there are no designated pastures allocated specifically for camels. As a result, camels are left to graze in open spaces, often encroaching on military firing ranges and crossing highways. Because camels roam freely in search of food and water, and because they are large animals, car accidents can cost the lives of both people and animals. In 2019, a new law entered into force, obliging camel owners to mark their camels.[8] Camels are supposed to carry an ID tag that identifies their owner in case they are lost or involved in a road accident. However, the Bedouins are in no rush to cooperate.

The camel races often take place in a military area in the Negev, regarded by the Bedouins as 'the best location for camels to run because it is flat and smooth'. But there is also another reason for selecting this site; it represents a political protest by Bedouins who view the military sites in the Negev as their own territory, taken from them by the Israeli state. They believe they have the right to hold camel races wherever they choose on their own land. Despite years of efforts by Bedouin representatives, attempts to regulate camel racing have been unsuccessful.

Camel races on TikTok: irony of secrecy

Various social media, particularly the popular TikTok platform, feature numerous images and videos of illegal camel races in the Negev. In September, coinciding with the beginning of the Bedouin wedding season, these races have become a regular occurrence in southern Israel. Hundreds of people gather to witness these events, recording them on their smartphones, and cars, jeeps, and motorbikes accompany the camels as they compete for first prize. One race posted on social media featured eight racing camels, but they were accompanied by many more vehicles, which kicked up a swirling dust storm, while enthusiastic onlookers shouted and fired guns from the open windows of their cars, all of which added to the intense excitement of the race. Loud Arabic background music made the video even more exciting.[9]

The images on social media, where participants are easily identifiable, are an example of what is known in criminology as the 'irony of secrecy' dilemma (Adler & Adler, 1980). In their study on drug trafficking in California, Adler & Adler showed that people involved in such activities struggled to gain respect and admiration as long as they couldn't share their secret with others. They called this the 'irony of secrecy', a dilemma characterised by the conflict between maintaining secrecy and the desire to reveal their hidden activities. This internal conflict involved a choice between rational goals and strategies on one hand and a hedonistic lifestyle and impulsive behaviour on the other (Adler & Adler, 1980). This dilemma is also evident among participants and organisers of camel races. On one hand, they try to avoid drawing attention from the police to these illegal events. On the other hand, the races unfold under the nose of law enforcement, with participants sharing extensive video footage and photos on various social media platforms. It serves as a form of boasting, a demonstration of success and power despite the potential risk of being caught and arrested.

The sharing of these images on social media platforms offers organisers and participants a channel to amplify their messages, presenting them as displays of power and protests against existing policies and legislation. Despite the risk of identification, they often opt to emphasise the importance of these events. Conversely, these messages from Bedouins are often downplayed. Police are unable to control every intersection in the Negev, leading them to claim that social media platforms exaggerate the scale of these events.

Conclusion

Once referred to as the 'ship of the desert', an integral part of Bedouin society, economy, and mobility in various Arab countries, the camel's role is gradually evolving. What has changed less, however, is the phenomenon of camel racing, which is flourishing as a sport and festive event in various countries around the world. However, camel racing has received contradictory interpretations when we look at the socio-political context of Israel and shed light on the perspectives of those directly involved in the event. The different meanings of the seemingly innocent, centuries-old event become apparent when comparing the examples of camel racing in Israel and Egypt, two neighbouring countries in the Middle East. In the Egyptian Sinai, camel races are being resumed after a period of violent struggle against terrorists and are now promoted as a message of normalisation, victory, and the unification of the tribes in the region. Camel racing serves as a political symbol, endorsed and celebrated by the Egyptian state.

In Israel, camel racing has also become a political symbol, but unlike in Egypt, it mirrors the Jewish–Arab relationship within the country. It serves as a symbol of protest against the discrimination and marginalisation faced by the Bedouin minority at the hands of the state. Through illegal camel racing, Bedouins seize an opportunity to assert their independence, uphold their own laws and traditions, and challenge the authority of the state. There is no consensus on camel riding in general, and on camel races in particular, among animal rights groups around the world. In a specific case of camel races in Israel and in Egypt, the question of the well-being of camels is almost totally disregarded.

Notes

1 https://ich.unesco.org/en/RL/camel-racing-a-social-practice-and-a-festive-heritage-associated-with-camels-01576.
2 Fieldwork notes, the Negev, 24 December 2022.

3 Fieldwork notes, the Negev, 24 December 2022.
4 https://www.peta.org.uk/features/egypt-camels/.
5 bornfree.org.uk.
6 Personal communication with Bedouin participants, 5 October 2022.
7 Fieldwork conversations with participants of camel races in the Negev, December 2022.
8 https://www.israelnationalnews.com/news/247553.
9 https://www.tv2000.co.il/article/29116.

References

Abu-Guider, A. (2018). Media representations of the Arab-Bedouin Minority within the green line in Israel from Young Arabs' perspectives. *Revue Européenne des Études Hébraïques*, (20), pp. 144–173.

Adler, P. A., & Adler, P. (1980). The irony of secrecy in the drug world. *Urban Life*, *8*(4), pp. 447–465.

Albanese, J. (2018). Illegal gambling business and organized crime: An analysis of federal convictions. *Trends in Organized Crime*, *12*(3), pp. 262–277.

Asgar, S, Farhat, S., & Niaz, S. (2005). *Camel Jockeys of Rahimyar Khan*, Report, Sweden: Save the Children, Pakistan Programme.

Degen, A., El-Meccawi, S., & Kam, M. (2019). The changing role of camels among the Bedouin of the Negev. *Human Ecology*, *47*, pp. 193–204.

Khalaf, S. (1999). Camel racing in the Gulf. Notes on the evolution of a traditional cultural sport. *Anthropos*, *94*, pp. 85–106.

Khalaf, S. (2000). Poetics and politics of newly invented traditions in the Gulf: Camel racing in the United Arab Emirates. *Ethnology*, *39*(3), pp. 243–261.

Khomeiri, M., & Yam, B. (2015). Introduction to camel origin, history, raising, characteristics, and wool, hair, and skin: A review. *Research Journal of Agricultural and Environmental Management*, *4*, pp. 496–508.

Kizar, O. (2018). The place of sports in the light of quran, hadiths and the opinions of the muslim scholar in islam. *Universal Journal of Educational Research*, *6*(11), pp. 2663–2668.

Marx, E. (1973). Circumcision feasts among the Negev Bedouins. *International Journal of Middle East Studies*, *4*(4), pp. 411–427.

Nurse, A., & Wyatt, T. (2020). *Wildlife Criminology*, Bristol: Bristol University Press.

Robinson, A. (1936). The camel in antiquity, in: *Sudan Notes and Records*, *19*(1), pp. 47–69.

Siegel, D., & Van Uhm, D. (2021). Illegal dogfighting: Sport or crime. *Trends in Organized Crime*, *24*(4), pp. 563–580.

South, N. (1998). Corporate and state crimes against the environment: Foundations for a green perspective in European criminology, in: Ruggiero, V., N. South & I. Taylor (eds.) *The New European Criminology*, London: Routledge, pp. 443–461.

Tuitel, R. (2014). The future of the Sinai Peninsula. *Connections*, *13*(2), pp. 79–92.

Wardeh, M. (2004). Classification of the dromedary camels. *Journal of Camel Science*, *1*, pp. 1–7.

White, R., & Heckenberg, D. (2014). *Green Criminology: An Introduction to the Study of Environmental Harm*. London: Routledge.

7
ORGANISED CRIME AND WILDLIFE TRAFFICKING

Daan van Uhm

Introduction

Traditionally, organised crime is linked to international smuggling activities, such as drugs, arms, and human trafficking, but the illegal trade in wildlife has increasingly become connected to the criminal portfolios of modern organised crime. Influenced by changing socio-economic, political, and ecological contexts, wildlife trafficking has become a significant new form of organised criminal activity, with disastrous impacts on the environment. Criminal networks have become involved in a wide range of high-value illegal wildlife markets, including rhino horn, tiger bone, ivory, and caviar, among others. This chapter will discuss the involvement of organised crime in the lucrative market of the illegal wildlife trade, partly based on interviews with people directly or indirectly involved in the illegal wildlife trade and fieldwork in key hotspots of wildlife trafficking (van Uhm, 2016, 2023).

Organised wildlife trafficking through history

Organised groups of wildlife smugglers and traffickers existed already in early civilisations across the world; some were entirely integrated, and others were banned from societies. A prime example is illustrated by the significantly increased illegal trade in exotic wildlife species with the emergence of the Roman games in the arena in ancient Rome (Belozerskaya, 2006; Whatmore, 2002). The animals were deployed to fight against gladiators or were slaughtered by the hundreds during celebrations of the Roman emperors.[1] For example, the events organised by Pompey around 55 BCE were well known for their large-scale massacres; at one of the events, 20 elephants, 600 lions, and 410 leopards were slaughtered, along with a Northern European Lynx and a rhinoceros. Pompey's rival, Caesar, had a similar number of wildlife slaughtered in the arena after celebrations or events (Wiedemann, 2002). The use of exotic animals during the Roman games was important for the rulers to demonstrate their status, power, worship, and popularity among the Roman social hierarchy. The demand for such a constant supply of exotic animals was huge and supported by special private enterprises. Even though a large part was controlled by private businesses

 DOI: 10.4324/9781003333906-8

on which the government levied an import tax of 2.5%, organised smuggling rings also emerged to avoid the taxes along the way (Hughes, 2003).[2]

Similarly, during the early modern period, characterised by European exploration, colonisation, and imperialism, large-scale trade with exotic areas in Africa, Asia, and the Americas took place. Imported exotic animals and products served as an endorsement of modern colonial power and became a symbolic representation of the conquest of all distant and exotic countries (Collard, 2020; Kalof, 2007). Large trading companies were established, including the British East India Company, the Dutch East India Company, and the Dutch West India Company around the 1600s, but in addition to the official trade, smuggling became a common issue in the decades that followed. Many employees received low salaries, and smuggling networks were established to earn extra money. Records of smuggled animal products and live animals illustrate the emergence of such animal smuggling rings. For example, the West India Company had a monopoly on the fur trade until 1640. Fur traders were required to sell the furs they acquired from trade to the Dutch West India Company (WIC) for a fixed price, resulting in minimal profits for fur traders. Therefore, illegal trade and smuggling of furs on the ships were flourishing. According to an ordinance of 1638, the illicit cargo took up so much room in the holds of the WIC ships that hardly any room was left for the legal goods (Jacobs, 2009).

Characterised by technological innovation as an outcome of the industrial revolution, large structural trade in wildlife emerged in the modern era as we know it today. Influenced by the post-war overall growth in the economy and prosperity, the trade in wildlife evolved as a trend-sensitive trading market that responded to consumer demand. The excessive and unsustainable exploitation of wildlife has reached alarming levels in our time (Moreto & Pires, 2018; Nurse, 2015; Sollund, 2019; van Uhm, 2016; Wong, 2019; Wyatt, 2013). Large trade flows of endangered species have become important to the global economy for a wide variety of purposes, such as medicine, food, pets, clothing, or status symbols (Hübschle, 2016; Kalof, 2007; van Uhm, 2018a). This ensured the emergence of sophisticated smuggling rings across the world to fuel the demand for illegal wildlife species. The large trade in wildlife had a devastating effect on species' populations, such as the trade in ivory that led to the extinction of elephant populations in many parts of Africa. Consequently, in the 20th century, the first international treaties to regulate and control the trade in wildlife came into force.

Box: Definitions: wildlife trafficking and organised crime

There is no universally accepted definition of wildlife trafficking. Wildlife trafficking involves the illegal trade, smuggling, poaching, capture, or collection of endangered species, protected wildlife, derivatives, or products thereof. There is also, no universally accepted definition of the term, and different jurisdictions and organisations employ different terminology (UNODC, 2024).[3] The United Nations Convention Against Transnational Organised Crime defined organised crime group as a

> structured group of three or more persons, existing for a period of time and acting in concert with the aim of committing one or more serious crimes or offences established in accordance with this Convention, in order to obtain, directly or indirectly, a financial or other material benefit.
>
> *(United Nations Office on Drugs and Crime 2004, p. 5)*

In this chapter, organised crime will be approached as organised, disciplined, and rational groups that may use violence or corruption to control illegal goods and services for profit for a significant length of time (van Uhm, 2023a; Wyatt et al., 2020). For instance, opportunistic poachers who sell protected animals, without having a buyer lined up, would not be classified as organised crime, while an established wildlife trade network that has existed for years with clearly agreed-upon duties and collaborating with corrupt government officials to smuggle illegal wildlife across borders could be considered organised crime by this definition.

Diversification of organised crime into the illegal wildlife trade

In the 21st century, organised crime groups rapidly diversify from "traditional" organised crime activities, such as drug or human trafficking, to the illegal trade in wildlife (van Uhm, 2023a). The developing interconnectedness between wildlife trafficking and other serious crimes shows that traditional lines of separation are no longer appropriate for understanding and dealing with the increasing complexities of organised crime (Martínez & Alonso, 2021; van Uhm et al., 2021). New illegal collaborations, alliances, and fluid criminal networks are developed in our globalised world to facilitate the illegal trade in wildlife. These criminal networks interact, cut across borders, infiltrate illicit markets, and penetrate fragile governments. Some sophisticated crime organisations can make a career shift to new businesses and infiltrate new markets in order to adapt to changing socioeconomic, political, and situational conditions, whereas other groups dominate new wildlife markets completely.

In order to understand the past, current, and potential future evolution of organised crime into the illegal trade in wildlife, Figure 7.1 presents the environmental crime continuum. Five points are situated on the environmental crime continuum. Each single point displays a form of convergence between organised crime (situated on the far left) and wildlife trafficking (situated on the far right). The level of diversification of organised crime into wildlife trafficking increases gradually during the stages: from minor diversification of organised crime groups, expanding their illegal activities with wildlife trafficking to a limited extent

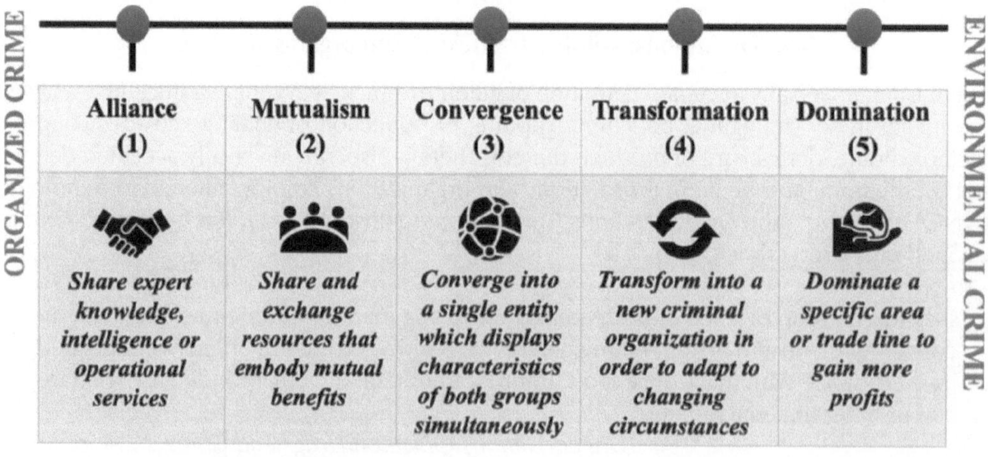

Figure 7.1 The environmental crime continuum.

based on alliances with wildlife trafficking groups (1) to overall domination over wildlife trafficking (5).

First, *alliances* between traditional crime groups and wildlife trafficking groups can be established for expert knowledge or operational services akin to social relationships within legitimate business environments (Makarenko, 2004; Ruggiero, 1996). The nature of alliances can vary; they can be one-off, short-term relationships, or long-term relationships (Morselli, 2009; Paoli, 2003). For instance, alliances arrange for wildlife to be smuggled along routes established for other types of contraband, such as drugs or arms (van Uhm et al., 2021). The first stage reflects minor diversification of organised crime groups, extending their illegal activities into wildlife trafficking to a limited degree, based on alliances with wildlife trafficking organisations. Second, *crime mutualism* between organised crime groups and wildlife trafficking groups can also be established in the long run to share resources, stop mutual adversaries and spread crime (Felson, 2006; van Duyne et al., 2002). Crime mutualism embodies barter trade, such as wildlife being exchanged for drugs and stolen cars, or camouflage, when a legitimate wildlife company is used to shield illegal trafficking (South & Wyatt, 2011). This second stage involves a higher degree of diversification, as the organised crime groups become increasingly involved in wildlife trafficking through crime mutualism. Eventually, they could merge into a single entity that simultaneously displays characteristics of both groups in the third stage; they *converge* entirely and become one and the same in the middle of the continuum (Miklaucic & Brewer, 2013). An example would be a criminal organisation involved in both human trafficking and fisheries, or a wildlife trader involved in the supply of weapons utilising multiple trade lines (Anagnostou & van Uhm, 2025). Stage three displays hybrid and multifaceted crime groups with the capability to diversify into wildlife trafficking and retain their traditional crime characteristics (Shelley & Kinnard, 2018). Fourth, organised crime groups can also decide to move away from their traditional activities and *transform* their characteristics to infiltrate the wildlife trafficking market; they become so fixated on wildlife trafficking that they drop the others altogether (van Uhm & Wong, 2021). In stage four, the ultimate aims and motivations of the crime group have actually changed in order to diversify into the illegal trade in wildlife. Fifth, organised crime can also *dominate* a specific area or line of wildlife trafficking (Varese, 2001). By monopolising elements of trade lines or by controlling the entire supply chain through the use of force, intimidation or extortion, in this final stage, criminal groups can effectively gain control, reputation and authority and dominate a wildlife trafficking market completely (van Uhm & Siegel, 2016).

Drug lords, crime syndicates, and mafia bosses

One of the most famous historical cases, reflecting the link between traditional organised crime and wildlife, may be Pablo Escobar, the notorious drug lord of the Medellín cartel. His brother, and co-founder of the cartel, Roberto Escobar, explained during an interview with him at his property in Medellín that they grew up together with a lot of affection for animals, because their father was a farmer with livestock and poultry. In the late 1970s, Pablo and his brother Roberto Escobar began to collect endangered species from all over the world, hiring expert advice on which species could survive and bribing customs officials to import illegal species by their small drug planes (Escobar, 2014; van Uhm, 2023a). They established Hacienda Nápoles, 530 ha of jungle and grassland along the Magdalena River in Colombia, with a collection of 1,900 animals from more than 100 species, including elephants, rhinos, hippos, and exotic birds; this became one of the most valuable animal collections in Latin America. Roberto Escobar explained:

> We had a special magazine that in chocolate bars there were little stickers, you get the magazine, and you put the sticker in each spot. They will teach you about the animals. So, Pablo had that magazine he filled it up and he said that eventually one day he will bring the opportunity to Colombia for them to see the animals. We had always the idea to have a zoo together. One day, we could buy cheap land between Bogota and Medellín. We brought animals from different places, and the most important ones were the elephant, the giraffe, the rhino and the hippo because everybody here saw them in magazines. We imported also white and black cockatoos from as far as Australia. Birds were big things for Pablo and me. (…) The animals in range Nápoles had everything they need; Pablo had the best way to treat the animals. And it was open to the public![4]

Another notorious drug lord, Mario Tabraue, demonstrated how animals not only play a role as symbols of status and power but can also be used as a business commodity for organised crime. Like Tony Montana from the movie *Scarface*, Mario Tabraue owned big cats and ran a powerful cocaine drug ring; the Tabraue Syndicate, worth $75 million at its height, was considered one of Florida's most violent drug organisations of the 1980s. The drug lord combined his cocaine trade from Colombia with an exotic animal business called Zoological Imports, one of the premier animal-importing companies in the United States. However, Tabraue was also involved in the trafficking of endangered species for profit, such as smuggling 35 endangered hyacinth macaws into the United States inside PVC pipes to avoid detection. Tabraue had his own animal store where he sold exotic animals, lent some of his animals to zoos, while others were kept as pets (van Uhm, 2018b).[5] Mario Tabraue explained during an interview at his private zoo in Miami, showing his favourite animal, the black panther:

> Before the drug business I had a zoo. When I had the drug business, I always had a zoo and import business. At 21 years old I had my first tiger. When I had the money for the drugs, of course, I invested in the most beautiful animals in the world. Everybody was in the business and the wild animals were in demand. I sold things cheap to make other people happy and zoos happy. Later, I invested drugs money into my animal business. Then the bullets started flying.

After an investigation by the FBI, the drug-wildlife kingpin was sentenced to 100 years imprisonment by a federal judge for racketeering, trafficking in cocaine, marijuana, and murder.[6] After 12 years, he had already been released after testifying against other members of the Miami underworld, and he founded a private zoo in 2000, where one can celebrate weddings, birthdays, corporate retreats, or getaway parties. "When I was arrested, I had to sell the animals. After 12 years in prison, I started from zero". By reflecting upon similarities with Tony Montana, he continued: "There is some similarity with Tony Montana, obviously I didn't get killed (laughs). But part of it yes, drugs, big cats, but they sensationalised it. Tony Montana wears the first initials of my name MT!"[7]

In addition to using exotic animals as status symbols or contraband, mafia bosses have also used dangerous animals to cement their grip on power or facilitate organised crime activities. A prime example is the notorious Italian American mobster Joseph Gallo of the Colombo crime family of New York City, who kept a pet lion named Cleo in his basement to frighten debtors into paying what they owed (van Uhm & Siegel, 2021). Resembling the James Bond scene at Kananga's farm, where crocodiles protected heroin processing plants, a crocodile on a rooftop was recently used by the Camorra mafia in Naples to intimidate business owners who were late paying their "pizzo" protection money. A pair of African grey parrots that had been trained to act as drug pushers for the mafia was also seized by the Neapolitan police. When the phone rang, the birds would respond 'Hello. How much do you need?' and in stressful situations they would say 'I'll shoot you.'

Wildlife trafficking and organised crime

Wildlife trafficking has been linked to several organised crime groups over the past decades, including Chinese organised crime, Colombian and Mexican drug cartels, Congolese militias, and Russian mafia. Oftentimes, wildlife trafficking is part of a wider network of illegal activities, and the criminal infrastructure provides opportunities to be involved in various illegal markets in order to spread the risks and increase both their control and profits.

Chinese triads and tiger bone and rhino horn

The combination of decentralisation, privatisation, and the opening of borders and the increasing economic prosperity in China that has allowed more people to afford expensive traditional Chinese medicines based on endangered species,[8] created new opportunities for Chinese organised crime in wildlife trafficking (van Uhm, 2020b; Wong, 2019). From the 1990s onwards, Chinese triads such as the 14K and Wo Shing Wo were linked to different illegal wildlife markets, including tiger bone and rhino horn. Even though Chinese triads were mainly active in the early days, nowadays a new Chinese underworld generation has emerged – some come from triad members, but they no longer consider themselves members (van Uhm, 2023a). The Chinese organised crime groups look across the borders to Africa, Europe, Latin America, and Southeast Asia.[9] For example, Chinese organised crime became highly involved in the illegal wildlife trade in the Golden Triangle, the border area between Laos, Myanmar, and Thailand, which gained notoriety in the 1920s as one of the world's most prolific opium producers. Since tiger skins and wines are highly profitable, the investment required to enter this trade is only available for a few. A famous tiger trader, who invested the profits from the methamphetamine business into the tiger business in Kengtung, the capital of Shan State (Myanmar), explained their powerful role during an interview at his grand villa:

> We know how to make things work. We have Chinese contacts in every small village here. I pay some of those guys with money you know, or we pay with poaching equipment, others know to find us when they have shot big cats. Then we arrange the transaction and further trade it to Jinghong [China], most is from Thailand and Chin state, but we know how to get it there.

Their network extends from local trade in Shan State to Thailand in the south for the import of captive-bred tiger parts, to Chin State in Western Myanmar and India to import wild tiger parts, and to the north with partners in Yunnan province in China, which illustrates their tentacles and transnational scope. After the tiger trader proudly showed more than 30 tiger skulls ready for export to China, he explained how he invested the profits from his methamphetamine business into his tiger business and highlighted:

> I am ethnic Chinese myself and speak the language. […] Otherwise, you will quickly be treated as an outsider. (…) However, Chinese or not Chinese, if you have no contacts with the militia groups, you have a small chance that your business will be successful.[10]

Informants explained how several Chinese business networks, some linked to the Chinese triads, have partnered with local armed groups and Burmese and Laotian elites who provide their services in exchange for a share of the profit; some of the Chinese organised crime groups even have their own tiger zoos (van Uhm, 2023a). An undercover agent in Bangkok echoed that the Golden Triangle is "a clear case where Chinese organised crime is active. Opium, methamphetamine, wildlife it's all part of their business (…) They can operate without intervention [here] and various gangsters from China come to visit and invest in the place."[11] The long history of Chinese migration to the borderlands, in combination with the flourishing opium and methamphetamine business, facilitates the increasing diversification of Chinese organised crime into wildlife trafficking in the Golden Triangle.

Colombian drug cartels and the bird trade

Both the famous Medellín and Calí drug cartels have been linked to the illegal trade in wildlife in the 1990s. The Medellín cartel would be involved in the illegal trade in parrots, mammals, and reptiles, and the Calí drug cartel would smuggle mixed shipments of wildlife and drugs by regional fishing fleets across the Caribbean into the United States and Europe; sometimes, the cartels would exchange drugs for endangered species, resulting in cashless transfers (van Uhm, 2016). The brother of Pablo, Roberto Escobar, explained:

> It was the secret passion of Pablo, he was a real animal lover! (…) Range Nápoles had it all and Pablo had two African jeeps, you could take the screen off and take friends on a tour and showing each section of the animals to friends and family.

In addition to the trade of wildlife for Hacienda Nápoles by Pablo and Roberto Escobar of the Medellín cartel in the 1980s and 1990s, Roberto underlined that: "Exotic animals were also important for other cartels. Another cartel had a property at the coast, and they also had giraffes and other big animals, but those people are still alive. The other cartels were inspired by us?!"[12] More recently, the sale of two rhino horns by a trader to a federal agent for $55,000 in a hotel in Las Vegas linked a former member of the Medellín drug cartel who was involved in the illegal trade in cocaine from the Medellín drug cartel to the United States. However, there is also a connection between the current drug cartels and the wildlife trade, explained respondents in Colombia. After the peace agreement between the Colombian government and the Fuerzas Armadas Revolucionarias de Colombia (FARC) in 2016, the Gulf Clan cartel – Clan del Golfo – considered to be the most powerful drug cartel in Colombia, extended its territories into the jungles of the Darién Gap at the border area with Panama (Ellis, 2018). For example, a large bird smuggling ring operated in and around the Katíos National Park in Colombia, home to about 450 species of birds, which represent, respectively, of the avian fauna of Colombia and Panama (UNESCO, 2009). The bird smuggling ring paid extortion taxes to the Gulf Clan cartel to transport the birds through the Gulf of Urabá, connecting the Darién jungles with port cities in Antioquia in Colombia (van Uhm, 2020a). A member of the Gulf Clan cartel explained in Acandí:

> The economic opportunities are very limited here. As you can see, we live in remote areas that have been forgotten by the Colombian state. The people need our support. (…) We provide them with services, and they pay us, we are together in this.[13]

According to bird traders at Playona beach, the birds were not destined for the Colombian market, but were exported abroad, to other countries. This illustrates how, with its highly strategic location and entangled network of rivers in the middle of the jungle, the Darién Gap has provided additional income to drug cartels for wildlife trafficking.

Congolese militias and the ivory trade

Located on the eastern edge of the Congo Basin, the border area between the Democratic Republic of Congo, Rwanda, and Uganda is one of the richest places on Earth in terms of biodiversity and natural resources, but at the same time it is one of the poorest regions in the world in terms of wealth, security, and peace, plagued by numerous militia groups. Throughout Eastern Congo's history, the interest in its natural wealth has fuelled decades of oppression, infighting, wars, and violence that continue to cause major environmental and human rights abuses (Verweijen, 2016; Vlassenroot & Huggins, 2005). During the extraction regime of the Congo Free State – then a private possession owned by King Leopold II of Belgium – characterised by systematic violence and brutality through the collection of ivory, Africans were forced to deliver ivory to the state, thus creating a *de facto* state-controlled monopoly over the ivory business (Conrad, 1899; Milburn, 2015). After decolonisation and the Congolese wars that officially ended in 2003, the conflicts between armed groups in Eastern Congo, as well as the ivory trade, continued. A militia member in North Kivu in Congo explained:

> The FDLR [militia] has several levels. (…) Some of the members have a crucial role in the ivory trade to neighbouring countries and facilitate this via their networks of their people and through their territories (…) Some substructures focus entirely on the charcoal trade, while other substructures focus on the military aspects and combatants, mining, or ivory trade[14]

The militia groups have been in and around the forests of eastern Congo for years and fund their activities via the trafficking of natural wealth, including ivory. Much like gold and coltan, ivory's profits have funded and enabled military and militant operations for years (van Uhm et al., 2022). An undercover agent from Uganda explained that Eastern Congo is a transit hub for the ivory trade, since several big ivory traders are part of the armed groups, which facilitates the smuggling through Eastern Congo into Uganda on its way to Asia.[15] A representative of a local community in Rutshuru explained:

> They [FDLR militia] stock the ivory in Rutshuru [90 USD per kg] and transport it in collaboration with the Chinese across the border into Uganda. They use secret roads,

and the price is good! The Chinese in Congo are also interested in the ivory, but most of it goes directly to Asia.[16]

This reflects how the ivory trade is often part of the broader network of activities by militia groups, including illegal mining, arms trafficking, extortion, and kidnapping for ransom, which fuels the ongoing conflict. These diverse criminal portfolios resemble organised crime groups more closely than a traditional rebel army, given the militia's financial lifelines, diverse illegal activities, and capitalisation facilitated by the regional extreme poverty and weak governance.

Russian mafia and the caviar trade

The caviar trade has a long history as a luxury food, from the ancient Greeks to the Russian tsars, but it became a trend in Western Europe in the 20th century. After the dissolution of the Soviet Union in the early 1990s, the Russian mafia became linked to the illegal fishing of sturgeons in the Caspian Sea for trade in the lucrative business of black caviar – from the poaching areas to major smuggling operations in collaboration with officials (Vaisman, 1997; Sellar, 2014; van Uhm & Siegel, 2016). An interviewed police spokesman in Makhachkala – the capital of Dagestan, a republic in southwestern Russia that lies on the flank of the Greater Caucasus mountain range, along the western shore of the Caspian Sea – highlighted at his office: "Sturgeon poachers used to travel in old rickety *Vikhr* boats, but these days the mafia is well organised and use state of the art means of transport, so that they are better equipped than the government guards."[17] Many informants along the Caspian Sea linked the power of the Russian mafia in the caviar business to the ongoing local wars over caviar, leading to bombings of buildings where coast guards lived and kidnappings and murders of officials after interventions, but also to offering protection services. A sturgeon poacher in a small village near the Caspian Sea explained the important role of *krysha* (literally a 'roof') by the mafia that is needed for the business:

> The middle enforcers, captains and lawyers, they all have a powerful person from the mafia standing behind their back and taking care of them. (…) If we have sturgeons on my boat and the police arrive, I call a certain person from the local mafia to say that the police have arrived. I give the phone to the police officer, he then talks to the person and they tell the police to calm down and then the police let me go.[18]

This reflects how *krysha* turns into a mutually beneficial arrangement between the mafia and sturgeon poachers; this not only covers a promise not to attack the business but would also defend it from other racketeers or law enforcers. Despite such protection schemes, respondents explained that it's crucial to protect themselves with arms. "Local poachers use hand grenades (…) others have automatic firearms to protect themselves from being shot," explained a caviar trader in the Astrakhan Oblast.[19] As a major outlet, West European countries also experienced violence and counterstrategies by the Russian mafia after seizures of caviar; enforcers, caviar dealers, CITES officials, and prosecutors were threatened from the 2000s onwards in Europe (van Uhm, 2016). The scarcity and extremely high prices paid for caviar, the social instability and corruption in the poaching regions, and a relative lack of control and loopholes in regulation attracted the involvement of Russian organised crime.

Mexican Sinaloa cartel and totoaba poaching

The Sea of Cortez, situated between the California peninsula and mainland Mexico, has become a notorious landscape for totoaba poaching. The totoaba, a marine fish endemic to the Gulf of California, is highly valued for its swim bladder, or "maw," which is considered a delicacy in Chinese cuisine and believed to have medicinal properties. Historically, Chinese traders controlled the totoaba maw trade, but over the past decade, the Sinaloa Cartel (Cártel de Sinaloa) has diversified into this lucrative business. Founded in the late 1960s in Sinaloa, the cartel initially focused on small-scale drug smuggling into the United States, but over time, it expanded its operations, becoming one of the world's most powerful drug trafficking organisations (Toro, 1995). The imprisonment of Joaquín "El Chapo" Guzmán, a former leader within the cartel, triggered a process of decentralisation and diversification among its factions. In recent years, the Sinaloa Cartel has extended its operations to Baja California, near the Gulf of California. The Sinaloa Cartel now plays a dominant role in brokering arrangements between Chinese buyers and local totoaba poachers, known as *totoabaeros* (Martínez & Alonso, 2021). An interviewed poacher in the poaching hotspot San Felipe explained that the cartel initially infiltrated and first targeted the fishers:

> The cartel took control by infiltrating the fishing cooperatives and initially offered us 40,000 pesos (+/-20,000 USD) per kilo. That was acceptable, but now they have lowered the price, started to extort us, imposed new rules, and forced us to sell exclusively to them.[20]

The cartel financed large mesh gillnets (*totoaberas)* and coerced several local fishermen into the drug trade, sometimes paying the totoaba poachers in narcotics. Another poacher noted the influx of poachers from Sinaloa, which disrupted local communities and established the cartel's dominance. "Poachers from Sinaloa have arrived. Some were 'imported' [by the Sinaloa Cartel], which has also impacted the local communities. We had to compete, you know!"[21] Previously, key Chinese traders in San Felipe directly interacted with Mexican fishers. However, the Sinaloa Cartel has pushed out several Chinese traders from direct purchases from local fishermen to monopolise the totoaba trade. Totoaba maw trade, storage, and processing are no longer the exclusive domain of the Chinese traders, but Chinese traders still play a crucial role in exporting totoaba maw from towns like Ensenada. An organised crime expert explained during a drive along the 'ruta narcobuchera' that the smuggling route is well known because both drugs and totoaba maws are transported along this chain, from the fishing town of San Felipe towards Mexicali, Tijuana, and Ensenada.[22] This illustrates how infrastructures from drugs are used for wildlife trafficking and how the diversification has ensured domination and altered the dynamics between the local fishers and the Chinese traders.

Discussion and conclusion

Organised crime groups across the world have been linked to the illegal trade in wildlife in recent decades. Alliances between organised crime and wildlife trafficking groups are established for expert knowledge or operational services, including access to smuggling routes. For instance, the Colombian Gulf Clan controls large parts of the Darién jungle, and therefore, bird smuggling rings in and around Katíos National Park operate under the control of the

Gulf Clan and pay extortion taxes to transport the birds through the Gulf of Urabá for export abroad. Crime mutualism is reflected by clear appointments between wildlife poachers and organised crime groups, illustrated by Burmese and Lao wildlife poachers who receive financial injections as well as poaching equipment from Chinese organised crime groups to facilitate the illegal trade in big cat bones. In addition, Chinese crime groups outsource not only the tasks of poaching but also product promotion or product smuggling to localised crime groups. Some organised crime groups and wildlife trafficking groups converge and form a single entity that displays characteristics of both groups simultaneously, illustrated by several militia groups in Eastern Congo that control diverse illegal trades, including minerals, metals, timber, arms, and ivory. Some of these groups consist of different substructures that control different illegal submarkets and collaborate with each other, while in other situations, the substructures are involved in the illegal activities simultaneously. In order to adapt to changing socioeconomic, political, or ecological circumstances, organised crime groups also transform their organisations, reflected by Chinese smuggling rings specialised in opium and methamphetamine, along the Laos and Myanmar borders, that adopted smuggling tiger bones and rhino horn in their criminal portfolios; they gradually transformed their organisations in order to specialise in illegal wildlife trade. Finally, organised crime groups may dominate a specific area or trade line, mirrored by Russian mafia groups that started to control the caviar trade from the late 1990s or the diversification of the Sinaloa cartel in the totoaba trade of San Felipe in Mexico, but also the territories controlled by militia groups in Eastern Congo reflect the role of domination over natural wealth, resulting in high-level collaboration with officials, but also violence in the respective areas.

This chapter has discussed how organised crime is involved in the lucrative market of wildlife trafficking. The analysed organised crime–wildlife connection is important to understand how organised crime infiltrates new wildlife markets as well as how it has been influenced by socioeconomic, geopolitical, and ecological changes. The relationships provided earlier illuminate some of the different dimensions of their involvement and show the complexity of the involvement of organised crime groups in the illegal wildlife trade, how they are embedded in the local context and how they interact with upperworld actors.

Notes

1 At that time, the animal trade was becoming a financially lucrative enterprise; a leopard cost 4,000 sesterces (£25,000) and an African lion was worth 600,000 sesterces (£4 million) (Bomgardner, 2000).

2 For example, the Roman vast mosaics in Cherchell in Algeria, former capital of Mauritania, reflect the scenes of hunting in natural surroundings.

3 Since the conventional definition of wildlife trafficking is usually tied to the state, defining what is criminal and what is not, crime is usually defined in relation to the law. However, many harmful activities against wildlife are not criminalised by law (Brisman & South, 2020; Van Uhm, 2023b). The nature of what constitutes a criminal act and how the criminal is portrayed changes over time (Sellin, 1938). Therefore, in a more expansive definition, as used by green criminologists, wildlife trafficking also extends to harm, including offenses harmful to humans, the environment and non-human animals, regardless of legality per se (Lynch & Stretesky, 2014; South, 2014).

4 Interview with Roberto Escobar in Medellín, Colombia, February 2022.

5 The animal business turned out to be a perfect cover for drug trafficking and laundering money; they used codes such as "I need three cockatoos," meaning three kilos of cocaine or "three lesser-crested cockatoos," which meant three ounces.

6 United States of America 'Operation Cobra indictment,' 16 December 1987.

7 Interview in Miami, the United States, November 2023.

8 Tiger bones are used to cure ulcers, rat-bite sores, abdominal pain, typhoid fever, malaria, hydrophobia, and especially to combat muscle cramps, migratory joint pain, pain in the bones, and rheumatism, while rhino horns are used as a substitute to cool the blood, alleviate fever (especially children's high fever), erythema, purpura, nosebleeds, convulsions, and cramps, as well as heart disease. Tiger bones are used in medicinal wines, as powder, or in pills, and rhino horn is prescribed as a powder or is ground into a "juice." Consuming medicine derived from endangered species would also be a cultural symbol of status and pride.
9 During the 2000s, investigations revealed how the Chinese triads became active in the trafficking of South African abalone, a highly priced delicacy and aphrodisiac, when the price of abalone spiralled. The triads brokered arrangements with local gangs—such as the Cape Flats gangs— over the poaching of abalone as well as the manufacture and distribution of the methamphetamine-based drug, *tik*. More recently, in Latin America, the Putian gang involved in jaguar products in Bolivia has been linked to the notorious Fujian triad.
10 Interview with tiger trader in Kengtung, Myanmar, March 2020.
11 Interview with undercover agent in Bangkok, Thailand, March 2020.
12 Interview with Roberto Escobar in Medellín, Colombia, February 2022.
13 Interview with Gulf Clan member in Acandí, Colombia, April 2019.
14 Interview with militia member in North Kivu, DR Congo, February 2021.
15 Interview with undercover agent in Kampala, Uganda, April 2022.
16 Interview with representative of a local community in Rutshuru, DR Congo, February 2021.
17 Interview with police officer in Makhachkala, Russia, September 2024.
18 Interview with local poacher near the Caspian Sea, March 2014.
19 Interview with local poacher in the Astrakhan Oblast, Russia, March 2014.
20 Interview with totoaba poacher in San Felipe at the Sea of Cortez, Mexico, November 2024.
21 Interview with totoaba poacher in the north of San Felipe, Mexico, November 2024.
22 Interview with organised crime expert during a drive along the 'ruta narcobuchera,' Mexico, November 2024.

References

Anagnostou, M. & van Uhm, D.P. (2025). The Links Between Human Trafficking and Wildlife Trafficking, In: Stoett, P. & Omrow, D. (Eds.), *Ecoviolence Studies: Human Exploitation and Environmental Crime*. Cambridge: Cambridge University Press.

Belozerskaya, M. (2006). *The Medici Giraffe: And Other Tales of Exotic Animals and Power*. New York: Little, Brown & Company.

Brisman, A., & South, N. (2020). *Routledge International Handbook of Green Criminology*. Routledge.

Collard, R. (2020). *Animal Traffic: Lively Capital in the Global Exotic Pet Trade*. Durham: Duke University Press.

Conrad, J. (1899). *Heart of Darkness*. London: Willan Blackwoord and Sons.

Ellis, R. (2018). *Transnational Organised Crime in Latin America and the Caribbean*. MarylandL Lexington Books.

Escobar, J. (2014). *Pablo Escobar: My Father*. New York: Penguin Random House.

Felson, M. (2006). *Crime and Nature*. London: Sage.

Hübschle, A. (2016). *A Game of Horns: Transnational Flows of Rhino Horn*. Universität Köln.

Hughes, J. D. (2003). Europe as consumer of exotic biodiversity: Greek and Roman times. *Landscape Research*, 28(1), 21–31.

Jacobs, J. (2009). *The Colony of New Netherland: A Dutch Settlement in Seventeenth-Century America*. New York: Cornell University Press.

Kalof, L. (2007). *Looking at Animals in Human History*. London: Reaktion Books.

Lynch, M. J., & Stretesky, P. B. (2014). *Exploring Green Criminology: Toward a Green Criminological Revolution*. London: Ashgate.

Makarenko, T. (2004). The crime-terror continuum: Tracing the interplay between transnational organised crime and terrorism. *Global Crime*, 6(1), 129–145.

Martínez, I. A., & Alonso, A. I. (2021). Mexican organised crime and the illegal trade in totoaba maw. *Trends in Organised Crime*, 24(4), 526–546.

Miklaucic, M., & Brewer, J. (2013). *Convergence: Illicit Networks and National Security in the Age of Globalisation*. Government Printing Office.
Milburn, R. (2015). Gorillas and guerrillas: Environment and conflict in the democratic republic of Congo. In *Environmental Crime and Social Conflict*. London: Routledge.
Moreto, W. D., & Pires, S. (2018). *Wildlife trafficking: An Environmental Criminology and Crime Science Perspective*. Durham: Carolina Academic Press. 57–74.
Morselli, C. (2009). *Inside Criminal Networks*. New York: Springer.
Nurse, A. (2015). *Policing Wildlife*. London: Palgrave Macmillan UK.
Paoli, L. (2003). *Mafia Brotherhoods: Organised Crime, Italian Style*. Oxford University Press.
Ruggiero, V. (1996). *Organised and Corporate Crime in Europe: Offers that can't be Refused*. Dartmouth.
Sellar, J. M. (2014). *The UN'S lone ranger: Combating international wildlife crime*. Whittles Publishing.
Shelley, L., & Kinnard, K. (2018). Convergence of rhino horn and ivory trade with other
forms of criminalist. In W. Moreto (Ed.), *Wildlife Crime* (pp.109–134). Philadelphia: Temple University Press.
Sollund, R. (2019). *The Crimes of Wildlife Trafficking: Issues of Justice, Legality and Morality*. New York: Routledge.
South, N. (2014). Green criminology: Reflections, connections, horizons. *International Journal for Crime, Justice and Social Democracy*, 3(2), 6–21.
South, N., & Wyatt, T. (2011). Comparing illicit trades in wildlife and drugs: An exploratory study. *Deviant Behavior*, 32(6), 538–561.
Toro, M. C. (1995). *Mexico's" war" on Drugs: Causes and Consequences*. Lynne Rienner Publishers.
UNESCO. (2009). Los Katios National Park (Colombia).
UNODC. (2024). World WIldlife Crime Report 2024. Vienna: UNODC.
Vaisman, A. (1997). *Sturgeon Catch and Trade in the Russian Part of the Caspian Sea*. Traffic Europe.
van Duyne, P., von Lampe, K., & Passas, N. (2002). *Upperworld and Underworld in Cross-border Crime*. Nijmegen: Wolf Legal Publishers.
van Uhm, D. P. (2016). *The Illegal Wildlife Trade: Inside the World of Poachers, Smugglers and Traders*. Switzerland: Springer International Publishing.
van Uhm, D. P. (2018a). The social construction of the value of wildlife: A green cultural criminological perspective. *Theoretical Criminology*, 22(3), 384–401.
van Uhm, D. P. (2018b). Wildlife and laundering. In Spapens, T., White, R., Van Uhm, D.P. & Huisman, W (eds), *Green Crimes and Dirty Money* (pp. 197–214). London: Routledge.
van Uhm, D. P. (2020a). *Crimen Organisado Ambiental en el Parque Nacional Los Katíos*. Utrecht: Utrecht University.
van Uhm, D. P. (2020b). Wildlife trafficking and criminogenic asymmetries in a globalised world. In Brisman, A. & South, N. (eds), *Routledge International Handbook of Green Criminology* (pp. 529–542). London: Routledge.
van Uhm, D. P. (2023a). *Organised Environmental Crime: Black Markets in Gold, Wildlife, and Timber.*. London: Bloomsbury.
van Uhm, D. P. (2023b). The Criminalisation of the Trade in Wildlife. In Nelen, H. & Siegel, D. (eds), *Organised Crime in the 21st Century* (pp. 155–169). Switzerland: Springer International Publishing.
van Uhm, D. P., & Siegel, D. (2016). The illegal trade in black caviar. *Trends in Organised Crime*, 19(1), 67–87.
van Uhm, D. P., & Siegel, D. (2021). Organised crime and animals. *Trends in Organised Crime*, 24(4), 419–424.
van Uhm, D. P., South, N., & Wyatt, T. (2021). Connections between trades and trafficking in wildlife and drugs. *Trends in Organised Crime*, 24(4), 425–446.
van Uhm, D. P., Tjoonk, M. M., & Bakole, E. Y. (2022). Organised Forest Crimes: Charcoal and Timber Trade in the Democratic Republic of Congo. In Zabyelina, Y. & Thachuk, K.L. (eds). *The Private Sector and Organised Crime* (pp. 149–162). London: Routledge.
van Uhm, D. P., & Wong, R. W. Y. (2021). Chinese organised crime and the illegal wildlife trade: Diversification and outsourcing in the golden triangle. *Trends in Organised Crime*, 24(4), 486–505.
Varese, F. (2001). *The Russian Mafia*. Oxford: Oxford University Press.
Verweijen, J. (2016). *Stable Instability: Political Settlements and Armed Groups in Congo*. Rift Valley Institute.

Vlassenroot, K., & Huggins, C. (2005). Land, migration and conflict in Eastern DR Congo. *Eco-Conflicts*, 3(4), 1–4.
Whatmore, S. (2002). *Hybrid Geographies: Natures Cultures Spaces*. London: Sage.
Wiedemann, T. (2002). *Emperors and gladiators*. Routledge.
Wong, R. W. Y. (2019). *The Illegal Wildlife Trade in China*. Switzerland: Springer International Publishing.
Wyatt, T. (2013). *Wildlife Trafficking: A Deconstruction of the Crime, the Victims, and the Offenders*. London: Palgrave Macmillan UK.
Wyatt, T., van Uhm, D. P., & Nurse, A. (2020). Differentiating criminal networks in the illegal wildlife trade: Organised, corporate and disorganised crime. *Trends in Organised Crime*, 23(4), 350–366.

PART II

Critical approaches, methods, and reflections on researching wildlife crime

8

ZOOMING-IN, ZOOMING-OUT

Closing the gap between practices in the wildlife trade and associated theories

Janine Heim

Introduction

The complex, scattered, and decentralised nature of the wildlife trade (Collard, 2020, p. 26; Duffy, 2022, pp. 46–51; Nadini et al., 2022) poses significant difficulties for scholars seeking to thoroughly investigate this issue. The situation is further complicated by the lack of data for some species, leading to instances of misunderstandings and incorrect representations of the trade (Challender et al., 2022). I propose the zooming-in, zooming-out approach as a practical and effective method for capturing the intricate complexities of the wildlife trade, whether it is legal or illegal. To illustrate its application, this chapter references my ongoing research on the trade in *Bitis*, a genus of African venomous vipers, though the primary emphasis will be on the methodology utilised and the rationale behind it. Species of the *Bitis* genus face threats like habitat loss, exploitation for food, and international pet trade, leading to population fragmentation ("The IUCN Red List of Threatened Species", 2024). Their conservation status varies from Least Concern to Endangered. The legality of their trade depends on the species, its conservation status, country of origin, market country, and the route it has taken there. For example, in South Africa, home to 12 of the 22 *Bitis* species, stringent permit regulations prohibit the keeping and collection of indigenous species in most provinces and mandate extensive permits for exports in provinces where these are legal, while countries like Togo, home to three *Bitis* species, have more relaxed rules (D'Cruze et al., 2020; Harrington et al., 2021; Wyk, 2023). In Europe, where this research is primarily focused, regulations vary extensively both within and across countries (Amidzich, 2021; Gebhardt-Brinkhaus, 2023).

To explore this trade, I employed the zooming-in, zooming-out approach. This method, which has its origins in organisational research (Nicolini, 2009), emphasises the need to trace the interconnectedness between various practices (Nicolini, 2009, p. 1408). Hence, Nicolini (2009) advocates for oscillating between the execution of specific practices in distinct locations (zooming-in) and exploring their associations and effects when connected in different ways (zooming-out). These connections are upheld through the coordination of both human and nonhuman mediators (Nicolini, 2009, p. 1409). This continuous zooming-in and zooming-out activity is crucial for recontextualising aspects of the practice in question

DOI: 10.4324/9781003333906-10

and highlighting different dimensions of the practice (Nicolini, 2009, pp. 1399–1402). I adapted the zooming-in, zooming-out method to focus on three main areas to make it effective for the study of the wildlife trade: exploring connections across the virtual and physical realms, studying the trade as a whole rather than isolating legal or illegal aspects, and employing multiple theoretical perspectives to describe the practice of trading wildlife. This approach does require the researcher to give considerable thought to the ethics of observing practices, both virtual and offline.

The virtual and the physical

Gorbis (2013) observed that social media was creating new types of networks that form an economy driven by personal relationships and knowledge sharing that was previously gatekept by corporations, governments, or education establishments. This observation took place among the move of brick-and-mortar businesses to digital and e-commerce platforms, which was then expedited by the COVID-19 pandemic (Sirimanne, 2021). This shift, as well as the formation of informal networks, was also apparent in the wildlife trade, which had become a notable presence on online markets well ahead of COVID-19 lockdowns (Krishnasamy & Stoner, 2016; Siriwat & Nijman, 2020; Vietnamese, 2016). While pandemic-induced supply chain issues caused delays in the online trade of exotic pets, it has not been found to curb the trade, with lockdown being advertised as an ideal time to acquire infant animals that require more attention (Morcatty et al., 2021). This development has made online observations a necessity when trying to understand the trade in any wildlife. Research on social media platforms has shown how exotic pet influencers affect the trade (Anagnostou & Doberstein, 2024), as well as how they can form desire for exotic pet ownership (Clarke et al., 2019). However, it is important to remember that what people say and do online isn't an exact reflection of their behaviour in their offline lives (Boellstorff, 2012; Boellstorff & Rysuľová, 2022). This makes following what is observed online into the physical world necessary to produce an accurate picture of the trade. Following practices across both realms will result in a more detailed picture of the trade.

Online observations are often a practical place to start observing if the practitioners of the practice in question are members of public online communities and advertise their expertise, goods, and/or services. This allows researchers to gain insights into individual attitudes, beliefs, and behaviours without direct interference. I also found that algorithms trying to push engagement and the panopticon-like set-ups of many social media platforms further aid in finding relevant groups and individuals engaging in the same practice. As the algorithm recognises these interests after joining a few relevant groups, it will recommend similar groups relevant to the research, ensuring the researcher is at least aware of these communities. Within these groups, the most engaged practitioners typically emerge as administrators, founders, or regular contributors who link to further social media platforms.

Zooming-in on these online groups and their members creates a picture of the interpersonal interactions among the exotic pet trade practitioners, as well as their relationships with the wildlife they trade. The latter is often multifaceted and complex, and understanding whether wildlife is viewed as a product, companion, status symbol, friend, or souvenir provides valuable insights into the practice of trading them. Zooming-out and trailing connections across other online platforms reveals additional explanations of the practice. I zoomed-in on these online groups and individuals and zoomed-out to follow them onto other platforms such as Instagram, which offers more visual material, as well as forums, which offer longer, more detailed discussions among practitioners. Due to the international nature of the trade in exotic pets, online observations also help elucidate the different moral codes each participant operates under, which they interpret and act upon in their unique ways.

I frequently follow connections from the offline to the virtual realm and vice versa during the zooming-in and zooming-out process, as practitioners of the wildlife trade draw on both realms to accomplish the practice of trading *Bitis*. Neglecting to trace these connections and observing only one realm of trade would result in a substantial omission of critical elements that go into accomplishing the practice of trading *Bitis*. What goes into the practice of trading *Bitis* was not always depicted online or was potentially inaccurate. Through online observations, I was able to identify relevant in-person events such as exotic pet expos, as people frequently showcased the animals that they would bring to these events. These events are an illustration of the virtual and offline directly connecting and present an opportunity to experience in-person knowledge exchanges, as well as nuances and tools integral to a practice that may not be discernible online. This is often due to these elements either being hidden from view on social media or being inherently understood without explanations by the practitioners being studied but not by the researcher, as I experienced it.

For instance, despite the widespread popularity of colour morphs in the reptile market (Valdez, 2021), posts about the successful breeding of albino *Bitis rhinoceros* sparked interest but, oddly, did not garner the level of admiration and sharing that other *Bitis* breeding successes or news often did in the *Bitis* community. Later, in-person interviews revealed a partial disapproval and disregard for colour morphs within sections of the *Bitis* community. This disapproval stems from a preference for natural colours and patterns linked to certain localities. This combination of online and in-person observations offered more profound insights into how the value system related to *Bitis* might differ from the overall reptile market.

Some individuals also displayed different practices in the offline world, like showcasing different animals at expos than advertised online or reserving animals for specific buyers, which was not mentioned online. The extension of online observations into in-person observations is crucial not only for this reason but also for collecting material artefacts. These can include magazines, books, handbooks, flyers, and business cards, all of which provide additional context for practices and discourses related to the trade in *Bitis*. These contain information on localised practices that "enter as elements, ingredients, or resources in other activities" (Nicolini, 2009, p. 24).

Comparing how far practices in the trade observed in the virtual realm transfer to practices in the offline realm is another reason why in-person observations still play an important role. What people say and do online is not an exact reflection of their offline lives (Boellstorff & Rysuľová, 2022). In this context, it is helpful to consider research that has shown that the digital and physical worlds are not converging but interacting, meaning that the differentiation between the virtual and offline is crucial to human conditions and social practice (Boellstorff, 2012; Boellstorff & Rysuľová, 2022). In my research, individuals are more inclined to express themselves without self-censorship in person rather than online. I found that this online self-censorship often appeared to stem from a fear of breaching the morals or unspoken rules of the communities within which practitioners of the *Bitis* trade operate. It is important to note that this does not necessarily imply illegal activities, although it can encompass such practices. I, however, found that more often, posters are concerned about contradicting prevalent views held by their community members. These views are sometimes more openly expressed in interviews, another important component to include in the zooming-in process, as having practitioners explain their practice adds further valuable data.

Beyond legality and illegality

Just as online observations or offline observations alone do not paint a complete picture of a practice, the legal and illegal parts of a trade cannot always be neatly separated (Gregson & Crang, 2017; Nijman et al., 2022; Wyatt, 2009). Investigating the legal market in live reptiles can provide insights into trends in the illicit market. For instance, the legal wildlife trade in a country like the United States can strongly predict which species are being smuggled illegally into another Western country, like Australia (Stringham et al., 2021). I, however, found that the observable trade in *Bitis* cannot easily be categorised into clearly legal and illegal elements. For example, a legally traded *Bitis* was found in breach of laws due to the mode of transportation, not because the act of selling or acquiring the *Bitis* was illegal, giving the transaction both legal and illegal aspects. *Bitis*, illegally procured in one nation, may also be legally sold in Germany, where another complicating factor is the significantly different venomous animal ownership laws among federal states (Gebhardt-Brinkhaus, 2023). Hence, most of my observations can be better explained by what scholars in the field of human geography have termed 'customary illegality', a degree of accepted or tolerated illegality in otherwise lawful business, as prevalent in most economies, particularly those involving cross-border transactions (Gregson & Crang, 2017; Gutierrez & Duffy, 2023). This includes practices such as the sale of counterfeit goods, money laundering, the use of forced labour in production, or the acquisition of raw materials in violation of laws, whether deliberately or inadvertently, which are often associated with otherwise legal industries (Gregson & Crang, 2017).

When looking at legality, researchers also need to critically examine the history of legality in the wildlife trade. The mirror thesis describes the theory that the law serves as a formalised reflection of informal social norms, essentially suggesting that legal frameworks embody prevalent societal conventions (Tamanaha, 2001). However, Tamanaha (2001) argued that there exists a fundamental disconnect between law and society for various reasons, such as legal systems representing the interests of those who control political and economic systems, legal systems being imposed by foreign cultures, and legal systems being maintained by legal technocrats. The concept of contested illegality in the wildlife trade can mean actions are not always motivated by profit but rather are a statement to display discontent with certain rules and laws, which can be traced back to colonial times in some parts of the world such as South Africa (Dore et al., 2022; Hübschle, 2017a), where a range of Dwarf *Bitis* originate from. In my interviews, I found that the efficiency, purpose, and fairness of laws surrounding exotic pet ownership and trade were frequently questioned by keepers and breeders. The terrarium size requirements for *Bitis* have frequently been cited as an example of excessive regulation in central Europe. Many have argued that these size requirements, especially the height stipulations, fail to account for some of the species' ground-dwelling and sedentary nature. Despite these criticisms, the majority of individuals appear to comply with these regulations.

In addition to the spectrum of legality and illegality in which trades operate, moral economies, meaning the societal acceptability surrounding economic practices within a 'sphere of circulation' (Gregson & Crang, 2017, p. 207), are an important factor influencing trades. The acceptance of keeping exotic animals as pets, for example, differs depending on the culture and society of the communities involved (Kieswetter, 2017, pp. 5–6). Exotic pet sub-communities influence each other's buying behaviour even when their reasoning for acquiring exotic pets varies, including those who buy exotic pets in bad health out of pity, collectors, those who seek out an unusual pet and those who acquired an exotic pet through circumstance without intending to adopt one (Shukhova & MacMillan, 2020, pp. 392–395). In the *Bitis* trade, keepers and traders interact with their animals differently. Some forge emotional bonds, preserving the remains of deceased *Bitis*, such as the skin or bones, as mementoes. Others exhibit a collector's mentality, seeking various *Bitis* species or focusing on specific ones, alongside other snake genera or animals. Conversely, general exotic pet importers engage with *Bitis* intermittently, showing only a casual appreciation of these snakes.

In addition to the difficulty of separating legal and illegal strands of the trade and changing regulations, individual practitioners have varying awareness and interpretations of the legality of trading and keeping animals (Anagnostou & Doberstein, 2024). This awareness can be observed in online discussions and comments (Anagnostou & Doberstein, 2024, p. 4) as well as in interviews with practitioners of the trade (Shukhova & MacMillan, 2020, p. 941). I found that the more immersed any practitioner was in the trade, the stronger was their awareness of the laws surrounding the trade of *Bitis*. This does not mean there was always agreement with the laws, which were often perceived to be counterproductive in terms of protecting *Bitis* or simply perceived as unnecessary due to success in captive breeding.

Il/legal are also inadequate descriptors when considering that outright bans on trading specific species are rare. This is because zoos and research institutions frequently have the option to apply for exemptions (Melisch et al., 2022; van Uhm & Spapens; Wyatt, 2021). It is also critical to acknowledge that policy shifts can swiftly render the possession of certain animals' illegal, putting existing ownership and animals in transit in a grey area. An illustrative example from the exotic pet trade is the 2022 European Union ban on several animals, including the Kingsnake (*Lampropeltis getula*), due to its status as an invasive species (European Union, 2022). The Kingsnake, available in a variety of artificially bred colours, referred to as 'traits', is a popular exotic pet (van Doorn et al., 2021). I observed the frustration this elicited from the community, who previously were able to keep this snake legally, particularly among those in central, eastern, and northern Europe who argue the improbability of this snake becoming an invasive species in their climate.

Future research on tracing the practice of trading *Bitis* to their countries of origin, which have varying laws around their export, will most likely reveal further views, interpretations, and opinions affecting the practice of trading *Bitis*.

Multi-theoretical examinations

To describe the intricacies of the practices related to the exotic pet trade without relying on, for example, binary concepts such as il/legal, I posit that it is necessary to build what Nicolini (2009, p. 1395) termed a "a toolkit-logic", with which he built the zooming-in, zooming-out framework. Several analytical frameworks are needed to build the vocabulary to both describe and analyse the practice in a way that is still reflective of the reality of what is being done. I found critical criminology approaches (DeKeseredy, 2021; Sollund, 2021) and constructivist security (Buzan et al., 1998; Kirk & McDonald, 2021; Massé & Lunstrum, 2016) to be particularly helpful when examining the trade in exotic pets. Both concepts, respectively, challenge traditional ideas of crime and security and permit the examination of the practice of trading wildlife from both within and outside traditional conceptions of crime and security, as well as the nexus between crime and security.

The research I carried out on the *Bitis* trade was significantly informed by a variety of examples that illustrated the value of critically examining prevailing narratives. As Duffy (2022) illustrates, existing research on wildlife trade can yield considerable insights, especially in light of the growing criminalisation and securitisation of this sector over the past decade. This is evident in South Africa, home of many *Bitis* species, where narratives linking organised crime and wildlife poaching have influenced both national and international security discourses (Duffy et al., 2019; Hübschle, 2011; Wittig, 2017). This dominant narrative on the wildlife trade in Africa was felt during my research, as most interviewees assumed I would exclusively be interested in the illegal trade or even looking for organised crime links.

The role of colonial powers in the wildlife trade, particularly in South Africa, has also been a significant consideration in my research. The intertwining of anti-poaching laws with colonialism and the exploitation of African peoples (Hübschle, 2017b) has been a recurring theme. The practices of colonial powers, such as the extraction of wildlife from their reserves for display as status symbols among European middle and upper classes (Aravamudan, 1999; Robbins, 2002; Simons, 2014), have been critical to my understanding of the historical context of the *Bitis* trade, with evidence I found in the Zoological Society of London archive of Western zoos requesting *Bitis* for their collection dating back at least to the 1940s. Indeed, in my observations, the most prominent members in the *Bitis* trade still appear to be comparatively affluent, white, male Europeans.

Conservation laws have been another critical area of focus in my research. For instance, conservation measures in the Maya Biosphere Reserve in Guatemala have been criticised for prioritising conservation issues over human rights and using conservation as a tool for social control (Lunstrum & Ybarra, 2018). This has been a particularly useful insight in my research, encouraging a more critical examination of conservation concerns surrounding the *Bitis* trade and clashes between private breeders and NGOs.

Drawing on these diverse examples and critical approaches, I have been able to develop a comprehensive understanding of the *Bitis* trade from multiple perspectives. This has enabled me to engage in meaningful discussions with participants and has facilitated the collection of diverse perspectives on existing issues. Additionally, these discussions have provided valuable recommendations for first-hand observations, which have been instrumental in informing my primary research.

This critical examination should also be directed inwards, towards the researcher's own preconceived notions about the trade they examine. Researchers themselves are not immune to the politicisation, criminalisation, and securitisation of many aspects of trading wildlife, which has caused tension between researchers and practitioners in the exotic pet trade, the latter of whom I found will protect themselves through opaqueness. Some academic discourse in this area has been implicated in the misinterpretation of wildlife trade data due to bias in data sets or sources, resulting in misrepresentations, especially pertaining to illegal activities in the wildlife trade space (Challender et al., 2022). Part of the reason for misrepresentation is because of a lack of care when it comes to using the appropriate language and labels to describe a trade (Challender et al., 2022). Bruskotter et al. (2019) also found that group identity played a significant role in social conflict in the conservation space, especially among those who consider themselves animal rights advocates or hunters, with both groups also forming vocal parties in conservation discourse. These issues illustrate not only the importance of multi-theoretical examinations to break out of binary frameworks that are counterproductive to accurately describing the trade but also the need for researchers to examine their own perspectives and allegiances.

The ethics of observing practices

How biases might frame a researcher's perceptions of practices in the wildlife trade, both when zooming-in on practices and when zooming-out to examine the wider network and effects of individual practices, is an important consideration. The risk of bias in research is not restricted to misrepresentations of the trade but may also impact ethical decision-making in the field. While researchers tied to organisations generally have some ethical standards to abide by, organisations are not always quick to catch up to the variety of social media and technology available now (Thompson et al., 2021). Whether online or offline, they do not necessarily consider the rights of groups that could be regarded as vulnerable, which is a subjective concept that can change depending on the context, situation and individual characteristics (Shaw et al., 2020).

Irrespective of an individual's perceived vulnerability, whether a person can reasonably expect privacy should be an important consideration that should not be impacted by the researcher's personal views on the practices observed. For example, at markets or other public events, sellers would not expect to know who is browsing their stands, and researchers do not necessarily have to disclose their identity when looking at what is displayed. But posing as a buyer to initiate interactions would interfere with informed consent. The same considerations should be applied to observations in the virtual world. When people actively advertise their products, such as wildlife, and share their own social media content widely on public platforms, there is not necessarily an expectation of privacy (Kozinets, 2023). However, when joining closed forums, groups, or chats, this expectation of privacy must be a consideration (Kozinets, 2023). While researchers may perceive all publicly accessible information as authored works under public scrutiny, this ignores the necessity of an online presence for many people, as non-participation in online activities can lead to social exclusion and even economic suppression (Bossio & McCosker, 2021; Büttner & Rudert, 2022). In the case of any form of closed group, I found announcing your presence to moderators or the group when there is any doubt is a good practice that demonstrates respect and transparency. This recommendation presumes that the researcher does not enter situations in which they could face danger for revealing who they are.

In my experience, while many practitioners will decline any interaction once you disclose your identity as a researcher, there will still be people willing to let you observe, consent to interviews or informal chats and even suggest other interviewees or avenues for observation. This even applies to not strictly legal activities. Disclosing your identity is also preferable to practitioners theorising about who you are. Unless the researcher is already a member of the groups they study, anything from the way they dress, talk and/or the kind of questions they ask will quickly reveal themselves as an outsider and without any explanation of this interest in the trade by an outsider, this will quickly elicit distrust. I found that it is preferable to be known as an academic researcher than being suspected of being a journalist, activist, or law enforcement in disguise. It is important to acknowledge that, even with the best intentions, explaining what you are studying can be difficult, and the truth is that it can be vague until researchers find what exactly they are looking for. In addition that "everything is capable of multiple interpretations and misunderstandings" (Fine & Shulman, 2009, p. 184). Because of the possibility of these unintentional deceptions, aiming for transparency is even more important. I found that transparency in all interactions, including those on social media accounts about the researcher's identity and the research itself, helps in building relationships.

Conclusion

The zooming-in and zooming-out process provides a promising approach to a more accurate representation of practices in the wildlife trade. By zooming-in, we can examine individual attitudes, beliefs, and behaviours within communities, gaining insights into the multifaceted relationships between practitioners and their practices. Conversely, zooming-out enables us to trace connections across different locations (virtual and physical) and explore broader themes and trends within the practice. The integration of online observations with physical interactions is key to understanding the wildlife trade's ecosystem and reveals the complexity and motivations of the trade participants. Acknowledging the blurred lines between legal and illegal trade, influenced by factors like historical legality, motivations, and 'customary illegality' is essential. A multi-theoretical approach is beneficial, inviting critical criminology and constructivist security perspectives. This approach challenges existing notions of illegality, morality, and security threats. During the research, self-reflection on biases and preconceptions is crucial, especially considering the need for transparency and respect for privacy.

While this approach means that there is a lot of data to digest for just one species, there is a deluge of data found following this approach. The result, however, promises to be holistic and as close as possible to the reality of the trade. This approach not only uncovers the implicit rules and norms within these communities but also gathers concrete evidence that augments our understanding of the trade in specific species. This enhances the accuracy and depth of our understanding of practices, providing a more comprehensive picture of the complex dynamics at play. Based on my preliminary findings, this method has yielded tangible insights into how the practice of trading *Bitis* is accomplished, including insights into not just what goes into the practice but also who is part of the practice. Future endeavours to further test this method will include connecting localised practices across different geographic regions in the trade in *Bitis* through in-person observations in source countries. The resulting findings aim to further contribute to the tools available in wildlife trade research and bolster the depth of our understanding in this field of study.

References

Amidzich, M. R. (2021). Stop pur-petuating the norm: Amending the lacey act to include a "dangerous or potentially dangerous wildlife" definition for exotic pet protection. *Vermont Journal of Environmental law, 23*, 56.

Anagnostou, M., & Doberstein, B. (2024). Exotic pet trade in Canada: The influence of social media on public sentiment and behaviour. *Journal for Nature Conservation, 77*, 126522. https://doi.org/10.1016/j.jnc.2023.126522

Aravamudan, S. (1999). *Tropicopolitans: Colonialism and Agency, 1688–1804*. Duke University Press.

Boellstorff, T. (2012). Rethinking digital anthropology. *Digital Anthropology, 39*, 60.

Boellstorff, T., & Rysuľová, A. (2022). The perception of culture in the virtual worlde. *Acta Ludologica, 5*(2), 104–107.

Bossio, D., & McCosker, A. (2021). Reluctant selfies: Older people, social media sharing and digital inclusion. *Continuum, 35*(4), 634–647.

Bruskotter, J. T., Vucetich, J. A., Dietsch, A., Slagle, K. M., Brooks, J. S., & Nelson, M. P. (2019). Conservationists' moral obligations toward wildlife: Values and identity promote conservation conflict. *Biological Conservation, 240,* 108296. https://doi.org/10.1016/j.biocon.2019.108296

Büttner, C. M., & Rudert, S. C. (2022). Why didn't you tag me?!: Social exclusion from instagram posts hurts, especially those with a high need to belong. *Computers in Human Behavior, 127,* 107062. https://doi.org/10.1016/j.chb.2021.107062

Buzan, B., Wæver, O., & Wilde, J. d. (1998). *Security: A New Framework for Analysis*. Lynne Rienner Pub.

Challender, D. W. S., Brockington, D., Hinsley, A., Hoffmann, M., Kolby, J. E., Massé, F., Natusch, D. J. D., Oldfield, T. E. E., Outhwaite, W., 't Sas-Rolfes, M., & Milner-Gulland, E. j. (2022). Mischaracterizing wildlife trade and its impacts may mislead policy processes. *Conservation Letters, 15*(1), e12832. https://doi.org/10.1111/conl.12832

Clarke, T. A., Reuter, K. E., LaFleur, M., & Schaefer, M. S. (2019). A viral video and pet lemurs on Twitter. *PLoS One, 14*(1), e0208577.

Collard, R.-C. (2020). *Animal Traffic: Lively Capital in the Global Exotic Pet Trade*. Duke University Press.

D'Cruze, N., Harrington, L. A., Assou, D., Green, J., Macdonald, D. W., Ronfot, D., Segniagbeto, G. H., & Auliya, M. (2020). Betting the farm: A review of ball python and other reptile trade from Togo, West Africa. *Nature Conservation, 40.* https://ora.ox.ac.uk/objects/uuid:4eda3cbb-e875-47f2-bdb1-383ad5789219

DeKeseredy, W. S. (2021). *Contemporary Critical Criminology*. Routledge.

Dore, A., Hübschle, A., & Batley, M. (2022). Towards environmental restorative justice in South Africa: How to understand and address wildlife offences. In Pali, B., Forsyth, M., Tepper, F. (eds.), In *The Palgrave Handbook of Environmental Restorative Justice* (pp. 333–359). Springer.

Duffy, R. (2022). Crime, security, and illegal wildlife trade: Political ecologies of international conservation. *Global Environmental Politics, 22*(2), 23–44. https://doi.org/10.1162/glep_a_00645

Duffy, R., Massé, F., Smidt, E., Marijnen, E., Büscher, B., Verweijen, J., Ramutsindela, M., Simlai, T., Joanny, L., & Lunstrum, E. (2019). Why we must question the militarisation of conservation. *Biological Conservation, 232,* 66–73.

European Union (2022). *Amending Implementing Regulation (EU) 2016/1141 to update the list of invasive alien species of Union concern*. https://eur-lex.europa.eu/legal-content/EN/TXT/?uri=OJ%3AL%3A2022%3A186%3ATOCfiles/292/TXT.html

Fine, G. A., & Shulman, D. (2009). *Lies From the Field: Ethical Issues in Organizational Ethnography*. Organizational ethnography: Studying the complexities of everyday life, 177–195.

Gebhardt-Brinkhaus, R. (2023). *Überblick über die rechtlichen Regelungen zur Gift- und Gefahrtierhaltung in den Ländern der Bundesrepublik*. ASPE-Institut GmbH.

Gorbis, M. (2013). *The Nature of the Future: Dispatches from the Socialstructed World*. Simon and Schuster.

Gregson, N., & Crang, M. (2017). Illicit economies: Customary illegality, moral economies and circulation. *Transactions of the Institute of British Geographers, 42*(2), 206–219.

Gutierrez, L., & Duffy, R. (2023). Harms and the illegal wildlife trade: Political ecology, green criminology and the european eel. *Critical Criminology*, 32, 1–16.

Harrington, L. A., Auliya, M., Eckman, H., Harrington, A. P., Macdonald, D. W., & D'Cruze, N. (2021). Live wild animal exports to supply the exotic pet trade: A case study from Togo using publicly available social media data. *Conservation Science and Practice, 3*(7), e430.

Hübschle, A. (2011). From theory to practice: Exploring the organised crime-terror Nexus in Sub-Saharan Africa. *Perspectives on Terrorism, 5*(3/4), 81–95. https://www.jstor.org/stable/26298525?seq=1

Hübschle, A. (2017a). Contested illegality. *The Architecture of Illegal Markets: Towards an Economic Sociology of Illegality in the Economy*, 177–197. Oxford University Press.

Hübschle, A. (2017b). The social economy of rhino poaching: Of economic freedom fighters, professional hunters and marginalized local people. *Current Sociology, 65*(3), 427–447.

Kieswetter, S. (2017). *The Motivations Behind Obtaining Exotic Pets: A Discussion Paper*. Zoocheck inc.

Kirk, J., & McDonald, M. (2021). The politics of exceptionalism: Securitization and COVID-19. *Global Studies Quarterly, 1*(3), ksab024.

Kozinets, R. V. (2023). Immersive netnography: A novel method for service experience research in virtual reality, augmented reality and metaverse contexts. *Journal of Service Management, 34*(1), 100–125.
Krishnasamy, K., & Stoner, S. (2016). TRADING FACES: A rapid assessment on the use of Facebook to trade wildlife in Peninsular Malaysia (PDF, 2.3 MB).
Lunstrum, E., & Ybarra, M. (2018). Deploying difference: Security threat narratives and state displacement from protected areas. *Conservation and Society, 16*(2), 114–124.
Massé, F., & Lunstrum, E. (2016). Accumulation by securitization: Commercial poaching, neoliberal conservation, and the creation of new wildlife frontiers. *Geoforum, 69*, 227–237.
Melisch, R., Xu, L., Badola, S., Compton, J., & Ziegler, S. (2022). *An Overview of the Threats to Red Pandas Posed by Domestic and International Trade*. Red Panda, 489–494.
Morcatty, T. Q., Feddema, K., Nekaris, K. A. I., & Nijman, V. (2021). Online trade in wildlife and the lack of response to COVID-19. *Environmental Research, 193*, 110439.
Nadini, M., Bracci, A., ElBahrawy, A., Gradwell, P., Teytelboym, A., & Baronchelli, A. (2022). Emergence and structure of decentralised trade networks around dark web marketplaces. *Scientific Reports, 12*(1), 5425.
Nicolini, D. (2009). Zooming in and out: Studying practices by switching theoretical lenses and trailing connections. *Organization studies, 30*(12), 1391–1418.
Nijman, V., Morcatty, T. Q., Feddema, K., Campera, M., & Nekaris, K. A. I. (2022). Disentangling the legal and illegal wildlife trade–insights from Indonesian wildlife market surveys. *Animals, 12*(5), 628. https://doi.org/10.3390/ani12050628
Robbins, L. E. (2002). *Elephant Slaves and Pampered Parrots: Exotic Animals in Eighteenth-Century Paris*. JHU Press.
Shaw, R. M., Howe, J., Beazer, J., & Carr, T. (2020). Ethics and positionality in qualitative research with vulnerable and marginal groups. *Qualitative Research, 20*(3), 277–293. https://doi.org/10.1177/1468794119841839
Shukhova, S., & MacMillan, D. C. (2020). From tigers to axolotls: Why people keep exotic pets in Russia. *People and Nature, 2*(4), 940–949.
Simons, J. (2014). The scramble for elephants: Exotic animals and the imperial economy. In M. Boyde (ed.), *Captured: The Animal within Culture* (pp. 26–42). Springer.
Sirimanne, S. (2021). *COVID-19 and e-commerce: A global review*. UNCTAD. https://unctad.org/publication/covid-19-and-e-commerce-global-review
Siriwat, P., & Nijman, V. (2020). Wildlife trade shifts from brick-and-mortar markets to virtual marketplaces: A case study of birds of prey trade in Thailand. *Journal of Asia-Pacific Biodiversity, 13*(3), 454–461. https://doi.org/10.1016/j.japb.2020.03.012
Sollund, R. (2021). Green criminology: Its foundation in critical criminology and the way forward. *The Howard Journal of Crime and Justice, 60*(3), 304–322.
Stringham, O. C., García-Díaz, P., Toomes, A., Mitchell, L., Ross, J. V., & Cassey, P. (2021). Live reptile smuggling is predicted by trends in the legal exotic pet trade. *Conservation Letters, 14*(6), e12833.
Tamanaha, B. Z. (2001). *A General Jurisprudence of Law and Society*. Oxford University Press. https://www.google.co.uk/books?id=g6Dn5jVTaY0C
The IUCN Red List of Threatened Species. (2024). *IUCN Red List of Threatened Species*. https://www.iucnredlist.org/search?query=Bitis&searchType=species
Thompson, R. M., Hall, J., Morrison, C., Palmer, N. R., & Roberts, D. L. (2021). Ethics and governance for internet-based conservation science research. *Conservation Biology, 35*(6), 1747–1754. https://doi.org/10.1111/cobi.13778
Valdez, J. W. (2021). Using Google trends to determine current, past, and future trends in the reptile pet trade. *Animals, 11*(3), 676.
van Doorn, L., Speybroeck, J., Brys, R., Halfmaerten, D., Neyrinck, S., Engelen, P., & Adriaens, T. (2021). Aesthetic aliens: Invasion of the beauty rat snake, Elaphe taeniura Cope, 1861 in Belgium, Europe. *BioInvasions Records, 10*(3), 741–754.
van Uhm, D., & Spapens, T. (2020). *Illegal Trade in Protected Birds in the Netherlands*. Boom uitgevers Den Haag.
Vietnamese, E. (2016). A rapid assessment of e-commerce wildlife trade in Viet Nam. *TRAFFIC Bulletin, 28*(2), 53.

Wittig, T. (2017). *Poaching, wildlife trafficking and organised crime. In Poaching, Wildlife Trafficking and Security in Africa* (pp. 77–101). Routledge.
Wyatt, T. (2009). Exploring the organization of Russia Far East's illegal wildlife trade: Two case studies of the illegal fur and illegal falcon trades. *Global Crime, 10*(1–2), 144–154.
Wyatt, T. (2021). *Is CITES Protecting Wildlife? Assessing Implementation and Compliance*. Routledge.
Wyk, S. V. (2023). *Snake Conservation. African Snakebite Institute*. https://www.africansnakebiteinstitute.com/articles/snake-conservation-2/

9

A FRAMEWORK FOR CONDUCTING RESEARCH ON CONSUMER DEMAND FOR HIGH-VALUE WILDLIFE PRODUCTS IN ASIA

Dang Vu Hoai Nam

Introduction

The consumption of high-value wildlife products, including those illegally harvested from endangered species, is deeply rooted in Asian culture and traditions and often involves little or no stigma (Cheung et al., 2021a; Truong, Dang, and Hall, 2015; Swan and Conrad, 2014). Consumer demand for these products has increased significantly over the past two decades due to economic development, leading to the emergence of a new affluent class in some Asian countries (Dang et al., 2022). Asian consumers use exotic wildlife products for different purposes, notably traditional medicine, food, decoration, cultural, or spiritual purposes (Knight, 2004). For instance, tiger bone has been used in traditional medicine in many Asian countries (e.g. China, Japan, South Korea, Vietnam) for centuries to treat bone-related diseases such as arthritis and rheumatism (Dinerstein et al., 2007). Rhino horn is used to treat a number of diseases and health conditions, such as high fever, body detoxification (Dang and Nielsen, 2018), and it is considered one of the collectible items in the Chinese art and antiques market (Gao et al., 2016). Chinese consumers also collect decorative items from ivory, mila amber, agarwood, and rosewood (Zhu, 2020; Gao et al., 2016). Pangolin is served as a delicacy food in wildmeat restaurants, while pangolin scales are used in traditional medicine with perceived benefits in stimulating human milk secretion, treating infertility, liver diseases, and tumours (Wang, Turvey, and Leader-Williams, 2020). Besides utilitarian purposes, high-value wildlife products are also used for hedonic purposes, for instance, showing wealth and social status or sharing within social networks and clubs to search for pleasures among wealthy males (Dang and Nielsen, 2018; Truong, Dang, and Hall, 2015). Despite the expansion of wildlife farms to meet demand, wealthy consumers have a strong preference for wild-sourced products, including those from illegal supplies (Dang et al., 2022; Dang, Nielsen, and Jacobsen, 2022). The escalating demand and evolving consumer preferences have exacerbated the poaching crisis in both Asia and Africa, posing a significant threat to species conservation (Moorhouse et al., 2020; 't Sas-Rolfes et al., 2019).

DOI: 10.4324/9781003333906-11

Managing demand has therefore been identified as an urgent conservation priority, yet it requires a thorough understanding of consumers and the context in which wildlife consumption is embedded (Thomas-Walters et al., 2021). Various approaches have been employed to influence demand, but evidence of success is limited (Veríssimo et al., 2020). Demand reduction interventions have been criticised for being based on delusion rather than evidence, and campaign designs often do not enable rigorous impact assessments (Dang and Nielsen, 2021; Greenfield and Veríssimo, 2019; Olmedo, 2018). Understanding consumer demand for high-value wildlife products represents a significant challenge. Despite numerous studies aimed at gaining insights into demand, each study typically relies on a specific theoretical framework and data collection method, such as qualitative or quantitative approaches. There is a lack of a cohesive methodological approach to conducting research on wealthy consumers and their demand.

Drawing from my extensive experience conducting interviews with over 1,000 wealthy consumers of rhino horn and tiger bone in Vietnam (Dang, 2021), this chapter is dedicated to developing a framework for studying consumer demand for high-value wildlife products in Asia. It proceeds as follows: the next section sets the scene by providing insights into the complex and multifaceted concept of demand for high-value wildlife products. Then, I discuss demand reduction approaches and challenges in understanding high-income wildlife consumers. This chapter is followed by a framework for generating insights into demand, which includes both qualitative and quantitative techniques. I conclude this chapter with some thoughts about potential applications and further development of the framework.

Demand for high-value wildlife products

Consumer demand is an integral part of wildlife markets. Addressing the illegal and unsustainable wildlife trade (henceforth IUWT) requires an in-depth understanding of demand and the context in which wildlife consumption occurs (Veríssimo et al., 2020; Veríssimo, 't Sas-Rolfes, and Glikman, 2020). The markets for high-value wildlife products are often characterised by a lack of information about supply, demand, and prices (Nadal and Aguayo, 2014). Demand is a complex and multifaceted concept, which is defined with variations across disciplines and over time. In classical economics, demand is defined as the amount of a product that consumers are willing and able to purchase over a price range in a given time period, which is often referred to as the 'law of demand' (Hildenbrand, 1983). This definition regards demand as a sequence of rational choices, rather than encapsulating the entire decision-making process (Veríssimo et al., 2020). Besides price, demand is influenced by an array of factors that are internal or external to consumers, as well as irrational choices driven by status symbols, social pressure, misinformation, superstition and beliefs (Hübschle, 2016). For instance, the demand for tiger bone in Vietnam is driven by the health condition of consumers and their relatives, as well as the influence of peer users (Dang et al., 2022). Demand for agarwood and rosewood in China is driven by traditions and cultural factors (Zhu, 2020). In Yemen, the jambiya, a traditional ceremonial dagger, carved out of rhino horn, has historically been used as a status symbol and cultural emblem among Yemeni men of high social status (Hübschle, 2016).

Fischer (2004) identified two demand externalities – stigma and outrage. The former arises from the illegal and inhumane nature of the wildlife trade, while the latter originates in altruism and depends on the size of the illegal activity (Fischer, 2004). Other factors that may affect high-value wildlife demand include the product's rarity, i.e. the anthropogenic Allee effect (Hausmann, Cortés-Capano, and Di Minin, 2023; Hall, Milner-Gulland, and Courchamp, 2008), product availability and its substitutes (Chen and 't Sas-Rolfes, 2021), consumer preferences related to legality, quality, and harvesting methods (Dang et al., 2022; Dang, Nielsen, and Jacobsen, 2022; Wyatt et al., 2021; Hanley et al., 2018), alongside income levels and perceived benefits associated with a particular product (Cheung et al., 2021b, c; Hinsley and 't Sas-Rolfes, 2020; Dang and Nielsen, 2018).

The distinction between high-value and regular or low-value wildlife products can be fluid, as the luxury status and pricing of a product often fluctuate over time, with bear bile as an example (Davis et al., 2022). Once highly coveted as a precious traditional medicine, bear bile has seen a drastic decline in value owing to the proliferation of bear farms in Vietnam. Currently, farmed bear bile is considered a normal product, while wealthy consumers still pay a premium for bile extracted from Siberian or polar bears. The primary distinction between high-value and low-value products lies in the fact that high-value items are not essential for subsistence purposes, such as food or other necessities, while low-value products lack the status-conferring function (Ayling, 2016).

Determining the demand of a specific product can become intricate when the product serves different purposes and caters to consumers of different income groups. To exemplify this complexity, I consider rhino horn and tiger products as two examples. The rhino horn market comprises multiple products with varying prices, and it can be segmented into three categories – top price, middle price, and low price (Patton and Ammann, 2016). Very wealthy consumers often acquire whole rhino horns, rhino horn antiques and crafts, or horns from Asian rhinos, whereas those of considerable means opt for smaller portions or several hundred grams of African rhino horns to share with others in special events to confer status and search for pleasure (Truong, Dang, and Hall, 2015). Rhino horn powder and off-cuts are supplied at reduced prices to the traditional medicine market, catering to regular consumers seeking personal use or aiding relatives who are suffering from terminal illnesses (Dang and Nielsen, 2018). In the case of tigers, every part and derivative of the animal can be traded within the black market. Tiger skin is used for home decoration, while tiger bone is processed into glue, a traditional medicine to treat bone-related ailments (Dang et al., 2022). Hence, it becomes imperative to comprehend the distinct consumer groups, their demand, and the socioeconomic and cultural contexts that underpin this demand.

An essential aspect of high-value wildlife markets is that demand is not solely driven by consumers (Ayling, 2016). Traders may play a significant role in shaping demand. For instance, the spread of a rumour exemplifies the tactics employed by traders to stimulate demand. Traders always want to sell their products at higher prices, but this is not an easy task. To bolster demand, they frequently embellish the product's advantages, constructing narratives highlighting its scarcity or endorsement by influential figures. Demand is also different between law-abiding and law-ignoring consumers. Law-abiding consumers only buy wildlife products from legal sources, whereas law-ignoring consumers buy from both legal and illegal sources (Fischer, 2004).

Approaches to reducing demand

Demand-side strategies are increasingly recognised as indispensable to tackle the IUWT. Previous studies have categorised approaches to reducing demand for illegal and unsustainable wildlife products ('t Sas-Rolfes et al., 2019; Veríssimo and Wan, 2019; Wallen and Daut,

2018). Here, I do not aim to conduct a systematic review of demand reduction approaches. I discuss the main approaches that have been employed to influence demand for high-value wildlife products, which comprise legal measures and behaviour change interventions. Legal measures include bans and restrictions on the purchase, consumption, or possession of wildlife products ('t Sas-Rolfes et al., 2019). For instance, Vietnam has increased legal sanctions on the consumption or possession of rhino horn, ivory, and tiger products since 2018 (Vietnam's Penal Code 2018, amended in 2017). China banned the auction of rhino horn and ivory products in 2011. Following a national crackdown on corruption and extravagance in late 2012, government officials in China were not allowed to receive luxury gifts (Smith, 2016). Shark fin soup, bird's nest soup, and other high-value wild animal products were banned from official events of the Chinese government (Gao, 2013). Nevertheless, the effectiveness of legal measures and regulations in curbing the demand for high-value wildlife products has faced scepticism due to limited evidence demonstrating their impact. Critics often label these measures as "eye watering" or "window dressing," suggesting they are more for show than practical implementation or substantial change (Ammann, 2012).

In addition to legal measures, efforts have been made to change consumer behaviour voluntarily to manage demand. These behavioural change interventions draw from social and behavioural sciences, employing diverse approaches, methods, and tools. They encompass various labels, including awareness-raising activities, outreach initiatives, social influence strategies, environmental education programmes, behavioural economics interventions, and social marketing campaigns (Thomas-Walters et al., 2020; Wallen and Daut, 2018). However, evidence of success is scant given that these activities often do not follow rigorous impact assessment methods (Dang, Nielsen, and Jacobsen, 2020; Veríssimo and Wan, 2019). Among the very few examples of behaviour change interventions with certain success is the Rare Pride campaign on the island of Bonaire to reduce the local demand for pet Lora (*Amazona barbadensis*), a threatened parrot (Salazar, Mills, and Veríssimo, 2019). Whereas campaigns targeting high-value wildlife consumers have been criticised for being based on unreliable insights into consumers, not using behavioural theories, and using messages and message-deliverers that are not evidence-based (Dang and Nielsen, 2021; Olmedo et al., 2018; Duthie et al., 2017). Therefore, success is often questionable (Vu, 2023; MacFarlane et al., 2022; Margulies, Wong, and Duffy, 2019; Roberton, 2014), and interventions sometimes involve unintended consequences (Thomas-Walters et al., 2020). For instance, the framing of rhino horn as fingernail in demand reduction campaigns has infuriated the Vietnamese government and consumers because they see it as untrue and consider it an oversimplification of the science (Smith, 2018).

An alternative approach to reducing wildlife demand involves addressing the supply side of the wildlife trade chain. This strategy focuses on creating synthetic products, such as synthetic rhino horn, that mimic the biological or chemical properties of those derived from wild fauna and flora (Chen and 't Sas-Rolfes, 2021). It aims to offer consumers comparable substitutes without relying on products sourced from endangered or threatened species. Despite its potential, the strategy of creating synthetic alternatives is contentious. Proponents argue that flooding the market with affordable synthetic products could meet consumer demand, reducing reliance on wild-sourced products and deterring poaching (Mi, Shao, and Vollrath, 2019). Conversely, opponents express concern that introducing synthetic alternatives would reduce the stigma and amplify demand for wild-sourced products. A legal supply of synthetic alternatives could also facilitate the laundering of products illegally harvested from the wild, exacerbating poaching (Chen, 2017).

Studies investigating the impact of synthetic products on wildlife demand mostly draw data from the general public or university students rather than actual consumers (e.g. Rock and MacMillan, 2022; Davis et al., 2016). These studies provide inconsistent results, and limited implications can be developed to inform policies or behaviour change interventions. It is also assumed that synthetic products and genuine ones are perfect substitutes (Mi, Shao, and Vollrath, 2019). This assumption lacks empirical evidence. For instance, if consumers try a synthetic rhino horn but find it lacks medicinal effects, the initiative might not succeed. Further research is essential to understand the potential impact of synthetic alternatives on demand, employing robust methods and collecting data directly from real consumers to gauge their preferences and behaviour more accurately.

Other market-based interventions including the introduction of a legally controlled trade for sustainably farmed products, the sale of large stockpiles of confiscated products or dead plants or animals, incremental releases of small batches, or episodic auctions are based on the same argument that these would devalue wildlife products and reduce poachers' incentives (Moorhouse et al., 2020; 't Sas-Rolfes et al., 2019). Again, evidence supporting these initiatives is very limited. Dang, Nielsen, and Jacobsen (2022) and Dang et al. (2022) conducted choice experiments with high-income consumers of rhino horn and tiger bone in Vietnam and found a large heterogeneity in consumer preferences. While consumers generally showed support for a legal trade, higher-income consumers strongly preferred wild-sourced products, including those from illegal supplies. Behaviour change interventions and market-based initiatives should rely on empirical evidence, which has not been adequately collected thus far.

Understanding consumers of high-value wildlife products

Designing effective demand management strategies necessitates a profound comprehension of consumers, their attitudes, behaviours, beliefs, preferences, trade-offs, as well as the socioeconomic, cultural, and historical contexts that influence the consumption of high-value wildlife products (Veríssimo et al., 2020). Demand for high-value wildlife products mainly comes from a specific group of wealthy consumers, who are often averse to investigations of their sensitive behaviours (Dang, 2021). Numerous surveys coordinated by conservation organisations have contributed to informing demand reduction campaigns, yet a predominant portion of the data originates from the general public. A minute fraction of respondents in these surveys actually comprises buyers or consumers (e.g. Cheung et al., 2021a, b, c; Davis et al., 2019, 2016; Olmedo et al., 2021; USAID Vietnam, 2018; USAID Wildlife Asia, 2018; TRAFFIC, 2017). Consequently, these surveys mostly target individuals without a demand for high-value wildlife products. Insights from these surveys are often perceived as speculative rather than evidential (Dang and Nielsen, 2021).

Collecting data from wealthy consumers represents a significant challenge to researchers. Although specialised questioning techniques and bias-mitigation tools have been developed and empirically tested (Arias et al., 2020; Nuno and St. John, 2015), they prove to be less effective in interviewing wealthy consumers (Dang, 2021). Data collection requires, in the first instance, identifying potential respondents who match the profile of current or future consumers and persuading them to participate in interviews. Yet, wealthy consumers are unlikely to spend time filling out online questionnaires or talking to enumerators over the phone. Neither are they motivated by small gifts or abstract benefits such as species conservation. Making an appointment with them is usually the only way to collect data via face-to-face interviews. In Confucian societies such as China or Vietnam, seniors and high-income individuals often do not want to keep company with younger people or those of lower income groups. And staging such interviews requires a certain level of trust (Dang, 2021; Wong, 2019; Hübschle, 2016).

A framework for conducting research to inform demand management interventions

Changing consumer behaviours to manage demand is the focus of many conservation organisations and projects (Wallen and Daut, 2018). Several behaviour change methods and frameworks have been established to guide these efforts, such as the Wildlife Consumer Behaviour Change Toolkit developed by TRAFFIC. While these frameworks draw from various disciplines and encompass illicit wildlife products broadly, none specifically targets high-value wildlife products. The challenge in understanding wealthy consumers of these products remains unresolved. This section proposes a framework for researching high-value wildlife consumers in Asia to inform demand management strategies. My research on over 1,000 wealthy consumers of rhino horn and tiger bone in Vietnam (Dang, 2021) serves as a practical illustration showcasing the application of this framework. The framework comprises three main stages: (i) literature review and key informants, (ii) qualitative interviews, and (iii) surveys (Figure 9.1).

The initial stage entails conducting a comprehensive literature review to establish the foundational understanding of the trade and consumption related to a particular species. This involves delving into existing studies exploring socioeconomic, cultural, and historical aspects, aiming to comprehend consumer behaviours and demand while pinpointing research gaps. For instance, reviewing the literature about the rhino horn trade and consumption, we found that motivations for this product are different between China and Vietnam. Additionally, multiple clinical studies investigating rhino horn's medicinal benefits as a traditional medicine have been documented (Dang and Nielsen, 2021). To supplement preliminary insights garnered from the literature review, conducting semi-structured or in-depth interviews with key informants, including experts and actual consumers who have engaged with the product, offers valuable perspectives and insights into consumer motivations, behaviours, beliefs, preferences, and the contextual factors influencing consumption (Nguyen et al., 2020). The research conducted by Rebecca Wong or Annette Hübschle serves as a prime illustration (Wong, 2019; Hübschle, 2016). The first stage lays the groundwork for designing consumer research, shaping research inquiries, establishing theoretical

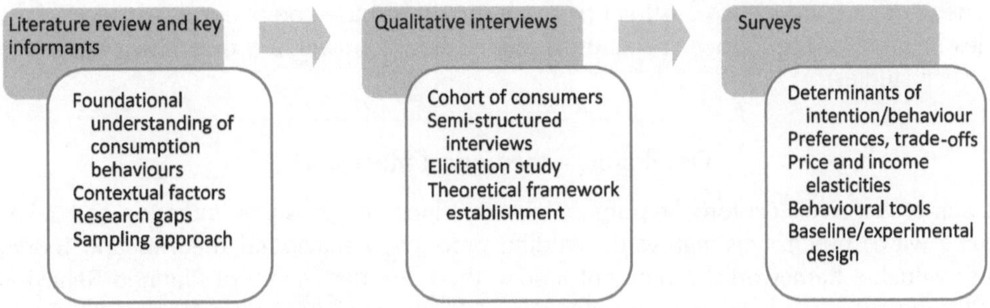

Figure 9.1 A framework for conducting research on high-value wildlife consumers.

frameworks, determining methods, and selecting sampling approaches. Qualitative insights from this stage also serve as potent resources for informing targeted demand management interventions and serve as foundational material for designing subsequent quantitative studies with larger sample sizes (Dang, Nielsen, and Jacobsen, 2020).

During the second stage, collaborating with researchers or research assistants entrenched within the same networks as the consumers becomes essential. This facilitates the identification of potential respondents and the establishment of an initial cohort for interviews. However, recruiting suitable research assistants presents challenges, particularly for researchers situated in different countries and unfamiliar with the local language. My own efforts led to the recruitment of ten research assistants, each with diverse levels of research experience, to conduct interviews with wealthy consumers in Vietnam, but only three displayed strong performance (Dang, 2021). Here, elicitation studies can be conducted to explore the underlying factors shaping demand and to construct a relevant theoretical framework alongside quantitative methods. I have conducted two elicitation studies, which include in-depth interviews and focus groups with 10–15 consumers and former traders of rhino horn and tiger bone. Participants were asked open-ended questions about what factors affect, facilitate, and influence their intention to buy these products, and prominent factors were recorded (Dang et al., 2025; Dang and Nielsen, 2022).

The third stage centres on broadening the initial sample size through methods like convenience sampling or snowballing, enabling the collection of statistically inferable results. Random sampling might prove impractical or less effective given the challenges highlighted earlier in engaging wealthy consumers, as mentioned above. Outsourcing data collection to a local research institute or company could pose challenges in ensuring data quality control. Moreover, panel samples sourced from market research firms might not be applicable (e.g. Cheung et al., 2021b, c), as wealthy consumers are unlikely to participate in these panels for nominal fees or benefits. Past surveys organised by conservation organisations, through contracted market research firms, have predominantly captured data from the general public rather than specifically targeting wealthy consumers of high-value wildlife products (e.g. USAID Vietnam, 2018; USAID Wildlife Asia, 2018; TRAFFIC, 2017). Reaching a large-enough sample requires time, and the initial cohort holds significance in directing strategies aimed at engaging more consumers for survey participation. This sample could potentially contribute to shaping demand management strategies, developing campaign messages, and determining effective dissemination channels. Key insights to gather in this stage encompass a wide array of consumer aspects: attitudes, motivations, beliefs, behaviours, preferences, trade-offs, influential individuals or organisations, preferred communications channels, as well as price and income elasticities. Importantly, this three-stage framework for conducting consumer research should be based on relevant behavioural theories or models. The following section discusses how to develop a theoretical framework for such studies.

Developing a theoretical framework

To comprehensively explore the primary factors influencing consumer intention and behaviour towards purchasing high-value wildlife products, behavioural theories and models offer valuable frameworks. Prominent among these are the Theory of Planned Behaviour (TPB), the Theory of Interpersonal Behaviour (TIB), the Social Cognitive Theory (SCT), and the Health Belief Model (HBM) (Doughty et al., 2021). Yet, many applications are isolated in disjunct disciplines, and each theory involves its own assumptions and limitations (Eyster, Satterfield, and Chan, 2022). The TPB, for instance, has been widely used to study wildlife-related behaviours (Miller, 2017), but it is criticised for not including non-cognitive factors such as emotions or habits (Russell et al., 2017). The TPB may be useful for predicting individual behaviour in the short term and constrained decision spaces. But it is unlikely to predict long-term behaviour in broader contexts (Sniehotta et al., 2014; McEachan et al., 2011). Many other theories are relevant for certain behaviours and contexts but may not apply to others. Fragmentation of theories poses a further critical challenge to developing an overarching solution to a problem (Eyster, Satterfield, and Chan, 2022). Also, socioeconomic aspects interact in complex ways with the specific context in which the consumption is embedded (Olmedo et al., 2021).

To better predict human behaviours, scholars have started combining theories and models that complement each other (e.g. Klöckner, 2013; Pee, Woon, and Kankanhalli, 2008). These combinations can yield a more comprehensive theoretical framework, providing a nuanced understanding of demand determinants, albeit demanding extensive theoretical groundwork. I have merged the TPB and the TIB to examine determinants of the intention to buy rhino horn among wealthy consumers in Vietnam (Dang and Nielsen, 2022). I have also employed and compared the TPB and the SCT in a survey involving tiger bone consumers (Dang et al., 2025). The selection of theories should be guided by insights collected from Stage 1 and Stage 2 of the three-stage framework described above, particularly results of the literature review and qualitative interviews with consumers and other key informants. Despite potential benefits in combining theories, this requires intensive theoretical legwork.

Understanding consumer preferences and trade-offs

In addition to behavioural theories and models, stated preference methods such as contingent valuation and choice modelling serve as valuable tools to glean insights into consumer preferences, trade-offs, and elasticities (Mariel et al., 2021). These methods, employing hypothetical scenarios, facilitate the exploration of policy impacts and behaviour interventions on consumer decision-making. For instance, our research investigated the effects of legally controlled trade and peer influence on the preferences and trade-offs of rhino horn consumers (Dang, Nielsen, and Jacobsen, 2022). Stated preference methods can also examine consumer preferences for substitutes such as biosynthetic horns (Coals et al., 2020) or herbal substitutes to tiger bone wine, bear bile, and pangolin scale (Moorhouse et al., 2020). Choice experiments furthermore provide information about price and income elasticities. For instance, we found that the demand for a small piece of rhino horn (100 g), which is often used to reduce hangover, fever, and to share with others in the same social network or club, is elastic to price changes (Dang, Nielsen, and Jacobsen, 2022). This contradicts the long-standing assumption of price-inelastic demand for rhino horn (Milner-Gulland et al., 1993). In another choice experiment, Davis et al. (2022) have investigated consumer preference for synthetic bear bile product. Including behavioural tools as attributes in choice sets allows testing the effectiveness of those tools in the design of demand management strategies. I have used peer reference, i.e. experience using rhino horn of people influential to respondents, as a fixed attribute in choice experiments and observed its effect on consumer choices (Dang, Nielsen, and Jacobsen, 2022). Also, surveys should include questions about consumer activities, key interests, networks, communication channels, and internet platforms, so that the most effective channels to disseminate messages and to influence demand can be identified for interventions. In a recent study, I have added a construct to the SCT investigating the perceived influence of mass media channels, i.e. the extent to which specific messages disseminated in each media channel influence the intention to buy tiger bone (Dang et al., 2025).

Although my studies on rhino horn and tiger bone consumers provide concrete examples, each wildlife product has a specific consumer group with distinct characteristics, motivations, beliefs, and preferences. Demand, therefore, is intricate and idiosyncratic. The consumption context also differs from product to product and across geographical areas, which requires a nuanced approach to research and demand management interventions (Thomas-Walters et al., 2020). The proposed framework allows generating sufficient insights into consumer motivations and demand, while collecting baseline data to measure the impact of demand management interventions and facilitating the design of field experiments using treatment and control groups.

Conclusion

Understanding demand for high-value wildlife products represents a major research challenge. I discuss the intricacy of demand and approaches to influencing demand. I propose a framework for conducting consumer research, building upon our experience studying the consumption of rhino horn and tiger bone in Vietnam. Collecting reliable insights into consumers and the consumption context is critical to designing effective demand management interventions, yet conventional data collection methods may not work. The proposed framework is built upon a mixed-methods approach, in which recruiting the right researchers plays an important role. Qualitative interviews with experts and consumers could be helpful, providing initial and contextual insights into consumer attitudes, motivations, beliefs, and behaviours, which feed into the design of surveys on larger samples to understand their preferences, trade-offs, determinants of demand, and demand elasticities. I propose stated preference methods and the application of broader theoretical frameworks to generate detailed insights into demand. There are opportunities to test policy options, alternative products, and behavioural tools in managing demand. As demand is idiosyncratic and contextual, it is unlikely a one-size-fits-all approach to designing demand management interventions. Thus, an effective framework for consumer research would provide reliable insights to develop workable strategies.

References

Ammann, K. (2012). The Vietnamese and rhino horn - A dealer speaks. *Swara, 3*, 32–37.

Arias, M., Hinsley, A., & Milner-Gulland, E. J. (2020). *Implementing the Ballot Box Method to Reduce Social Desirability Bias When Researching Sensitive Behaviours in Conservation. Tools and Guidance for Illegal Wildlife Trade*. Oxford Martin Programme on the Illegal Wildlife Trade.

Ayling, J. (2016). Reducing demand for illicit wildlife products: Crafting a 'whole-of-society' response. In L. Elliott & W. H. Schaedla (Eds.), *Handbook of Transnational Environmental Crime* (Chapter 17, pp. 346–368). Edward Elgar Publishing.

Chen, F. (2017). The economics of synthetic rhino horns. *Ecological Economics, 141*, 180–189.

Chen, F., & 't Sas-Rolfes, M. (2021). Theoretical analysis of a simple permit system for selling synthetic wildlife. *Ecological Economics. 180*, e106873.

Cheung, H., Mazerolle, L., Possingham, H. P., & Biggs, D. (2021a). A survey of traditional Chinese medicine consumers to investigate the impact of China's legalization of rhino horn trade on stigmatization and likelihood of use. *Conservation Science and Practice, 3*(11), e536 https://doi.org/10.1111/csp2.536.

Cheung, H., Mazerolle, L., Possingham, H. P., & Biggs, D. (2021b). Rhino horn use by consumers of traditional Chinese medicine in China. *Conservation Science and Practice, 3*(5), e365 https://doi.org/10.1111/csp2.365.

Cheung, H., Mazerolle, L., Possingham, H. P., & Biggs, D. (2021c). China's legalization of domestic rhino horn trade: Traditional Chinese medicine practitioner perspectives and the likelihood of prescription. *Frontiers in Ecology and Evolution, 9*, 607660. https://doi.org/10.3389/fevo.2021.607660.

Coals, P., Moorhouse, T. P., D'Cruze, N. C., Macdonald, D. W., & Loveridge, A. J. (2020). Preferences for lion and tiger bone wines amongst the urban public in China and Vietnam. *Journal for Nature Conservation, 57*, 125874.

Dang, V. H. N. (2021). When cheap talk is not that cheap - interviewing the super-rich about illegal wildlife consumption. *International Journal of Social Research Methodology*, 1–6. https://doi.org/10.1080/13645579.2021.1904117

Dang, V. H. N., Gadbert, K., Nielsen, J. V., Nielsen, M. R., & Jacobsen, J. B. (2022). The impact of a legal trade in farmed tigers on consumer preferences for tiger bone glue - Evidence from a choice experiment in Vietnam. *Journal for Nature Conservation, 65*, e126088. https://doi.org/10.1016/j.jnc.2021.126088

Dang, V. H. N., & Nielsen, M. R. (2018). Understanding utilitarian and hedonic values determining the demand for rhino horn in Vietnam. *Human Dimensions of Wildlife, 23*(5), 417–432.

Dang, V. H. N., & Nielsen, M. R. (2021). Evidence or delusion: A critique of contemporary rhino horn demand reduction strategies. *Human Dimensions of Wildlife, 26*(4), 390–400.

Dang, V. H. N., & Nielsen, M. R. (2022). Understanding determinants of demand for rhino horn through the theory of planned behaviour and the theory of interpersonal behaviour. *Ecological Economics, 195*, e107361. https://doi.org/10.1016/j.ecolecon.2021.107361

Dang, V. H. N., Nielsen, M. R., & Jacobsen, J. B. (2020). Reference group influences and campaign exposure effects on rhino horn demand: Qualitative insights from Vietnam. *People and Nature, 2*(4), 923–939.

Dang, V. H. N., Nielsen, M. R., & Jacobsen, J. B. (2022). Conserving rhinos by legal trade: Insights from a choice experiment on rhino horn consumers. *Ecological Economics, 193*, e107287. https://doi.org/10.1016/j.ecolecon.2021.107287

Dang Vu, H. N., Taylor, L., Haysom, S., & Nielsen, M. R. (2025). Predicting Consumer Intention to Buy Tiger Bone Glue in Vietnam: A Comparison Between the Theory of Planned Behavior and the Social Cognitive Theory. Conservation Science and Practice, 7(9), e70130.

Davis, E. O., Glikman, J. A., Crudge, B., Dang, V., Willemsen, M., Nguyen, T., O'Connor, D., & Bendixsen, T. (2019). Consumer demand and traditional medicine prescription of bear products in Vietnam. *Biological Conservation, 235*, 119–127. https://doi.org/10.1016/j.biocon.2019.04.003

Davis, E. O., O'Connor, D., Crudge, B., et al. (2016). Understanding public perceptions and motivations around bear part use: A study in northern Laos of attitudes of Chinese tourists and Lao PDR nationals. *Biological Conservation, 203*(2016), 282–289.

Davis, E. O., Veríssimo, D., Crudge, B., Sam, S. H., Cao, D. T., Ho, P. V., … & Glikman, J. A. (2022). How will the end of bear bile farming in Vietnam influence consumer choice? *Conservation & Society, 20*(1), 1–11.

Dinerstein, E., Loucks, C., Wikramanayake, E., Ginsberg, J., et al. (2007). The fate of wild tigers. *BioScience, 57*(6), 508–514.

Doughty, H., Oliver, K., Veríssimo, D., Ser Huay Lee, J., & Milner-Gulland, E. J. (2021). Using theory and evidence to design behaviour change interventions for reducing unsustainable wildlife consumption. *People and Nature, 2*, 469–483.

Duthie, E., Veríssimo, D., Keane, A., & Knight, A. T. (2017). The effectiveness of celebrities in conservation marketing. *PloS One, 12*(7), e0180027.

Eyster, H. N., Satterfield, T., & Chan, K. M. (2022). Why people do what they do: An interdisciplinary synthesis of human action theories. *Annual Review of Environment and Resources, 47*(1), 725–751.

Fischer, C. (2004). The complex interactions of markets for endangered species products. *Journal of Environmental Economics and Management, 48*(2), 926–953.

Gao, P. J. (2013). *Chinese Anti-Corruption, Luxury Goods Hit. Elite Reference*. China Youth Daily.

Gao, Y., Stoner, K. J., Lee, A. T. L., & Clark, S. G. (2016). Rhino horn trade in China: An analysis of the art and antiques market. *Biological Conservation, 201*, 343–347. https://doi.org/10.1016/j.biocon.2016.08.001

Greenfield, S., & Veríssimo, D. (2019). To what extent is social marketing used in demand reduction campaigns for illegal wildlife products? Insights from elephant ivory and rhino horn. *Social Marketing Quarterly, 25*(1), 40–54.
Hall, R. J., Milner-Gulland, E. J., & Courchamp, F. (2008). Endangering the endangered: The effects of perceived rarity on species exploitation. *Conservation Letters, 1*(2), 75–81.
Hanley, N., Sheremet, O., Bozzola, M., & MacMillan, D. C. (2018). The allure of the illegal: Choice modeling of rhino horn demand in Vietnam. *Conservation Letters, 11*(3), e12417.
Hausmann, A., Cortés-Capano, G., Fraser, I., & Di Minin, E. (2023). Assessing preferences and motivations for owning exotic pets: Care matters. *Biological Conservation, 281*, 110007.
Hildenbrand, W. (1983). On the"law of demand". *Econometrica: Journal of the Econometric Society, 51*(4), 997–1019.
Hinsley, A., & 't Sas-Rolfes, M. (2020). Wild assumptions? Questioning simplistic narratives about consumer preferences for wildlife products. *People and Nature, 2*(4), 972–979.
Hübschle, A. (2016). *A Game of Horns: Transnational Flows of Rhino Horn*. International Max Planck Research School on the Social and Political Constitution of the Economy.
Klöckner, C. A. (2013). A comprehensive model of the psychology of environmental behaviour – a meta-analysis. *Global Environmental Change, 23*(5), 1028–1038.
Knight, J. (2004). *Wildlife in Asia*. RoutledgeCurzon.
MacFarlane, D., Hurlstone, M. J., Ecker, U. K., Ferraro, P. J., et al. (2022). Reducing demand for overexploited wildlife products: Lessons from systematic reviews from outside conservation science. *Conservation Science and Practice, 4*(3), e627.
Margulies, J. D., Wong, R. W., & Duffy, R. (2019). The imaginary 'Asian super consumer': A critique of demand reduction campaigns for the illegal wildlife trade. *Geoforum, 107*, 216–219.
Mariel, P., Hoyos, D., Meyerhoff, J., Czajkowski, M., et al. (2021). *Environmental Valuation with Discrete Choice Experiments: Guidance on Design, Implementation and Data Analysis*. Springer.
McEachan, R. R. C., Conner, M., Taylor, N. J., & Lawton, R. J. (2011). Prospective prediction of health-related behaviours with the theory of planned behaviour: A meta-analysis. *Health Psychology Review, 5*(2), 97–144.
Mi, R., Shao, Z. Z., & Vollrath, F. (2019). Creating artificial rhino horns from horsehair. *Scientific Reports, 9*(1), 1–6.
Miller, Z. D. (2017). The enduring use of the theory of planned behaviour. *Human Dimensions of Wildlife, 22*(6), 583–590.
Milner-Gulland, E. J. (1993). An econometric analysis of consumer demand for ivory and rhino horn. *Environmental and Resource Economics, 3*, 73–95.
Moorhouse, T. P., Coals, P. G., D'Cruze, N. C., & Macdonald, D. W. (2020). Reduce or redirect? Which social marketing interventions could influence demand for traditional medicines? *Biological Conservation, 242*, 108391.
Nadal, A., & Aguayo, F. (2014). *Leonardo's Sailors: A Review of the Economic Analysis of Wildlife Trade*. The Leverhulme Centre for the Study of Value.
Nguyen, H. P., Nguyen, H. T. M., & Pham, H. T. (2020). The price of hope—insights into rhino horn consumption in health-related contexts in Vietnam. *Journal of Consumer Affairs, 55*, 1249–1273. https://doi.org/10.1111/joca.12342.
Nuno, A., & St. John, F. A. V. (2015). How to ask sensitive questions in conservation: A review of specialized questioning techniques. *Biological Conservation, 189*, 5–15. https://doi.org/10.1016/j.biocon.2014.09.047
Olmedo, A., Sharif, V., & Milner-Gulland, E. J. (2018). Evaluating the design of behavior change interventions: A case study of rhino horn in Vietnam. *Conservation Letters, 11*(1), e12365.
Olmedo, A., Veríssimo, D., Milner-Gulland, E. J., Hinsley, A., Dao, T. T. H., & Challender, D. W. S. (2021). Uncovering prevalence of pangolin consumption using a technique for investigating sensitive behaviour. *Oryx*, 56, 1–9. https://doi.org/10.1017/S0030605320001040
Patton, F., & Ammann, K. (2016). From health to wealth – understanding the trade in rhino horn in the age of affluence. *Pachyderm, 57*, 122–124.
Pee, L. G., Woon, I. M. Y., & Kankanhalli, A. (2008). Explaining non-work-related computing in the workplace: A comparison of alternative models. *Information Management. 45*, 120–130.
Roberton, S. (2014). *Has demand for rhino horn truly dropped in Vietnam?* Retrieved from https://blog.nationalgeographic.org/2014/11/03/has-demand-for-rhino-horn-truly-dropped-in-vietnam/

Rock, K. I., & MacMillan, D. C. (2022). Can substitutes reduce future demand for wildlife products: A case study of China's millennial generation. *Human Ecology, 50*(1), 91–111.
Russell, S. V., Young, C. W., Unsworth, K. L., & Robinson, C. (2017). Bringing habits and emotions into food waste behaviour. *Resources Conservation Recycling, 125*, 107–114.
Salazar, G., Mills, M., & Veríssimo, D. (2019). Qualitative impact evaluation of a social marketing campaign for conservation. *Conservation Biology, 33*(3), 634–644.
Smith, K. M. (2016). *Tigers, Flies, and Longines: The Chinese Corruption Crackdown and Its Effect on Conspicuous Consumption*. The University of Mississippi. Retrieved from https://egrove.olemiss.edu/hon_thesis/530/
Smith, M. S. (2018). Framing rhino horn demand reduction in Vietnam: Dismissing medical use as voodoo. *Pacific Journalism Review, 24*(2), 241–256.
Sniehotta, F. F., Presseau, J., & Araújo-Soares, V. (2014). Time to retire the theory of planned behaviour. *Health Psychology Review, 8*(1), 1–7.
Swan, K., & Conrad, K. (2014). *The Conflict between Chinese Cultural and Environmental Values in Wildlife Consumption*. Routledge handbook of environment and society in Asia, 321–335.
Thomas-Walters, L., Hinsley, A., Bergin, D., et al. (2021). Motivations for the use and consumption of wildlife products. *Conservation Biology, 35*(2), 483–491. https://doi.org/10.1111/cobi.13578.
Thomas-Walters, L., Veríssimo, D., Gadsby, E., Roberts, D., & Smith, R. J. (2020). Taking a more nuanced look at behavior change for demand reduction in the illegal wildlife trade. *Conservation Science and Practice, 2*(9), e248.
TRAFFIC. (2017). *Chi initiative briefing paper*. Retrieved from https://www.traffic.org/site/assets/files/8811/chi-initiative-briefing-paper.pdf
Truong, V. D., Dang, V. H. N., & Hall, C. M. (2015). The marketplace management of illegal elixirs: Illicit consumption of rhino horn. *Consumption Markets and Culture, 19*(4), 353–369.
't Sas-Rolfes, M., Challender, D. W., Hinsley, A., Veríssimo, D., & Milner-Gulland, E. J. (2019). Illegal wildlife trade: Scale, processes, and governance. *Annual Review of Environment and Resources, 44*, 201–228.
USAID Vietnam. (2018). *Research Study on Consumer Demand for Elephant, Rhino and Pangolin Parts and Products in Vietnam*. Retrieved from. https://www.usaidwildlifeasia.org/resources/reports/inbox/ussv-quant-report-saving-elephants-pangolins- and-rhinos-20181105.pdf.
USAID Wildlife Asia. (2018). *Research Study on Consumer Demand for Elephant, Pangolin, Rhino and Tiger Parts and Products in China*. Retrieved from. https://www.usaidwildlifeasia.org/resources/reports/inbox/usaid_china_wildlife-demand-reduction_en glish_presentation_june12_2018_final.pdf/view.
Veríssimo, D., Vieira, S., Monteiro, D., et al. (2020). Audience research as a cornerstone of demand management interventions for illegal wildlife products: Demarketing sea turtle meat and eggs. *Conservation Science and Practice, 2*, e164. https://doi.org/10.1111/csp2.164
Veríssimo, D., 't Sas-Rolfes, M., & Glikman, J. A. (2020). Influencing consumer demand is vital for tackling the illegal wildlife trade. *People and Nature, 2*(4), 972–976.
Veríssimo, D., & Wan, A. K. Y. (2019). Characterizing efforts to reduce consumer demand for wildlife products. *Conservation Biology, 33*(3), 623–633.
Vu, A. N. (2023). Demand reduction campaigns for the illegal wildlife trade in authoritarian Vietnam: Ungrounded environmentalism. *World Development, 164*, 106150.
Wallen, K. E., & Daut, E. (2018). The challenge and opportunity of behaviour change methods and frameworks to reduce demand for illegal wildlife. *Nature Conservation, 26*, 55–75.
Wang, Y., Turvey, S. T., & Leader-Williams, N. (2020). Knowledge and attitudes about the use of pangolin scale products in traditional Chinese medicine (TCM) within China. *People and Nature, 2*(4), 903–912.
Wong, R. W. Y. (2019). *The Illegal Wildlife Trade in China: Understanding the Distribution Networks*. Palgrave MacMillan.
Wyatt, T., Maher, J., Allen, D., Clarke, N., & Rook, D. (2021). The welfare of wildlife: An interdisciplinary analysis of harm in the legal and illegal wildlife trades and possible ways forward. *Crime, Law and Social Change, 77*(1), 68–89.
Zhu, A. L. (2020). China's rosewood boom: A cultural fix to capital overaccumulation. *Annals of the American Association of Geographers, 110*(1), 277–296, https://doi.org/10.1080/24694452.2019.1613955

10

ESTIMATING WILDLIFE CRIME PREVALENCE

Quantitative methods and ethical considerations

Harriet Ibbett, Julia P.G. Jones, and Freya A.V. St. John

Introduction

So far, the chapters in this book have drawn on a range of methods, from ethnographic observation to forensic science, to present detailed accounts and case studies of wildlife crime. Access to accurate data on wildlife crime is essential for the development of appropriate policies and law enforcement strategies, as well as to assess their effectiveness once implemented. Yet, obtaining insights into wildlife crime is often challenging; those involved are rarely willing to reveal their involvement, activities are usually conducted in secret to minimise detection (Keane et al., 2008), and the direct victims (i.e. wild fauna and flora) are unable to self-report crimes. Given these challenges, how do researchers gain reliable, accurate data on the extent of wildlife crime? In this chapter, we briefly outline some of the methods commonly used to measure and monitor the extent of wildlife crime, before focusing on quantitative social science methods, including specialised questioning techniques (SQTs), used to collect data from people about wildlife crime. We then explore some of the ethical challenges associated with this type of research and provide advice for those looking to design questionnaires to collect data from potential offenders.

Approaches for estimating wildlife crime

Law enforcement records are a useful source of data that can be used to measure and monitor certain types of wildlife crime. During their duties, law enforcement officers (including protected area rangers, park guards, police, border forces, army, and navy) may record details on detected infractions (e.g. animal carcasses, tree stumps, sawmills, traps, poisoning events, egg collecting, illegal logging, hunting, fishing, or mining camps, fly tipping) (Table 10.1). Often, these data are compiled into databases (such as the Spatial Monitoring and Reporting Tool (SMART)), with arbitrary units, such as Catch per Unit of Effort (e.g. number of incidences detected per number of days or km patrolled) calculated to assess seasonal peaks in activity, while spatial data (e.g. GPS coordinates) can be used to map crime hotspots (Dobson et al., 2019). In some instances, data are compiled at national, regional, and even international levels to monitor global trends in wildlife crime; for example, ranger-collected

 DOI: 10.4324/9781003333906-12

Table 10.1 Approaches used to estimate the prevalence of wildlife crime, with examples of their application. Copyright: Adapted from Gavin et al. 2010; Bergseth et al., 2013; Arias, 2015.

Method	*Description*	*Provides information about:*				
		Crime type	*Timing of crime*	*Crime location*	*Perpetrator*	*Drivers of crime*
Enforcement records	Incidents of wildlife crime (e.g. traps, carcasses, tree stumps, camps, sawmills) detected and recorded by law enforcers while on foot, vehicle, boat, or air patrol.	✓	✓	✓	✓	✓
	Judicial records or legal proceedings (e.g. arrest and prosecution records).	✓	✓	✓	✓	✓
	Seizure data, such as records of wildlife products seized by law enforcers at air, rail, and seaports.[1]	✓	✓	✓	✓	✓
Observations made in absentia	Bio-surveillance technologies such as camera traps,[2] bio-acoustic monitoring, or drones used to record incidences of rule breaking.	✓	✓	✓	✓	✓
	Use of satellite imagery (remote sensing) to detect crime (e.g. elephant carcasses, illegal deforestation, agricultural expansion, or mining).	✓	✓	✓	✓	✓
	Online surveillance (e.g. through search of online marketplaces or social media platforms).[1]	✓	✓	✓	✓	✓

(*Continued*)

Table 10.1 (Continued)

Method	*Description*	*Provides information about:*				
		Crime type	*Timing of crime*	*Crime location*	*Perpetrator*	*Drivers of crime*
Observations made *in person*	Surveys of products available for sale in markets.[1]	✓	✓	✓	✓	✓
	Researchers observe crime covertly (i.e. without the consent of the person being observed).	✓	✓	✓	✓	✓
	Observations of people conducted in person with their consent (e.g. hunter follows or ethnographic studies).	✓	✓	✓	✓	✓
Interviews/ questionnaires	Data collected directly from people about wildlife crime using interviews, group exercises, or questionnaires (these could be collected in person or online).	✓	✓	✓	✓	✓
Personal records	Individuals provide information about wildlife crime (e.g. via hunter diaries or fisheries logbooks).	✓	✓	✓	✓	✓

1 If the crime of interest is trafficking, this approach can provide additional information such as the crime location, perpetrator, and timing. However, this approach does not necessarily reveal these details associated with the extraction of the wildlife product at source.
2 Camera traps may capture images of perpetrators. However, see guidelines on the ethical use of surveillance technologies (Sandbrook et al., 2018; 2021).

data on elephant poaching incidents are submitted to MIKE, an international database Monitoring the Illegal Killing of Elephants (CITES, 2022) (Table 10.1).

In addition to ranger-collected data, criminal justice administrative data, such as arrest and prosecution records, can provide insights into types of wildlife crime committed, where, when and by whom, as well as the punishments distributed (Gavin et al., 2010; Paudel et al., 2020). Seizure data, collected by law enforcers at transport hubs such as air and seaports, can also be analysed to assess wildlife trafficking trends (e.g. Kurland & Pires, 2017), with data accumulated and monitored by international conservation organisations (e.g. TRAFFIC) as well as anti-crime agencies (e.g. United Nations Office on Drugs and Crime World WISE database (UNODC, 2024)). Encouragingly, forensic advances mean that the geographic origin of seized wildlife products, such as elephant ivory, can increasingly be identified (Wasser et al., 2015).

In recent decades, the emergence of novel surveillance technologies has further advanced our ability to monitor and quantify the extent of wildlife crime remotely. While initially developed to survey wildlife populations, technologies such as camera traps, drones, and acoustic monitoring equipment have been used to study wildlife crimes (Table 10.1). For example, drones, camera traps and towers mounted with infrared cameras have been used to survey unauthorised entry into National Parks in Northern India (Simlai, 2021), while passive acoustic monitoring has been used to detect hunting with guns in Loango National Park, Republic of Gabon (Wrege et al., 2017). In addition, the development and increased affordability of high-resolution satellite imagery mean that crimes can be detected and monitored remotely; for example, in the Brazilian Amazon, researchers have used such data to assess compliance with deforestation embargoes (da Silva et al., 2022). While these emerging surveillance technologies present exciting opportunities to improve understanding of wildlife crime, their deployment requires careful consideration. If used carelessly, such technologies could alienate local people who may resent surveillance and even contribute to increased crime (e.g. as a form of social defiance) (Simlai, 2021), while the use of high-resolution satellite imagery raises concerns around data justice (Pritchard et al., 2022). Guidelines exist to support researchers to minimise ethical implications from the use of surveillance technologies (e.g. Sandbrook et al., 2021).

Data on incidents of wildlife crime can also be collected by researchers making observations either in person or remotely (Table 10.1). Conducting in-person surveys of products for sale at markets, for example, can help identify price changes and demand for specific wildlife products (Gomez & Shepherd, 2018; Milner-Gulland & Clayton, 2002), while monitoring of online sales platforms, social media accounts or even the dark web can also determine crime trends (e.g. Hinsley et al., 2016; Wyatt et al., 2022).

Limitations of these approaches

Despite the widespread application of the approaches listed above, they have many limitations, not least, potentially large and unknown measurement errors (Gavin et al., 2010). The challenging terrain of many places where wildlife crimes are committed, and the often subsistence nature of much wildlife crime (particularly in the Global South) means that incidents are often hard to detect (Robbins et al., 2005). For example, an experimental study conducted in the forests of eastern Cambodia found that despite significant search effort, rangers found only a small proportion of artificial snares that had been set (Ibbett et al., 2020). Even if incidents of crime are detected, they may remain unreported; particularly if law enforcers are involved in the crime themselves (Gavin et al., 2010). The unstructured, incidental way in which data are collected also makes the types of observational data described above inherently difficult to analyse (Dobson et al., 2020; Keane, Jones, et al., 2011), and such data rarely provide a complete picture of the crime. For example, seizure data do not reveal where resources originate and may only provide information about the trafficker, rather than the person responsible for extracting the wildlife product;

satellite imagery may not distinguish between illegal activity and natural processes; and market surveys do not account for items not reaching the market, consequently underestimating offtake (Gavin et al., 2010) and similarly to seizure data, do not reveal details associated with the extraction of the wildlife product at source. Arguably, the key limitation of these approaches is that they tell us little about the characteristics or motivations of those committing wildlife crimes (Nuno & St John, 2015).

Asking people directly about wildlife crime

Speaking to people directly can provide much richer insights into the types of people committing wildlife crimes, as well as their motivations. The use of qualitative methods, including ethnographic observation and in-depth interviews, can produce rich understandings of the processes underpinning wildlife crimes, yet perhaps the most commonly used quantitative tool for understanding drivers of crime is the questionnaire (Ibbett & Brittain, 2019; von Essen et al., 2014). Typically comprised of lists of closed questions (whereby respondents choose a response from a limited number of options), questionnaires can be delivered at multiple scales, enabling researchers to obtain small (fewer than 100 responses), medium (hundreds of responses) and large samples (many thousands of responses). Such data can be analysed statistically and even combined with ecological and environmental variables to determine trends, with results used to evidence specific policy recommendations.

Limitations of collecting data directly from people

However, a critical limitation of asking people directly about their behaviour is that people are rarely willing to self-report their involvement or discuss their motivations freely (Solomon et al., 2007). Doing so could present a substantial threat to participants, potentially resulting in significant psychological (e.g. feelings of guilt, shame, or embarrassment), physical (e.g. violence), social (e.g. social ostracism) or punitive (e.g. formal sanctions such as arrest or imprisonment) costs (Lee & Renzetti, 1990). Such threats can arise both because of the content of a response (i.e. admission of a crime) or, in some situations, simply participating in the research may present risks to participants' and researchers' safety, regardless of the answer provided. This may be especially so if wildlife crimes involve the elite, if findings impinge on powerful interests, or if results highlight corruption or abuses of power (Lee & Renzetti, 1990; Robbins, 2000).

The likely sensitivity of a topic, and therefore peoples' willingness to provide information, is also influenced by the social context in which the research occurs; what might be innocuous in one context might be highly sensitive in another (Ibbett, Jones, et al., 2023b; Lee & Renzetti, 1990). For example, if laws are poorly enforced and wildlife crime is common practice, respondents may willingly report their involvement (Razafimanahaka et al., 2012). Further, people's propensity to answer questions is often influenced by social norms; the unwritten rules that prescribe and regulate how people behave (Hechtor & Opp, 2001). If a behaviour is perceived as socially unacceptable or undesirable, admittance could result in social stigmatisation or ostracism. For example, in Nigeria, social norms were found to be a stronger driver of compliance with conservation rules than national legislation (Atuo et al., 2020). While in Madagascar, communities may be more familiar with cultural prohibitions known as fady than national legislation (Keane, Ramarolahy, et al., 2011), meaning research about violating fady may be more sensitive than research on law breaking. Failure to identify and acknowledge topic sensitivity may cause offence, alienate, or endanger those involved (Sieber & Stanley, 1988). Alternatively, researchers may perceive some subjects to be more sensitive than they are, resulting in the use of inappropriate or unnecessarily complex methods, which may in turn exacerbate issues such as non-response and acquiescence bias (Ibbett et al., 2023b).

Combined, these factors can have significant implications for data quality. Samples can become biased, only capturing responses from those who are willing to participate, while specific questions about involvement in crime may suffer high levels of refusals (known as non-response bias) (Krumpal, 2013; Tourangeau & Yan, 2007). Data provided by those who consent to participate, but who harbour concerns about repercussions, may be affected by social desirability bias, whereby inaccurate responses are provided to conform to prevailing social norms, thus reducing the risk of potential repercussions (Blair et al., 2020; Krumpal, 2013). Moreover, if the study population has previously been exposed to extensive research, respondents may be suffering from survey fatigue, which can result in acquiescence bias, where respondents agree with non-sensitive questionnaire statements simply to end the survey sooner.

Ethical considerations when researching wildlife crime

Beyond data quality, asking people about their involvement in wildlife crime also raises a range of ethical concerns which, if poorly addressed, can have significant implications on research participants, the research process, as well as the success of interventions informed by the research (Brittain et al., 2020). All researchers, regardless of their discipline, have a moral, professional, and often legal duty to ensure that research does no harm. Yet, by its very nature, research on wildlife crime has the potential to do significant harm; the primary objective after all, is to uncover what is purposely hidden. When activities are conducted for subsistence or cultural purposes, research on wildlife crime can have significant impacts on already marginalised groups (Satizábal et al., 2021) and risks placing the blame on smaller actors, while those that are perhaps better resourced to hide their practices continue unabated (Sundström, 2016). It is therefore important to recognise that as a researcher, you hold significant power; not only do you define the questions asked, by whom, and of whom, but you also control how information is collected, stored, and shared. With this power comes a responsibility to consider ethical issues that may arise during research and to implement measures to minimise potential harms.

For many researchers, this is facilitated by an ethical review process, whereby research protocols are submitted to an Ethical Review Board (sometimes called an Institutional Review Board), who assess whether sufficient measures are in place to protect participants and researchers. Central to this process is the concept of consent (Guillemin & Gillam, 2004), which can be thought of as a contract formed between a researcher and participant (Brittain et al., 2020). Here, researchers explain the conditions the participant can expect them to abide by in exchange for their involvement (Dyer & Demeritt, 2009). Vitally, this process should be Free, Prior, and Informed, whereby the purpose, methods, and intended uses of research are described, and the risks and benefits of participation are discussed in a language understandable to participants *before* research commences (Guillemin & Gillam, 2004; Speiglman & Spear, 2008; Wax, 1980). If sought diligently, informed consent should ensure that people participate after being provided with all the information they need to make informed and independent decisions about whether to participate (Brittain et al., 2020).

Additionally, measures such as *confidentiality* (whereby identifiable information about the respondent is collected, but never shared or made public) and *anonymity* (whereby no identifiable information is collected) are important to consider, particularly when gathering sensitive data on illegal activities, including wildlife crime (Ibbett & Brittain, 2019). These considerations may determine how other measures, such as consent, are recorded. For example, written consent could undermine assurances of anonymity, as people may be recognisable from their handwriting or signature; instead, verbal consent may be more appropriate. Sometimes, it may be necessary to record the identity of the participant (e.g. if individuals are to be re-visited), in which case measures such as pseudo-anonymisation, whereby respondents' identifiable information is stored separately from their responses, are important. When speaking to people directly, it is also important to consider whether data can ever be truly anonymous. Even if personal identifiers (including names, dates of birth, addresses, identity numbers, photographs, voice recordings, GPS coordinates, and any handwritten items such as fingerprints, signatures) are not collected and interviews are conducted away from participants' homes, interviewers could later recognise respondents by sight and/or name, and there is a risk they will remember their response. Additionally, other people, who may or may not be directly involved in the research, may observe from whom data are collected. When the research topic is sensitive, these limits to anonymity must be recognised by researchers. Steps to minimise these risks could involve, for example, ensuring interviews are conducted in safe, private spaces and ensuring interviewers act with the utmost discretion.

How data are stored, both during data collection and afterwards, is also an important consideration, particularly if collected in person. Data may be captured physically, for example on paper forms, in notebooks, or using audio recordings or electronic devices such as tablets or phones (e.g. using software such as KoboToolbox), meaning they are particularly vulnerable to being misplaced or stolen. Ensuring data are stored securely both in the field and back at the research base is vital. At times, it may also be pertinent to consider whether in-person possession of data presents any specific risk to researchers. For example, if conducting research on topics that involve or conflict with the interests of the powerful elite (e.g. if investigating the involvement of government officials in wildlife crime), being found in possession of such data could result in serious implications. If collecting data electronically, steps such as encrypting data at the point of entry, ensuring data are stored on servers outside of the country of origin and are immediately deleted from handheld devices after upload, are practical measures that prevent data loss, undesired access, and protect researchers.

Looking forward, increased internet accessibility and improved global literacy may lead to a shift of survey effort away from face-to-face data collection towards online research (Wardropper et al., 2021). While this can reduce research costs, empower researchers to access new participants, give voice to new groups, and allow for significantly larger sample sizes, there are potential ethical and methodological implications for those studying wildlife crime. Generally, online survey administration does not enable participants to check their understanding of survey questions or methods. Meanwhile, special consideration needs to be given to the anonymisation of data, for example by ensuring URLs are not captured during online survey completion, as well as factors such as where the platform used to capture data complies with legislation and funder requirements relating to the storage of data. For example, any data collected in the United Kingdom and European Union must be compliant with General Data Protection Regulation legislation, while some funders may require data to be stored on servers located in certain countries. Detailed guidance on ethical considerations when collecting data online is increasingly emerging (Monkman et al., 2018; Thompson et al., 2021; Wardropper et al., 2021).

Aside from the practical measures outlined above, wider ethical issues can arise during wildlife crime research. As previously outlined, the nature of the research topic is ethically loaded, and researchers may at times find themselves in challenging situations, understanding how to navigate these situations if they arise is important. For example, what if a researcher witnesses incidents of wildlife crime? Or uncovers new intelligence about an upcoming poaching event? Sharing this knowledge with authorities or professional networks could lead to crime prevention. However, doing so may also result in repercussions for the research team, those who revealed the information, and future researchers who wish to work in that space, particularly if the information was obtained in confidence. Revealing the information could breach the research contract formed in trust between the researcher and the participant. Yet, doing nothing may also be untenable, particularly if the uncovered knowledge conflicts strongly with the researcher's own beliefs and values (Brittain et al., 2020). In such instances, viewing the situation from a wider perspective can be helpful; by overlooking small-scale or low-order offences, the researcher may be able to access information which casts an informative light on the wider structural forces enabling wildlife crime to occur (Duffy et al., 2015). Indeed, similar approaches are used in law enforcement and intelligence gathering operations; lower-level transgressions may be overlooked to tackle higher-level crimes. However, at other times, information uncovered during research may warrant reporting, regardless of the implications, for example, if it represents a threat to human life or breaches human rights. Additionally, researchers may have legal obligations to share certain types of information with appropriate authorities (e.g. relating to terrorism). It is essential to be aware of these responsibilities at the beginning of research. To avoid any breaches of confidence, researchers may wish to outline any legal obligations they are under to participants as part of the consent process.

When these types of ethical conundrums occur, it can be useful to remember one's role as a researcher, whose primary purpose is to further knowledge, rather than to investigate crime or enforce laws. However, even if the role as a researcher is clear, it may not always be easy to enact, particularly when collaborating with others who have different objectives. For example, if collaborating with a conservation NGO, to help inform their interventions, they may be particularly interested in knowing the locations where data were collected, or the identity of participants the researcher spoke to. Further, what if a researcher uncovers information that casts a critical light on the activities of the NGO, which could affect their reputation or future ability to access funding? How does a researcher share this information constructively? Moreover, much research is only possible with the support of gatekeepers who, under certain provisos, enable research. For example, research proposals are often approved and granted at national, regional, and local governmental levels, and permits may come with clauses such as the requirement for researchers to register at local police stations, to be accompanied to certain locations by government staff, or to share data. Similarly, collaborating with NGOs may facilitate access to specific groups, but may come with conditionality. Failure to meet these conditions may damage collaborations or even result in legal sanctions and can raise questions about research sovereignty. Yet, fulfilling these requirements may seem like an untenable ethical compromise, particularly if there are clauses mandating access to sensitive data; being accompanied during research; or censoring of findings. Discussing these issues with collaborators at the beginning of research, and drafting Memorandums of Understanding outlining the objectives, expectations, and responsibilities of each partner can be helpful steps to overcome some of these challenges.

Throughout research, there is a need to consider who is collecting data and asking questions. In some instances, being viewed as an 'outsider' (defined as someone external to the context where data are being collected) can be beneficial, as respondents may be more willing to reveal information about wildlife crime to those, they are unlikely to have further contact with. Conversely, outsiders may be less trusted and viewed with greater suspicion (Holmes, 2020). Working with people local to study areas, for example a trusted leader, can help encourage greater participation (Razafimanahaka et al., 2012). However, it can also provide those who facilitate research with new knowledge, which may situate them in positions of power over participants (Brittain et al., 2020) or present risk to their safety (Lee & Renzetti, 1990). Careful consideration of these factors is particularly pertinent when investigating wildlife crime.

One further important ethical consideration is giving thought to the narratives that are adopted when reporting research findings, particularly if data have been collected at the individual or community level. Discourses used to frame wildlife crime can have powerful implications on how people are perceived and may perpetuate negative connotations (von Essen et al., 2014). For example, the term 'unlawful hunting' can be perceived relatively neutrally, whereas describing hunters as poachers criminalises individuals and evokes images of thievery and dishonesty (von Essen et al., 2014). Wildlife crimes are committed across diverse contexts by equally diverse actors for a variety of reasons (including for thrill-seeking, sport, to fulfil cultural traditions, or to make political statements (Nurse, 2011; Wyatt, 2016)), and it is important to be mindful that in some situations, conations of illegality can be both unhelpful and socially unjust. For example, in their introduction to a journal issue dedicated to the study of non-compliance (Solomon et al., 2016) highlight that many who do not comply with conservation regulations do so for subsistence reasons and that labelling those who do so as criminals is ethically and morally problematic, particularly given that conservation has often been imposed within these contexts (Duffy, 2010).

Methods designed to provide respondents with a greater sense of protection

Responding to these ethical concerns and recognising the importance of obtaining reliable data on sensitive topics, for several decades, social scientists have been developing survey methods to overcome these challenges. Known as specialised questioning techniques (SQTs) (also referred to as indirect questioning methods, concealed response methods), these methods are hypothesised to provide respondents greater protection and encourage more honest reporting when answering sensitive questions, thereby increasing the accuracy of prevalence estimates (Chaudhuri & Christofides, 2013). Numerous SQTs exist, each developed to overcome the limitations of others (Cerri et al., 2021; Nuno & St John, 2015). We outline three below.

The unmatched count technique

Some methods provide protection by aggregating answers. For example, the unmatched count technique (UCT, also known as a list experiment) divides the sample in half. One-half are provided a list of innocuous items, and the other receives the same list with the addition of a sensitive item (Droitcour et al., 1991), e.g. the wildlife crime. Respondents are presented with the list and asked to identify how many items apply to them without revealing which (Figure 10.1). The prevalence of the sensitive behaviour can be calculated at the sample level by subtracting the mean score of all those who received the control list from the mean score of all those who received the treatment list. This approach has been used to assess behaviours such as the compliance of orchid traders with CITES regulations (Hinsley et al., 2017), the illegal consumption of turtle eggs in Sao Tome (Nuno et al., 2018), and hunting in Serengeti National Park, Tanzania (Nuno et al., 2013). Variations of this design are also available; for example, the double list method uses two sets of lists, and respondents are asked to answer twice. Here, the sensitive item is the same on both sets of lists, but the non-sensitive items differ. Respondents are shown a control list from one set and a treatment list from the other. Doing so helps to decrease the amount of variation in the estimate and can improve accuracy. Various design considerations exist including the length of lists, the choice of non-sensitive items, and how lists

Figure 10.1 Summary of some specialised questioning techniques that are available to help researchers offer participants greater levels of protection when asking them about their involvement in wildlife crime (full comic on methods for asking sensitive questions available here: https://doi.org/10.6084/m9.figshare.25164347.v1).

are administered. More information can be found in a review by Hinsley et al. (2018), as well as a toolkit produced by Olmedo et al. (2019).

The bean method

The bean method is straightforward to administer and works by asking respondents to secretly move specific-coloured beans (or counters, buttons, marbles, etc.) from one jar to another, depending on their answer (Figure 10.1) (Lau et al., 2011). The total number of beans in each jar is known at the start of the sampling period (this might be a day, week, or month), and then the number of beans in each jar is recounted at the end of the sampling period. The difference in the number of beans in each jar represents the number of people who gave a specific answer. This method is relatively easy to explain and administer and has been used successfully in Liberia to understand wild meat hunting (Jones et al., 2020).

Randomised response techniques

Of the available SQTs, randomised response techniques (RRTs) are most commonly used to ask questions about wildlife crime (Ibbett et al., 2021). The term RRT is used to describe a suite of designs that use probability to administer questions. One of the most used designs is the forced-response RRT. Here, respondents are provided with a randomising device, such as a die or a bag of coloured counters, and are asked to provide a specific answer, depending on the outcome of a randomising event (Figure 10.1). For example, a respondent may be given a die, told to roll it in secret, and then asked to answer according to the score; if the die lands on a 1, 2, 3, or 4, they may be asked to answer truthfully; if they land a 5 they may be instructed to give the prescribed answer 'no', regardless of whether this answer applies to them; and if they land a 6, again their answer would be prescribed, but they may be instructed to answer 'yes'. Using the known probability of those instructed to answer 'yes' truthfully (in this example, 4/6 = 0.667), and the probability of the answer being a prescribed 'yes' (in this example, 1/6 = 0.167), researchers can estimate the prevalence using the following equation:

$$\pi = \lambda - \theta / S$$

where π is the estimated proportion of the sample admitting to the behaviour, λ is the proportion of all answers that are 'yes', θ is the probability of the answer being a prescribed 'yes', and S is the probability of being asked to answer the question truthfully (St John et al., 2015).

The forced-response RRT has been used to estimate the prevalence of conservation non-compliance in Chile (Galvez et al., 2021; Oyanedel et al., 2020), Taiwan (St John et al., 2015), Indonesia (St John et al., 2018) and Madagascar (Razafimanahaka et al., 2012), showing that it can be adapted for a variety of contexts. However, one of the downfalls of the forced-response RRT is that some people do not like being 'forced' to provide specific answers. Another design that can be applied is the Unrelated Question RRT (Ibbett et al., 2021). Here, respondents use randomisation to determine which question they should answer: a question on an innocuous, unrelated topic, or a sensitive question.

Regardless of the question, both answer options are the same (i.e. yes or no). This method is particularly effective when researchers know the probability of answering 'yes' to the unrelated question. This can be achieved by asking a question where population-level data are available (e.g. were you born in a January or February? Does your phone number end in a 1 or 2?). Or, by introducing an additional randomisation event, the outcome of which forms the subject of the unrelated question. For example, in Uganda, Solomon et al. (2007) first asked respondents to flip a coin, then presented two identical envelopes. Respondents were asked to choose one. Inside, one envelope contained a card showing the 'head' side of a coin, the other included a photograph of a snare. If respondents selected the first card showing the coin, they were asked to answer "yes" if they had flipped a head. If they selected the card with the photograph of a snare, they were asked to answer honestly whether they had set snares. Solomon et al. knew that there was a 0.5 probability that respondents would open the envelope with the coin, and that if they did, they should answer 'yes'. Researchers interested in incorporating RRT into a study need to make various decisions about design and administration, including the probability of respondents answering truthfully and the type of randomising device used. Further guidance, including a decision tree on whether an RRT is likely appropriate, can be found in Ibbett et al. (2021).

Limitations of specialised questioning techniques

While SQTs make a valuable contribution to preserving anonymity and protecting participants, they do not provide a silver bullet when asking sensitive questions. They often require much larger sample sizes; it remains difficult to validate the accuracy of the data produced, and sometimes their use can raise ethical challenges, for example, if respondents experience uneasiness using equipment (Figure 10.2). Moreover, questions remain about their suitability and effectiveness; these methods were developed in Western, educated populations in industrialised, rich, and democratic contexts (so-called WEIRD populations), raising questions about their applicability when applied in alternate contexts (Razafimanahaka et al., 2012). To be successful, SQTs require respondents to understand what they must do and why, and be willing to follow procedures fully (Hoffmann et al., 2017). Yet, a series of experiments which aimed to assess the effectiveness of these methods when asking people about wildlife crime around protected areas in Indonesia and Tanzania, found understanding of the methods was significantly affected by a respondent's age and their education level; with those who were older or less educated more likely to answer incorrectly (Ibbett et al., 2022). Further experimental research designed to assess how topic sensitivity influences respondents' propensity to follow RRT instructions revealed that despite a good understanding of the method, people living around protected areas in central-southern Tanzania were reluctant to provide accurate answers about fictional characters breaking protected areas rules, particularly when required by RRT instructions to provide a 'yes' response (Ibbett et al., 2023a). In this case study, the behaviour of research interest was considered so sensitive that despite assurances of anonymity, the additional protection offered by RRT, and indeed the use of fictional characters, people remained concerned that their responses would be taken as admissions of non-compliance (Ibbett, Dorward, et al., 2023a).

Figure 10.2 Examples of challenges often experienced when using SQTs to ask questions about wildlife crime (adapted from Ibbett et al., 2024).

Conclusions

Conducting research on wildlife crime is methodologically and ethically challenging. In this chapter, we outlined different ways in which researchers attempt to measure the extent of wildlife crime using a range of methods, including gathering data from people who may have committed wildlife crimes. We also detailed some of the limitations of these approaches, and crucially, we discussed ethical challenges associated with research on wildlife crime and provided guidance on mitigating potential harms, both to participants and researchers. Quantitative social science methods, including questionnaires in all their guises, are a powerful means for gathering informative insights into the prevalence and drivers of wildlife crime. However, to be effective, questionnaire-based studies require robust and rigorous design (Newing, 2011; St John et al., 2014), including time spent at the outset immersed in the research context to ensure the study design is practical and locally appropriate. Indeed, collaborating and co-designing with local researchers and extensive piloting and refinement of survey instruments are essential to improving data quality and ensuring respondents feel comfortable participating in research. We outlined how SQTs can usefully increase respondents' sense of protection when providing potentially incriminating answers to survey questions, and how this can enhance data quality, making SQTs a useful addition to the toolbox for research on wildlife crime. However, we emphasised that SQTs require careful design and extensive piloting – typically above and beyond what is required for standard question-answer items regularly encountered in questionnaires. To avoid wasting respondents' time and precious research resources, we encourage researchers to incorporate iterative rounds of analysis early on in SQT deployment to help determine the utility of the chosen method and to unleash the full potential that SQTs have to offer in gathering robust, reliable data on wildlife crime.

References

Arias, A. (2015). Understanding and managing compliance in the nature conservation context. *Journal of Environmental Management, 153*, 134–143. https://doi.org/10.1016/j.jenvman.2015.02.013

Atuo, F. A., Fu, J., O'Connell, T. J., Agida, J. A., & Agaldo, J. A. (2020). Coupling law enforcement and community-based regulations in support of compliance with biodiversity conservation regulations. *Environmental Conservation, 47*(2), 104–112. https://doi.org/10.1017/S0376892920000107

Bergseth, B., Russ, G., & Cinner, J. (2013). Measuring and monitoring compliance in no-take marine reserves. *Fish and Fisheries, 16*, 240–258. https://doi.org/10.1111/faf.12051

Blair, G., Coppock, A., & Moor, M. (2020). When to worry about sensitivity bias: A social reference theory and evidence from 30 years of list experiments. *American Political Science Review, 114*(4), 1297–1315. https://doi.org/10.1017/S0003055420000374

Brittain, S., Ibbett, H., de Lange, E., Dorward, L., Hoyte, S., Marino, A., Milner-Gulland, E. J., Newth, J., Rakotonarivo, S., Veríssimo, D., & Lewis, J. (2020). Ethical considerations when conservation research involves people. *Conservation Biology, 34*(4), 925–933. https://doi.org/10.1111/cobi.13464

Cerri, J., Davis, E. O., Veríssimo, D., & Glikman, J. A. (2021). Specialized questioning techniques and their use in conservation: A review of available tools, with a focus on methodological advances. *Biological Conservation, 257*(April), 1–19. https://doi.org/10.1016/j.biocon.2021.109089

Chaudhuri, A., & Christofides, T. C. (2013). *Indirect Questioning in Sample Surveys*. https://doi.org/10.1007/978-3-642-36276-7

CITES. (2022). *Monitoring the illegal killing of elephants*. https://cites.org/eng/prog/mike/index.php/portal

da Silva, V. C. S., Guimaraes Vieira, I. C., Galbraith, D., Potapov, P., Rivero, S. L. de M., de Lima, A. M. M., da Silva Pimentel, M. A., & Adami, M. (2022). Marked non-compliance with deforestation embargoes in the Brazilian Amazon. *Environmental Research Letters, 17*(5), 054033. https://doi.org/10.1088/1748-9326/AC6398

Dobson, A. D. M., Milner-Gulland, E. J., Aebischer, N. J., Beale, C. M., Brozovic, R., Coals, P., Critchlow, R., Dancer, A., Greve, M., Hinsley, A., Ibbett, H., Johnston, A., Kuiper, T., Le Comber, S., Mahood, S. P., Moore, J. F., Nilsen, E. B., Pocock, M. J. O., Quinn, A., … Keane, A. (2020). Making messy data work for conservation. *One Earth, 2*(5), 455–465. https://doi.org/10.1016/j.oneear.2020.04.012

Dobson, A. D. M., Milner-Gulland, E. J., Beale, C. M., Ibbett, H., & Keane, A. (2019). Detecting deterrence from patrol data. *Conservation Biology, 33*(3), 665–675. https://doi.org/10.1111/cobi.13222

Droitcour, J., Caspar, R. A., Hubbard, M. L., Parsley, T. L., Visscher, W., & Ezzati, T. M. (1991). The item count technique as a method of indirect questioning: A review of its development and a case study application. In S. Biemer, P.P. Groves, R.M. Lyberg, L.E. Mathiowetz, N.A. Sudman (Eds.), *Measurement Errors in Surveys* (pp. 185–210). John Wiley & Sons, Inc.

Duffy, R. (2010). *Nature Crime: How We're Getting Conservation Wrong*. Yale University Press.

Duffy, R., St John, F. A. V., Büscher, B., & Brockington, D. (2015). Toward a new understanding of the links between poverty and illegal wildlife hunting. *Conservation Biology, 30*(1), 14–22. https://doi.org/10.1111/cobi.12622

Dyer, S., & Demeritt, D. (2009). Un-ethical review? Why it is wrong to apply the medical model of research governance to human geography. *Progress in Human Geography, 33*(1), 46–64. https://doi.org/10.1177/0309132508090475

Gálvez N, St. John FAV, Davies ZG. 2021. Drivers of predator killing by rural residents and recommendations for fostering coexistence in agricultural landscapes. *Frontiers in Conservation Science, 2.* https://www.frontiersin.org/journals/conservation-science/articles/10.3389/fcosc.2021.712044/full

Gavin, M. C., Solomon, J. N., & Blank, S. G. (2010). Measuring and monitoring illegal use of natural resources. *Conservation Biology, 24*(1), 89–100. https://doi.org/10.1111/j.1523-1739.2009.01387.x

Gomez, L., & Shepherd, C. (2018). Trade in bears in Lao PDR with observations from market surveys and seizure data. *Global Ecology and Conservation, 15*, e00415.

Guillemin, M., & Gillam, L. (2004). Ethics, reflexivity, and 'Ethically important moments' in research. *Qualitative Inquiry, 10*(2), 261–280. https://doi.org/10.1177/1077800403262360

Hechtor, M., & Opp, K. D. (2001). *Social Norms* (M. Hechtor & K. D. Opp, Eds.). The Russell Sage Foundation.

Hinsley, A., Keane, A. M., St John, F. A. V., Ibbett, H., & Nuno, A. (2018). Asking sensitive questions using the unmatched count technique: Applications and guidelines for conservation. *Methods in Ecology and Evolution, 00*, 1–12. https://doi.org/10.1111/2041-210X.13137

Hinsley, A., Lee, T. E., Harrison, J. R., & Roberts, D. L. (2016). Estimating the extent and structure of trade in horticultural orchids via social media. *Conservation Biology, 30*(5), 1038–1047. https://doi.org/10.1111/cobi.12721

Hinsley, A., Nuno, A., Ridout, M., St John, F. A. V., & Roberts, D. L. (2017). Estimating the extent of CITES noncompliance among traders and end-consumers; Lessons from the global orchid trade. *Conservation Letters, 10*(5), 602–609. https://doi.org/10.1111/conl.12316

Hoffmann, A., Waubert de Puiseau, B., Schmidt, A. F., & Musch, J. (2017). On the comprehensibility and perceived privacy protection of indirect questioning techniques. *Behaviour Research Methods, 49*(4), 1470–1483. https://doi.org/10.3758/s13428-016-0804-3

Holmes, A. G. D. (2020). Researcher positionality: A consideration of its influence and place in qualitative research. A new researcher guide. *Shanlax International Journal of Education, 8*(4), 1–10.

Ibbett, H., & Brittain, S. (2019). Conservation publications and their provisions to protect research participants. *Conservation Biology, 34*(1), 80–92. https://doi.org/10.1111/cobi.13337

Ibbett, H., Dorward, L., Dwiyahreni, A. A., Jones, J. P. G., Kaduma, J., Kohi, E. M., Mchomvu, J., Prayitno, K., Sabiladiyni, H., Sankeni, S., Saputra, A. W., Supriatna, J., & St John, F. A. V. (2022). Experimental validation of specialised questioning techniques in conservation. *Conservation Biology*, 36(5), e13908. https://doi.org/10.1111/cobi.13908

Ibbett, H., Dorward, L., Dwiyahreni, A. A., Kohi, E. M., & St John, F. A. V. (2024). *Understanding the Links between Well-Being and Rule Breaking around Protected Areas. A comic series.* Figshare. https://doi.org/10.6084/m9.figshare.25164347.v2

Ibbett, H., Dorward, L. J., Kohi, E. M., Jones, J. P. G., Sankeni, S., Kaduma, J., Mchomvu, J., Mawenya, R., & St. John, F. A. V. (2023a). Topic sensitivity still affects honest responding, even when specialized questioning techniques are used. *Conservation Science and Practice, 5*(6), e12927. https://doi.org/10.1111/csp2.12927

Ibbett, H., Jones, J. P. G., Dorward, L., Kohi, E. M., Dwiyahreni, A. A., Prayitno, K., Sankeni, S., Kaduma, J., Mchomvu, J., Wijaya Saputra, A., Sabiladiyni, H., Supriatna, J., & St John, F. A. V. (2023b). A mixed methods approach for measuring topic sensitivity in conservation. *People and Nature, 5*(4), 1245–1261. https://doi.org/10.1002/pan3.10501

Ibbett, H., Jones, J. P. G., & St John, F. A. V. (2021). Asking sensitive questions in conservation using randomised response techniques. *Biological Conservation, 260*(109191), 1–14.

Ibbett, H., Milner-Gulland, E. J., Beale, C., Dobson, A. D. M., Griffin, O., O'Kelly, H., & Keane, A. (2020). Experimentally assessing the effect of search effort on snare detectability. *Biological Conservation, 247*(March), 108581. https://doi.org/10.1016/j.biocon.2020.108581

Jones, S., Papworth, S., Keane, A. M., Vickery, J., & St John, F. A. V. (2020). The bean method as a tool to measure sensitive behaviour. *Conservation Biology*, *0*(0), 1–10. https://doi.org/10.1111/cobi.13607

Keane, A., Jones, J. P. G., Edward-Jones, G., & Milner-Gulland, E. J. (2008). The sleeping policeman: Understanding issues of enforcement and compliance in conservation. *Animal Conservation*, *11*(2), 75–82. https://doi.org/10.1111/j.1469-1795.2008.00170.x

Keane, A., Jones, J. P. G., & Milner-Gulland, E. J. (2011). Encounter data in resource management and ecology: Pitfalls and possibilities. *Journal of Applied Ecology*, *48*(5), 1164–1173. https://doi.org/10.1111/j.1365-2664.2011.02034.x

Keane, A., Ramarolahy, A. A., Jones, J. P. G., & Milner-Gulland, E. J. (2011). Evidence for the effects of environmental engagement and education on knowledge of wildlife laws in Madagascar. *Conservation Letters*, *4*(1), 55–63. https://doi.org/10.1111/j.1755-263X.2010.00144.x

Krumpal, I. (2013). Determinants of social desirability bias in sensitive surveys: A literature review. *Quality and Quantity*, *47*(4), 2025–2047. https://doi.org/10.1007/s11135-011-9640-9

Kurland, J., & Pires, S. F. (2017). Assessing U.S. wildlife trafficking patterns: How criminology and conservation science can guide strategies to reduce the illegal wildlife trade. *Deviant Behaviour*, *38*(4), 375–391. https://doi.org/10.1080/01639625.2016.1197009

Lau, J. T. F. F., Yeung, N. C. Y. Y., Mui, L. W. H. H., Tsui, H. Y., & Gu, J. (2011). A simple new method to triangulate self-reported risk behaviour data-the bean method. *Sexually Transmitted Diseases*, *38*(9), 788–792. https://doi.org/10.1097/OLQ.0b013e318218cc66

Lee, R. M., & Renzetti, C. M. (1990). The problems of researching sensitive topics. An Overview and Introduction. *American Behavioural Scientist*, *33*(5), 510–528. https://doi.org/10.1177/0002764290033005002

Milner-Gulland, E. J., & Clayton, L. (2002). The trade in babirusas and wild pigs in North Sulawesi, Indonesia. *Ecological Economics*, *42*(1–2), 165–183.

Monkman, G. G., Kaiser, M., & Hyder, K. (2018). The ethics of using social media in fisheries research. *Reviews in Fisheries Science & Aquaculture*, *26*(2), 235–242.

Newing, H. S. (2011). *Conducting Research in Conservation. A Social Science Perspective*. Taylor & Francis.

Nuno, A., Blumenthal, J.M., Austin, T.J., Bothwell, J., Ebanks-Petrie, G., Godley, B.J. and Broderick, A.C. (2018), Understanding implications of consumer behavior for wildlife farming and sustainable wildlife trade. Conservation Biology, 32: 390-400. https://doi.org/10.1111/cobi.12998

Nuno, A., & St John, F. A. V. (2015). How to ask sensitive questions in conservation: A review of specialized questioning techniques. *Biological Conservation*, *189*, 5–15. https://doi.org/10.1016/j.biocon.2014.09.047

Nurse, A. (2011). Policing wildlife: Perspectives on criminality in wildlife crime. In *Papers from the British Criminology Conference* (Vol. 11, pp. 38–53).

Olmedo, A., Davis, E., & Hinsley, A. (2019). *Asking sensitive questions in conservation using the Unmatched Count Technique* (Oxford Martin Programme in the Illegal Wildlife Trade, pp. 1–5). University of Oxford. https://doi.org/10.31235/osf.io/rcdbk

Oyanedel R., Gelcich S., & Milner-Gulland E. J. (2020). Motivations for (non-)compliance with conservation rules by small-scale resource users. *Conservation Letters*, 13, e12725. https://doi.org/10.1111/conl.12725

Paudel, P. K., Acharya, K. P., Baral, H. S., Heinen, J. T., & Jnawali, S. R. (2020). Trends, patterns, and networks of illicit wildlife trade in Nepal: A national synthesis. *Conservation Science and Practice*, *2*(9), 1–11. https://doi.org/10.1111/csp2.247

Pritchard, R., Sauls, L. A., Oldekop, J. A., Kiwango, W. A., & Brockington, D. (2022). Data justice and biodiversity conservation. *Conservation Biology*, *36*(5), e13919. https://doi.org/10.1111/cobi.13919

Razafimanahaka, J. H., Jenkins, R. K. B., Andriafidison, D., Randrianandrianina, F., Rakotomboavonjy, V., Keane, A., & Jones, J. P. G. (2012). Novel approach for quantifying illegal bushmeat consumption reveals high consumption of protected species in Madagascar. *Oryx*, *46*(4), 584–592. https://doi.org/10.1017/S0030605312000579

Robbins, P. (2000). The rotten institution: Corruption in natural resource management. *Political Geography*, *19*(4), 423–443. https://doi.org/10.1016/S0962-6298(99)00087-6

Robbins, P., Mcsweeney, K., Waite, T., & Rice, J. (2005). Even conservation rules are made to be broken: Implications for biodiversity. *Environmental Management*, *37*(2), 162–169. https://doi.org/10.1007/S00267-005-0009-5

Sandbrook, C., Luque-Lora, R., & Adams, W. M. (2018). Human bycatch: Conservation surveillance and the social implications of camera traps. *Conservation and Society, 16*(4), 493–504.
Sandbrook, C., Clark, D., Toivonen, T. Simla, T., O'Donnell, S., Cobbe, J., Adams, W. (2021). Principles for the socially responsible use of conservation monitoring technology and data. Conservation Science and Practice; 3:e374. https://doi.org/10.1111/csp2.374
Satizábal, P., Le Billon, P., Belhabib, D., Saavedra-Díaz, L. M., Figueroa, I., Noriega, G., & Bennett, N. J. (2021). Ethical considerations for research on small- scale fisheries and blue crimes. *Fish and Fisheries, 22*, 1160–1166. https://doi.org/10.1111/faf.12590
Sieber, J., & Stanley, B. (1988). Ethical and professional dimensions of socially sensitive research. *Psychology in the Public Forum, 43*(1), 49–55. https://doi.org/10.1037/0003-066X.43.1.49
Simlai, T. (2021). *Negotiating the Panoptic Gaze: People, Power and Conservation Surveillance in the Corbett Tiger Reserve* [University of Cambridge]. https://doi.org/10.17863/CAM.84136
Solomon, J., Jacobson, S. K., Wald, K. D., & Gavin, M. (2007). Estimating illegal resource use at a Ugandan Park with the randomized response technique. *Human Dimensions of Wildlife, 12*(2), 75–88. https://doi.org/10.1080/10871200701195365
Solomon, J. N., Gavin, M. C., & Conteh, A. (2016). Codes of ethics are critical for research on non-compliance with conservation rules and regulations. *Biological Conservation, 196*(November 2015), 210. https://doi.org/10.1016/j.biocon.2016.01.026
Speiglman, R., & Spear, P. (2008). The role of institutional review boards: Now you see them, now you don't. In D. M. Mertens & P. Ginsberg (Eds.), *The Handbook of Social Research Ethics* (pp. 121–134). Thousand Oaks, CA: Sage
St John, F. A. V., Keane, A. M., Jones, J. P. G., & Milner-Gulland, E. J. (2014). Robust study design is as important on the social as it is on the ecological side of applied ecological research. *Journal of Applied Ecology, 51*(6), 1479–1485. https://doi.org/10.1111/1365-2664.12352
St John, F. A. V., Linkie, M., Martyr, D.J., Milliyanawati, B., McKay, J.E., Mangunjaya, F.M., Leader-Williams, N. and Struebig, M.J., (2018). Intention to kill: Tolerance and illegal persecution of Sumatran tigers and sympatric species. *Conservation Letters, 11*(4), e12451.
St John, F. A. V., Mai, C.-H., & Pei, K. J. C. (2015). Evaluating deterrents of illegal behaviour in conservation: Carnivore killing in rural Taiwan. *Biological Conservation, 189*, 86–94. https://doi.org/10.1016/j.biocon.2014.08.019
Sundström, A. (2016). Understanding illegality and corruption in forest governance. *Journal of Environmental Management, 181*, 779–790. https://doi.org/10.1016/j.jenvman.2016.07.020
Thompson, R. M., Hall, J., Morrison, C., Palmer, N. R., & Roberts, D. L. (2021). Ethics and governance for internet-based conservation science research. *Conservation Biology, 35*(6), 1747–1754.
Tourangeau, R., & Yan, T. (2007). Sensitive questions in surveys. *Psychological Bulletin, 133*(5), 859–883. https://doi.org/10.1037/0033-2909.133.5.859
UNODC. (2024). *United Nations Office on Drugs and Crime* [dataset]. https://dataunodc.un.org
von Essen, E., Hansen, H. P., Nordström Källström, H., Peterson, M. N., & Peterson, T. R. (2014). Deconstructing the poaching Phenomenon. *British Journal of Criminology, 54*(4), 632–651. https://doi.org/10.1093/bjc/azu022
Wardropper, C. B., Dayer, A. A., Goebel, M. S., & Martin, V. Y. (2021). Conducting conservation social science surveys online. *Conservation Biology: The Journal of the Society for Conservation Biology, August 2020*, 1–9. https://doi.org/10.1111/cobi.13747
Wasser, S. K., Brown, L., Mailand, C., Mondol, S., Clark, W., Laurie, C., & Weir, B. S. (2015). Genetic assignment of large seizures of elephant ivory reveals Africa's major poaching hotspots. *Science, 349*(6243), 84–87. https://doi.org/10.1126/science.aaa2457
Wax, M. L. (1980). Paradoxes of 'consent' to the practice of fieldwork. *Social Problems, 27*(3), 272–283. https://doi.org/10.2307/800246
Wrege, P. H., Rowland, E. D., Keen, S., & Shiu, Y. (2017). Acoustic monitoring for conservation in tropical forests: Examples from forest elephants. *Methods in Ecology and Evolution, 8*(10), 1292–1301. https://doi.org/10.1111/2041-210X.12730
Wyatt, T. (2016). Victimless venison? Deer poaching and black-market meat in the UK. *Contemporary Justice Review, 19*(2), 188–200. https://doi.org/10.1080/10282580.2016.1169700
Wyatt, T., Miralles, O., Massé, F., Lima, R., da Costa, T. V., & Giovanini, D. (2022). Wildlife trafficking via social media in Brazil. *Biological Conservation, 265*, 109420. https://doi.org/10.1016/j.biocon.2021.109420

11

THE COMMERCIAL POACHING-GREEN MILITARISATION NEXUS AS UNSUSTAINABLE ENVIRONMENTAL CONFLICT

Elizabeth (Libby) Lunstrum, Nicia Givá, and Francis Massé

Introduction

One of the most high-profile forms of wildlife crime has been the increase in the commercial poaching of rhinos and elephants for their horns and tusks, respectively, over the last decade. The source of many wildlife products for the global illegal wildlife trade (IWT), commercial poaching stands out as a driver of the overexploitation of wildlife, putting pressure on biodiversity and contributing to what some refer to as the sixth mass extinction (Ceballos et al., 2020). IWT is defined as the "trade in wildlife or wildlife parts that violates either international legal frameworks or the legislation of one or several of the countries through which a wildlife product has passed" (OECD, 2019). How to stem commercial poaching and IWT has been on the global agenda for over a decade, even becoming incorporated into broader security and geopolitical discussions and policy arenas (Massé & Margulies, 2020). Importantly, IWT not only harms biodiversity but also increasingly drives a new form of *environmental conflict* as states and conservation NGOs respond to the pressing concern of IWT with a heavy and increasingly militarised hand, amounting to practices of green militarisation (Lunstrum, 2014), or the militarisation of conservation.

In this chapter, we demonstrate how political ecology approaches to environmental conflict provide a useful lens to understand the increase in commercial poaching of rhinos and the militarised response in South Africa and Mozambique. We illustrate how the poaching-green militarisation nexus provokes violent environmental conflict between conservation actors and ground-level hunters who are often organised by global criminal syndicates. We argue this is a conflict rooted in important ways in long-standing and deepening inequalities in the South Africa–Mozambican borderlands. This conflict directly threatens rhino populations through unsustainable illegal hunting, or what is commonly referred to as poaching. The militarised response, equally, may violate human rights, leads to further dispossession of already marginalised rural communities and, subsequently, threatens the long-term viability of conservation by alienating the very people whose support is needed

DOI: 10.4324/9781003333906-13

for conservation's success. The environmental conflict provoked by and that encompasses the poaching-green militarisation nexus is thus unsustainable both ecologically and socially. These dynamics, and an environmental conflict approach, ultimately demand a policy response that rejects militarised interventions to commercial poaching in favour of addressing underlying inequalities.

Rhino poaching, IWT, and green militarisation

To grasp the severity and complexity of IWT as a form of environmental conflict and seek out viable solutions, it is important to understand the dynamics of rhino poaching and the emergence of green militarisation as a response. This chapter draws on our collective research from one of the world's most intense sites of commercial poaching and of militarised responses, namely South Africa's Kruger National Park and the adjacent Limpopo National Park and surrounding areas in Mozambique.[1] Commercial poaching and IWT have a long history but saw a dramatic increase in 2008, with the supply-side of the trade focused heavily on sub-Saharan Africa and Asia. South Africa is a case in point. Home to the world's largest population of rhinoceros, the country saw only 13 rhinos killed illegally in 2007, a number that skyrocketed to 1,215 in 2014. Those numbers fell to 499 in 2023 but are still concerningly high and show an upward trend from 2021 (DFFE, 2024). Elephants and pangolin, currently the world's most trafficked mammal, have also seen drastic increases in illicit harvest over the last 15 years, although both numbers have fallen in recent years (Kuiper et al., 2023; UNODC, 2020). Demand for these and other wildlife species has been driven largely by consumers in Asia, but substantial demand also comes from North America, Europe, and Australia. This underscores the need to avoid stereotypes, for instance, of "Asian super consumers" (Margulies et al., 2019) and reinforces the fact that IWT is truly global in scope (UNODC, 2020).

The rhino horn trade is an extraordinarily lucrative enterprise that is a relatively low-risk, high-reward activity for the networks organising the trade (Interpol, 2022). To give a sense of the wealth at stake, at the height of the rhino horn trade, pound-by-pound rhino horn was more expensive than gold and cocaine (Ripple et al., 2015). The trade in rhino horn, valued at approximately $230 million per year gross income (UNODC, 2024), forms part of the global IWT, whose precise value is difficult to estimate given its clandestine nature. Commonly cited estimates range from USD $5 to 23 billion annually (Tow et al., 2021), reflecting its position as one of the world's most profitable illegal economies. It has even been suggested that IWT's precipitous rise was tied to the 2008 global financial crisis (Di Minin et al., 2022), where investors were searching out more secure, albeit illegal, opportunities to reap impressive returns on investments. This is juxtaposed against the often abject and relative poverty and marginalisation that characterise many areas that are the sources of wildlife products, such as rhino horn, which exists alongside some of the poorest rural populations in sub-Saharan Africa.

Compared to other conservation challenges, IWT is distinct in two ways. It has received critical attention in the context of the COVID-19 pandemic (UNEP, 2020), given speculation that coronavirus may have emerged from the wildlife trade and broader recognition that IWT is a dangerous vector for zoonosis (Cassey et al., n.d.; Massé & Gladkova, 2024). More central to this intervention, IWT is increasingly understood not merely as a conservation challenge but framed as an issue of transnational crime (Massé et al., 2020) and security (Duffy, 2021; Massé & Margulies, 2020), both in a national and global sense. In other words, IWT and poaching are not inherently security issues, but have been framed as such through what security studies scholars refer to as a process of securitisation (Buzan et al., 1998). This move has been driven in part by concerns that illicit hunters are heavily armed (Mogomotsi & Madigele, 2017) and organised, that the trade is driven by global criminal syndicates (Walker, 2017), and largely unsubstantiated (Maguire & Haenlein, 2015) and overblown (Duffy, 2022) claims that poaching fuels terrorist networks like Al Shabaab, Boko Haram and the Lord's Resistance Army (U.S. House Hearing, 2015). Rhino poaching has an added layer in that the killing of rhinos in South Africa, especially by Mozambicans, is framed as a cross-border attack on national and natural heritage (Griffiths, 2017; Lunstrum, 2014). Conservation organisations, which are notoriously inadequately funded, have strategically tapped into and reinforced the security angle to garner needed support for their work. This has paid off in the form of increased assistance from foreign governments (Massé & Margulies, 2020) and philanthropists (SANParks, 2014), along with defence and military corporations (Lunstrum, 2018).

The securitisation of IWT has led to a broader security and policing response (Massé & Margulies, 2020) that includes judicial reform, criminal investigation, and intelligence gathering (Massé, 2022). Specifically, within and near protected areas (PAs), i.e. areas where wildlife and wildlife parts are often illicitly acquired, the security response has become an increasingly *militarised* response. This militarised response is captured in the concept of green militarisation, defined as "the use of military and paramilitary personnel, training, technologies, and partnerships in the pursuit of conservation efforts" (Lunstrum, 2014, p. 817). These practices reflect global changes in conservation practice (Duffy et al., 2019) that include the para-militarisation of ranger corps, which involves new paramilitary training and diverting work away from traditional ecological monitoring and towards policing (Annecke & Masubelele, 2016); deployment of national armies to patrol PAs (Somerville, 2018); partnerships with military and defence corporations to gain access to hardware and personnel (Lunstrum, 2018); the related use of (para)military technology (Massé, 2018) including drones (Millner, 2020) for surveillance and pursuit; the recruitment of (retired) military personnel rather than conservation actors to lead anti-poaching missions (Lunstrum, 2014); the militarisation of language, e.g. referring to conservation as a "war" (Swails & Magnay, 2014); and militarised ecotourism (Marijnen, 2022). These, in turn, have far-reaching impacts beyond militarisation (Mushonga & Matose, 2020).

In many ways, green militarisation is not new. Western conservation is marred by a colonial history of deploying force against local and Indigenous communities to create PAs (Neumann, 1998; Enns & Bersaglio, 2024), with militaries playing an important role (Jacoby, 2001; Spence, 1999). Current military buildup within PAs, moreover, can often trace its roots back to broader military conflict (Ashaba, 2021). What is new with this latest wave of militarisation is its unprecedented scale and intensity (Duffy et al., 2019), with large-scale (para)military buildup unfolding across PAs and hundreds of suspected illicit hunters shot and killed over the last decade in the fight against rhino poaching alone. While we see these changes most explicitly around IWT, the militarised approach is expanding to reshape conservation, for instance, in the contexts of forest governance (Dutta, 2020; Mushonga & Matose, 2020), climate change mitigation (Asiyanbi, 2016), along with conservation in conflict zones (Marijnen, 2017).

Green militarisation has, in short, produced a pattern of human rights abuses. This includes the killing of suspected environmental transgressors, both in countries like Botswana that has a 'shoot-on-sight' policy (Mogomotsi & Madigele, 2017) and South Africa that does not (Burleigh, 2017), along with forced displacement of communities away from PAs (Lunstrum, 2016). The killing of suspected poachers and long prison sentences have also led to a "crisis of widows" (Massé et al., 2021), in which the women and families left behind are left more vulnerable without an economic provider and other forms of support, worsening already extreme poverty. These are underscored by broader interventions in which local people are seen not as needed partners in conservation but rather as conservation's enemy (Mahato, 2022). While human rights organisations and academics have been charting these abuses for the last decade, they hit the mainstream news in a 2019 BuzzFeed (Warren & Baker, 2019) exposé of the World Wildlife Fund (WWF), which led to US congressional hearings to review support for the organisation (House Natural Resources Committee Democrats, 2021).

It is this constellation of forces that set the stage for understanding the poaching crisis as a distinctly violent and unsustainable form of environmental conflict.

The rhino poaching crisis as environmental conflict

Environmental conflict is often understood as the process through which pressures on natural resources stemming from processes like population growth and environmental changes – including deforestation, desertification, drought, and water insecurity, among others – lead to conflict (Homer-Dixon, 2010). Political ecology, sometimes defined as "the study of ecological distribution conflicts" (Martinez-Alier, 2003, p. 54), pushes back against this naturalised, deterministic, and apolitical understanding of environmental conflict. It approaches conflict as emerging from broader struggles of land and environmental resources, and how these struggles are shaped by unequal processes of power that produce and intensify an unequal distribution of resources and related marginalisation, dispossession, and vulnerabilities (Asiyanbi, 2016; Le Billon & Duffy, 2018). In this way, environmental conflict is best understood as produced and shaped by unequal political-economic structures and relations in which some actors capture resources and their benefits while others lose out and are dispossessed, for instance, of land, resources, related value, and cultural connection. Important here is the understanding that poor people do not simply degrade their environments out of ignorance, irresponsibility, or desperation, as more traditional Malthusian approaches would have us assume (Ehrlich, 1975; Homer-Dixon, 2010). Rather, environmental conflict is a process whereby incompatible uses and values of natural resources and understandings of environmental change emerge over time and in specific sites between different groups; this, in turn, leads to an unequal distribution of and access to resources and resulting tensions and conflict. Environmental conflict is thus site-specific and involves competing values (both material and symbolic) and competing uses of specific natural resources within a context of unequal access.

Political ecology has robustly studied biodiversity conservation as a process of competing values and uses with regard to wild flora, fauna, and the spaces they occupy (Neumann, 1998). It has also documented how such tensions often result in elite actors securing control over land and resources, often leading to the accumulation of wealth and power in their hands and the dispossession and marginalisation of the rural poor who depend on land and resources for their livelihoods, survival, and well-being (Brockington, 2002). The militarised response to the poaching crisis – in defence of conservation areas and specific wildlife populations more broadly – has and continues to replicate similar patterns. Namely, the security buildup in PAs like Kruger and the Limpopo National Parks has seen an increase in the use of violence and the dispossession of marginalised communities within and around it in the name of combatting rhino poaching and the trade in rhino horn (Lunstrum & Ybarra, 2018; Witter & Satterfield, 2019).

Political ecology debates on conservation and green militarisation, nonetheless, have typically not engaged with environmental conflict as a lens to understand and approach conservation or forms of wildlife crime. *We show below how environmental conflict is a productive and useful lens for understanding the rhino poaching conflict*, namely as a conflict over rhinos as a natural and national treasure and resource and as an illegal economic commodity on the international illegal wildlife market. Such an approach begins with understanding the drivers of IWT, and particularly commercial poaching.

A broad array of research across various disciplines demonstrates that people engage in different forms of illegal hunting, from commercial poaching to subsistence hunting, for a variety of reasons. These include rationales attached to recreation and cultural practice; resistance against conservation, wildlife-use restrictions, and associated violence; and, of course, for livelihoods, food security, and economic gain (Hübschle, 2016; Kahler & Gore, 2012; Knapp, 2012; Lunstrum & Givá, 2020; Witter, 2021). The latter economic reasons are indeed a primary driver for commercial poaching of wildlife, both for meat and wildlife parts like horn and ivory. Rhino hunting in the Mozambique–South Africa borderlands reflects this. As we illustrate in detail elsewhere (Lunstrum & Givá, 2020), economic motivations and especially the lack of jobs and other economic opportunities are the most frequently cited motivations for young men entering into the trade (also see Hubschle & Jooste, 2017). As criminal syndicates organise the rhino horn trade, driven by global demand and consumers willing to pay exorbitant prices, they find recruits often from the poor but skilled communities in the South Africa–Mozambique borderlands adjacent to the largest concentration of rhinos in the world. Our research shows that even though recruits receive only a fraction of the overall rhino horn profit, it is substantial if not unprecedented compared to other available economic options (Lunstrum & Givá, 2020). Employing their animal tracking and bush skills gained from cattle rearing and subsistence hunting, small hunting teams can earn upwards of USD $5,000 for a successful rhino hunt (Interviews 2012–2016). If we contrast the wealth of rhino horn consumers with the poverty of recruits, we see how rhino poaching is not driven by absolute poverty but rather inequality. Here, wealth is also a primary driver (Price, 2017) as is the vast economic gap separating the poor and the wealthy (Hubschle & Shearing, 2018; Lunstrum et al., 2023).

The stark monetary incentives and motivations to hunt rhinos for the lucrative international black market can be contrasted with the protection or policing side of the rhino poaching conflict. Here we see South African and Mozambican conservation and security bodies provided with the mandate and resources to protect rhinos from would-be poachers. While the efforts to prevent and address poaching are driven by a variety of concerns – from poaching posing a threat to wild rhino populations and conservation efforts to being a transboundary security and criminal concern – economic concerns are also not lost here. Rhino poaching is seen as threatening South Africa's ecotourism revenues (Annecke & Masubelele, 2016), a position that reflects broader arguments about how poaching and IWT can be a detriment to ecotourism, wildlife, and conservation-based economies, and thus of economic security, especially in Africa (Holterman, 2022).

An environmental conflict lens thus draws attention to this mismatch between (1) impoverished, vulnerable young men drawn into the poaching trade by economic opportunity, itself driven by demand from wealthy consumers, and (2) the state and conservation sector endowed with the mandate and resources to protect rhino as a national, economic, and natural resource. The deadly poaching conflict unfolds as those protecting and hunting rhino collide in Kruger National Park and adjacent PAs in Mozambique (and other sites of rhino conservation).

The militarised buildup in Kruger National Park and in the borderlands of Mozambique to protect rhino and address poaching is resulting in the use of violence and heavy-handed tactics against not only illicit rhino hunters but also their communities through their physical and indirect displacement (Lunstrum & Ybarra, 2018; Witter & Satterfield, 2019). This further threatens the well-being of already vulnerable and marginalised communities. Consequential in its own right, this also leads to community alienation from and even defiance against conservation efforts (Duffy et al., 2019; Witter, 2021). This threatens the very survival of conservation and species protection efforts (Duffy et al., 2019), which need the buy-in and support of local people for their long-term success. In other words, while the rhino poaching threat is real, the militarised approach to conservation is further alienating local communities from their resources and livelihoods, which in turn actively disrupts conservation (Lunstrum, 2014).

Importantly, the inequality underpinning rhino poaching is not simply the result of contemporary conservation efforts, militarised or otherwise. While beyond the scope of this chapter, it is important to understand this inequality as produced through a long-term generation of unequal economic and environmental relations, largely beginning with colonialism and continuing through with South Africa's war of destabilisation in Mozambique and the implementation, again in Mozambique, of neoliberal economic policies that further entrenched rural poverty and produced limited livelihood and economic options for rural smallholders. While well-intentioned, conservation efforts in the form of Kruger and Limpopo National Parks and adjacent PAs have unfortunately exacerbated these inequalities and further alienated rural people from the land and resources they depend on. The militarised approach to combatting rhino poaching replicates and furthers these dynamics (Lunstrum et al., 2023).

An environmental conflict lens coming from a political ecology tradition thus helps clarify that the motivations for rhino poaching is not merely reducible to absolute wealth or poverty. Rather, these motivations emerge from inequalities over the use of and access to resources embodied in rhino horn and to land as PAs and that stem from long-term underdevelopment within these same areas. This environmental conflict approach, which sees *inequality* rather than merely poverty or economic gain as driving the rhino horn trade, offers several useful insights. First, a focus on economic inequality offers a more nuanced understanding of and is arguably a more accurate primary driver of commercial rhino poaching. Second, and drawing on the concept of relative poverty, inequality captures how economic drivers include more than simple material deprivation but embody multiple experiences and interpretations of poverty and desires to escape it. These motivations range from caring for family and community, supplementing agricultural livelihoods, and otherwise making ends meet, to partying, showing off, increasing social standing, and reaching a previously unattainable level of freedom that being a successful rhino poacher is perceived to provide. Third, the focus on inequality also accounts for the demand side of IWT, and how commercial poaching and IWT economies like rhino horn emerge from and are driven by the wealth of those consuming and buying rhino horn on the black market. Brought to the fore is thus the stark gap between poverty and wealth that underpins the rhino horn economy, fuels poaching and the illegal trade, and ultimately underpins the rhino poaching conflict (Lunstrum et al., 2023).

Conclusion: moving beyond conflict

The above insights gleaned from approaching the poaching and wildlife crime crisis as environmental conflict raise important policy lessons for how to address poaching and IWT (Fukushima et al., 2021) and stem the environmental conflict. The reality is that rhino poaching, and IWT more broadly, is a complex problem requiring a complex slate of policy interventions. While there has been robust debate on the legalisation of IWT (Rizzolo, 2021), it is unclear whether this would solve the problem or make it worse. More viable solutions target rather than legitimise demand. Paralleling the "war on drugs," culturally sensitive demand reduction (Dang Vu & Nielsen, 2020) is arguably the most important policy intervention because if there is no demand, there is no trade. Elsewhere along the supply chain, needed interventions include strengthening international legal cooperation (Jiao et al., 2021), addressing corruption (Camargo & Burgess, 2022), and strengthening policing while avoiding overly militarised approaches (Nurse, 2015). Equally, if not more important, solutions rest in stronger community engagement (Roe & Booker, 2019) and support (IIED, 2015) that treat communities not as *prima facie* threats or enemies but rather partners in environmental and wildlife protection. This requires conservation efforts to not merely avoid harm but also materially benefit communities. More than this, there is a need to reimagine and approach perceived environmental transgressors not as racialised enemies, which is inherent in the current approach, but rather as people worthy of dignity, necessitating a non-violent response to wildlife crime (Ramutsindela et al., 2022).

Two additional needed interventions are not simple policy responses but rather complex bodies of responses. First, although insufficiently acknowledged, effective "community engagement" demands firmer commitment to understand and respect how local worldviews may provide more sustainable models of human–environment and biocultural protection (Bridgewater & Rotherham, 2019) (i.e. the joint protection of people/culture and biodiversity). This also requires stronger protection of community and Indigenous rights as part of a broader rights-based approach (Tauli-Corpuz et al., 2020) and broader efforts to decolonise conservation (Domínguez & Luoma, 2020). These simply cannot happen in a militarised framework (Lunstrum et al., 2025). Second, and what the environmental conflict approach to rhino poaching outlined above illustrates is that it is vital to address poverty and inequality as underlying drivers of poaching and the broader illegal trade in rhino horn (Lunstrum & Givá, 2020). This takes poaching and IWT out of a simple conservation framework. Unlike green militarisation, however, this reframes rhino poaching, and similar commercial poaching economies in marginalised and vulnerable areas, not as a security or criminal issue, but rather one of human development and well-being.

Acknowledgements

This chapter draws heavily from our previous work (see footnote 1) and directly from the blog post "The Illegal Wildlife Trade-Green Militarisation Nexus Provokes Unsustainable Environmental Conflict" published by the *Georgetown Journal of International Affairs* (Lunstrum, 2023). We thank the journal for allowing us to reproduce much of the blogpost in this chapter. We also thank previous co-authors, particularly Filipe Mate and Paulo Lopes Jose, along with many of the people we interviewed in Mozambique and South Africa from 2009 to 2016 whose insights are reflected throughout our joint publications upon which this chapter is built.

Note

1 See the following articles for more detailed work on these issues that we draw on for this chapter: Lunstrum & Givá, 2020; Lunstrum et al., 2023; Massé et al., 2021.

References

Annecke, W., & Masubelele, M. (2016). A review of the impact of militarisation: The case of rhino poaching in Kruger National Park, South Africa. *Conservation and Society, 13*(4), 195–204.

Ashaba, I. (2021). Historical roots of militarised conservation: The case of Uganda. *Review of African Political Economy, 48*(168), 276–288. https://doi.org/10.1080/03056244.2020.1828052

Asiyanbi, A. P. (2016). A political ecology of REDD plus: Property rights, militarised protectionism, and carbonised exclusion in Cross River. *Geoforum, 77*, 146–156. https://doi.org/10.1016/j.geoforum.2016.10.016

Bridgewater, P., & Rotherham, I. D. (2019). A critical perspective on the concept of biocultural diversity and its emerging role in nature and heritage conservation. *People and Nature, 1*(3), 291–304. https://doi.org/10.1002/pan3.10040

Brockington, D. (2002). *Fortress Conservation: The Preservation of the Mkomazi Game Reserve, Tanzania*. Indiana University Press. Table of contents https://www.loc.gov/catdir/toc/fy034/2001051529.html

Burleigh, N. (2017). Kruger park, south africa: Where black poachers are hunted as much as their prey. *Newsweek Magazine*. https://www.newsweek.com/2017/08/18/trophy-hunting-poachers-rhinos-south-africa-647410.html

Buzan, B., Wæver, O., & Wilde, J. d. (1998). *Security: A New Framework for Analysis*. Lynne Rienner Pub.

Camargo, C. B., & Burgess, G. (2022). *Behavioural Drivers of Corruption Facilitating Illegal Widlife Trade: Problem Analysis and State of the Field Review*. https://www.traffic.org/publications/reports/behavioural-drivers-of-corruption-facilitating-illegal-wildlife-trade-problem-analysis-and-state-of-the-field/

Cassey, P., Wittmann, T., Toomes, A., & Stringham, O. (n.d.). Illegal Wildlife Trade (IWT): A Threat to Environmental Biosecurity, Biodiversity, Human Health and Wellbeing. In: Invasion Science & Wildlife Ecology; University of Adelaide.

Ceballos, G., Ehrlich, P. R., & Raven, P. H. (2020). Vertebrates on the brink as indicators of biological annihilation and the sixth mass extinction [Article]. *Proceedings of the National Academy of Sciences of the United States of America, 117*(24), 13596–13602. https://doi.org/10.1073/pnas.1922686117

Dang Vu, H. N., & Nielsen, M. R. (2020). Evidence or delusion: A critique of contemporary rhino horn demand reduction strategies. *Human Dimensions of Wildlife, 26*(4), 390–400. https://doi.org/10.1080/10871209.2020.1818896

DFFE. (2024). *KwaZulu-Natal carried the brunt of rhino poaching in 2023, says Creecy*. Retrieved 5 August from https://www.dffe.gov.za/mediareleases/creecy_relasesrhinopoachingstats2024feb27

Di Minin, E., t Sas-Rolfes, M., Selier, J., Louis, M., & Bradshaw, C. J. A. (2022). Dismantling the poachernomics of the illegal wildlife trade. *Biological Conservation, 265*, Article 109418. https://doi.org/10.1016/j.biocon.2021.109418

Domínguez, L., & Luoma, C. (2020). Decolonising conservation policy: How colonial land and conservation ideologies persist and perpetuate indigenous injustices at the expense of the environment. *Land, 9*(3), 65. https://doi.org/10.3390/land9030065

Duffy, R. (2021). *Security and Conservation*. Yale University Press.

Duffy, R. (2022). Crime, security, and illegal wildlife trade: Political ecologies of international conservation. *Global Environmental Politics, 22*(2), 23–44. https://doi.org/10.1162/glep_a_00645

Duffy, R., Massé, F., Smidt, E., Marijnen, E., Büscher, B., Verweijen, J., Ramutsindela, M., Simlai, T., Joanny, L., & Lunstrum, E. (2019). Why we must question the militarisation of conservation. *Biological Conservation, 232*, 66–73. https://doi.org/10.1016/j.biocon.2019.01.013

Dutta, A. (2020). Forest becomes frontline: Conservation and counter-insurgency in a space of violent conflict in Assam, Northeast India. *Political Geography, 77*, Article 102117. https://doi.org/10.1016/j.polgeo.2019.102117

Ehrlich, P. R. (1975). *The Population Bomb* (Rev. ed.). Rivercity Press.
Enns, C., & Bersaglio, B. (2024). *Settler Ecologies: The Enduring Nature of Settler Colonialism in Kenya*. University of Toronto Press.
Fukushima, C. S., Tricorache, P., Toomes, A., Stringham, O. C., Rivera-Téllez, E., Ripple, W. J., Peters, G., Orenstein, R. I., Morcatty, T. Q., Longhorn, S. J., Lee, C., Kumschick, S., Freitas, M. A. d., Duffy, R. V., Davies, A., Cheung, H., Cheyne, S. M., Bouhuys, J., Barreiros, J. P.,… Cardoso, P. (2021). Challenges and perspectives on tackling illegal or unsustainable wildlife trade. *Biological Conservation, 263*, 109342. https://www.sciencedirect.com/science/article/pii/S0006320721003943
Griffiths, M. L. (2017). Heritage Lost: The cultural impact of wildlife crime in South Africa. *South African Crime Quarterly, 60*, 45–50. https://doi.org/10.17159/2413-3108/2017/i60a1728
Holterman, D. (2022). Stopping the slaughter: The economic rationale to anti-poaching in Tanzania. *Geoforum, 135*, 106–117. https://doi.org/10.1016/j.geoforum.2022.08.003
Homer-Dixon, T. F. (2010). *Environment, Scarcity, and Violence*. Princeton University Press.
House Natural Resources Committee Democrats. (2021). *Protecting Human Rights in International Conservation*. https://www.youtube.com/watch?v=reezIOe8CbY&ab_channel=HouseNaturalResourcesCommitteeDemocrats
Hubschle, A., & Jooste, J. (2017). Interview with major general Johan Jooste (Retired), South African National Parks, head of special projects. *South African Crime Quarterly, 60*, 61–68. https://doi.org/10.17159/2413-3108/2017/v0n60a2776
Hubschle, A., & Shearing, C. (2018). *Ending wildlife trafficking: local communities as change agents*. https://traffickingtransformations.org/wp-content/uploads/2021/03/HubschlewithShearingWildlife-Trafficking-Report-WEB-4.pdf
Hübschle, A. M. (2016). The social economy of rhino poaching: Of economic freedom fighters, professional hunters and marginalized local people. *Current Sociology, 65*(3), 427–447. https://doi.org/10.1177/0011392116673210
IIED. (2015). *Community solutions are the key to tackling wildlife crime* https://www.iied.org/community-solutions-are-key-tackling-wildlife-crime
Interpol. (2022). Stepping up the global fight against wildlife crime through a united response. https://www.interpol.int/en/News-and-Events/News/2022/Stepping-up-the-global-fight-against-wildlife-crime-through-a-united-response
Jacoby, K. (2001). *Crimes Against Nature: Squatters, Poachers, Thieves, and the Hidden History of American Conservation*. University of California Press.
Jiao, Y., Yeophantong, P., & Lee, T. M. (2021). Strengthening international legal cooperation to combat the illegal wildlife trade between Southeast Asia and China. *Frontiers in Conservation Science, 9*(645427), 1–15. https://doi.org/10.3389/fevo.2021.645427
Kahler, J. S., & Gore, M. L. (2012). Beyond the cooking pot and pocket book: Factors influencing noncompliance with wildlife poaching rules. *International Journal of Comparative and Applied Criminal Justice, 36*(2), 103–120. https://doi.org/10.1080/01924036.2012.669913
Knapp, E. (2012). Why poaching pays: A summary of risks and benefits illegal hunters face in Western Serengeti, Tanzania. *Tropical Conservation Science, 5*(4), 434–445.
Kuiper, T., Altwegg, R., Beale, C., Carroll, T., Dublin, H. T., Hauenstein, S., Kshatriya, M., Schwarz, C., Thouless, C. R., Royle, A., & Milner-Gulland, E. J. (2023). Drivers and facilitators of the illegal killing of elephants across 64 African sites. *Proc Biol Sci, 290*(1990), 20222270. https://doi.org/10.1098/rspb.2022.2270
Le Billon, P., & Duffy, R. (2018). Conflict ecologies: Connecting political ecology and peace and conflict studies. *Journal of Political Ecology, 25*, 239–260. https://doi.org/10.2458/v25i1.22704
Lunstrum, E. (2014). Green militarization: Anti-poaching efforts and the spatial contours of Kruger National Park. *Annals of the American Association of Geographers, 104*(4), 816–832.
Lunstrum, E. (2016). Green grabs, land grabs and the spatiality of displacement: Eviction from Mozambique's Limpopo National Park. *Area, 48*(2), 142–152. https://doi.org/10.1111/area.12121
Lunstrum, E. (2018). Capitalism, wealth, and conservation in the age of security: The vitalization of the state. *Annals of the American Association of Geographers, 108*(4), 1022–1037. https://doi.org/10.1080/24694452.2017.1407629
Lunstrum, E. (2023). The illegal wildlife trade-green militarization nexus provokes unsustainable environmental conflict. *Georgetown Journal of International Affairs*. https://gjia.georgetown.edu/2023/01/09/the-illegal-wildlife-trade-green-militarization-nexus-provokes-unsustainable-environmental-conflict/

Lunstrum, E., & Givá, N. (2020). What drives commercial poaching? From poverty to economic inequality. *Biological Conservation, 245,* 1–8. https://doi.org/10.1016/j.biocon.2020.108505

Lunstrum, E., Giva, N., Massé, F., Mate, F., & Jose, P. L. (2023). The rhino horn trade and radical inequality as environmental conflict. *Journal of Peasant Studies. 50*(3), 1085–1105. https://doi.org/10.1080/03066150.2021.1961130

Lunstrum, E., & Ybarra, M. (2018). Deploying difference: Security threat narratives and state displacement from protected areas. *Conservation and Society, 16*(2), 114–124. https://doi.org/10.4103/cs.cs_16_119

Lunstrum, E., Massé, F., Ashaba, I. M., Dutta, A., Marijnen, E., Mushonga, T., & Matose, F. (2025). Demilitarizing conservation. *Annals of the American Association of Geographers,* 1–13. https://doi.org/10.1080/24694452.2025.2534550.

Maguire, T., & Haenlein, C. (2015). *An Illusion of Complicity Terrorism and the Illegal Ivory Trade in East Africa.* Royal United Services Institute for Defence and Security Studies. https://static.rusi.org/201509_an_illusion_of_complicity_0.pdf

Mahato, B. (2022). 'People as the Enemy of Conservation': The Militarization of Conservation in Nepal. *Medium.* https://survivalinternational.medium.com/people-as-the-enemy-of-conservation-the-militarization-of-conservation-in-nepal-1769f4c069bd

Margulies, J., Wong, W. Y. R., & Duffy, R. (2019). The imaginary "Asian super consumer": A critique of demand reduction campaigns for the illegal wildlife trade. *Geoforum, 107,* 216–219. https://doi.org/10.1016/j.geoforum.2019.10.005

Marijnen, E. (2017). The 'green militarisation' of development aid: The European commission and the Virunga National Park, DR Congo. *Third World Quarterly, 38*(7), 1566–1582. https://doi.org/10.1080/01436597.2017.1282815

Marijnen, E. (2022). Eco-war tourism: Affective geographies, colonial durabilities and the militarisation of conservation. *Security Dialogue, 53*(6), 550–566. https://doi.org/10.1177/09670106221112215

Martinez-Alier, J. (2003). *The Environmentalism of the Poor: A Study of Ecological Conflicts and Valuation.* Edward Elgar Publishing.

Massé, F. (2018). Topographies of security and the multiple spatialities of (conservation) power: Verticality, surveillance, and space-time compression in the bush. *Political Geography, 67,* 56–64. https://doi.org/10.1016/j.polgeo.2018.10.001

Massé, F. (2022). Police power in green: Furthering political ecologies of the state. *Political Geography, 97,* 102627. https://doi.org/10.1016/j.polgeo.2022.102627

Massé, F., Dickinson, H., Margulies, J., Joanny, L., Lappe-Osthege, T., & Duffy, R. (2020). Conservation and crime convergence? Situating the 2018 London illegal wildlife trade conference. *Journal of Political Ecology, 27,* 23–42. <Go to ISI>://WOS:000582466300002

Massé, F., Givá, N., & Lunstrum, E. (2021). A feminist political ecology of wildlife crime: The gendered dimensions of a poaching economy and its impacts in Southern Africa. *Geoforum, 126,* 205–214. https://doi.org/10.1016/j.geoforum.2021.07.031

Massé, F., & Gladkova, E. (2025). Spatializing zoonotic disease dynamics from a political ecology perspective: Reconceptualizing spillover as structure. *Journal of Political Ecology. 32*(1), 5810. https://doi.org/10.2458/jpe.5810.

Massé, F., & Margulies, J. (2020). The geopolitical ecology of conservation: The emergence of illegal wildlife trade as a national security interest and the re-shaping of US foreign conservation assistance. *World Development, 132,* Article 104958. https://doi.org/10.1016/j.worlddev.2020.104958

Millner, N. (2020). As the drone flies: Configuring a vertical politics of contestation within forest conservation. *Political Geography, 80*(102163), 1–13. https://doi.org/10.1016/j.polgeo.2020.102163

Mogomotsi, G. E. J., & Madigele, P. K. (2017). Botswana's 'shoot-to-kill' policy as an anti-poaching strategy. *South African Crime Quarterly, 60,* 51–59. https://doi.org/10.17159/2413-3108/2017/v0n60a1787

Mushonga, T., & Matose, F. (2020). Dimensions and corollaries of violence in Zimbabwe's protected forests. *Geoforum, 117,* 216–224. https://doi.org/10.1016/j.geoforum.2020.10.005

Neumann, R. (1998). *Imposing Wilderness: Struggles Over Livelihood and Nature Preservation in Africa.* University of California Press.

Nurse, A. (2015). *Policing Wildlife: Perspectives on the Enforcement of Wildlife Legislation.* Springer.

OECD. (2019). *The Illegal Wildlife Trade in Southeast Asia: Institutional Capacities in Indonesia, Singapore, Thailand and Vietnam*. OECD Publishing. https://doi.org/10.1787/14fe3297-en

Price, R. A. (2017). *Economic drivers and effects of the illegal wildlife trade in Sub Saharan Africa* (K4D Helpdesk Report, Issue. https://opendocs.ids.ac.uk/opendocs/bitstream/handle/20.500.12413/14234/144_Economic_Drivers_of_IWT_in_Sub-Saharan_Africa.pdf?sequence=1&isAllowed=y

Ramutsindela, M., Matose, F., & Mushonga, T. (2022). *The Vlence of Conservation in Africa: State, Militarization and Alternatives*. Edward Elgar.

Ripple, W. J., Newsome, T. M., Wolf, C., Dirzo, R., Everatt, K. T., Galetti, M., Hayward, M. W., Kerley, G. I. H., Painter, L. E., Sandom, C. J., Terborgh, J. W., & Valkenburgh, B. (2015). Collapse of the world's largest herbivores. *Science Advances, 1*(4), 1–12. https://doi.org/10.1126/sciadv.1400103

Rizzolo, J. B. (2021). Effects of legalization and wildlife farming on conservation. *Global Ecology and Conservation, 25*, e01390. https://doi.org/10.1016/j.gecco.2020.e01390

Roe, D., & Booker, F. (2019). Engaging local communities in tackling illegal wildlife trade: A synthesis of approaches and lessons for best practice. *Conservation Science and Practice, 1*(5), e26. https://doi.org/10.1111/csp2.26

SANParks. (2014). *Media Release: Howard Buffett R255 million donation to rhino conservation*. https://www.sanparks.org/about/news/news.php?id=56022

Somerville, K. (2018). Militarization of conservation: An army of occupation, not protection *Global Geneva*. https://global-geneva.com/militarization-of-conservation-an-army-of-occupation-not-protection/

Spence, M. D. (1999). *Dispossessing the Wilderness: Indian Removal and the Making of the National Parks*. Oxford University Press.

Swails, B., & Magnay, D. (2014). Anti-poaching 'war' to save rhinos whose horns are more valuable than gold. *CNN, November 30*. https://www.cnn.com/2014/11/30/world/africa/rhino-poaching-kruger-national-park/index.html

Tauli-Corpuz, V., Alcorn, J., Molnar, A., Healy, C., & Barrow, E. (2020). Cornered by PAs: Adopting rights-based approaches to enable cost-effective conservation and climate action. *World Development, 130*, Article 104923. https://doi.org/10.1016/j.worlddev.2020.104923

Tow, J. H., Symes, W. S., & Carrasco, L. R. (2021). Economic value of illegal wildlife trade entering the USA. *PLoS One, 16*(10), e0258523. https://doi.org/10.1371/journal.pone.0258523

U.S. House Hearing, C. (2015). Poaching and Terrorism: A National Security Challenge: Hearing Before the Subcommittee on Terrorism, Nonproliferation, and Trade of the Committee on Foreign Affairs House of Representatives, 114th Congress. In.

UNEP. (2020). *Preventing the Next Pandemic: Zoonotic Diseases and How to Break the Chain of Transmission*. U. N. E. P. (UNEP). https://www.unenvironment.org/resources/report/preventing-future-zoonotic-disease-outbreaks-protecting-environment-animals-and

UNODC. (2020). *World Wildlife Crime Report: Trafficking in protected species*. https://www.unodc.org/unodc/en/data-and-analysis/wildlife.html

UNODC. (2024). *World Wildlife Crime Report 2024* https://www.unodc.org/documents/data-and-analysis/wildlife/2024/Wildlife2024_Final.pdf

Walker, D. (2017). *IWT and Its Relation to Criminal Organizations: An African Case Study with Global Implications*. California State University Maritime Academy]. https://scholarworks.calstate.edu/downloads/9593tw03b?locale=en

Warren, T., & Baker, K. (2019). WWF Funds Guards Who Have Tortured And Killed People. *Buzzfeed News*. https://www.buzzfeednews.com/article/tomwarren/wwf-world-wide-fund-nature-parks-torture-death

Witter, R. (2021). Why militarized conservation may be counter-productive: Illegal wildlife hunting as defiance. *Journal of Political Ecology, 28*(1), 175–192.

Witter, R., & Satterfield, T. (2019). Rhino poaching and the "slow violence" of conservation-related resettlement in Mozambique's Limpopo National Park. *Geoforum, 101*, 275–284. https://doi.org/10.1016/j.geoforum.2018.06.003

PART III

Emerging and alternative responses to wildlife crime

12
ECOCIDE AS WILDLIFE CRIME

Nick Fromherz and Erica Lyman

Introduction

The crime of ecocide increasingly finds purchase across domestic legal systems, just as it gains momentum for adoption internationally. In parallel, laws and judicial decisions recognising the rights of nature multiply. Against the backdrop of the twin crises of mass extinction and climate change, humanity's need for new and potentially transformational means of legal recourse has never been clearer. Accordingly, the growth of these two concepts—ecocide and the rights of nature—is at once unsurprising and encouraging. While this chapter focuses on ecocide, it also offers a brief reflection on the complementarity of rights-of-nature approaches.

As discussed below, no single definition of ecocide exists; however, among the many proposed and well-established definitions in national law and international fora, commonalities suggest the following basic definition: a natural or legal person commits ecocide when they commit unlawful or wanton acts or omissions with the actual or constructive knowledge that those acts or omissions entail a substantial likelihood of severe and either widespread or long-term damage to the environment, wherein "environment" includes both ecosystems and their organic and inorganic components.[1]

For its part, the rights-of-nature concept can be understood as operationalising the notion that nature is a holder of certain fundamental rights, including the right to exist free of pollution, depredation, degradation, and anthropogenic harm to ecological processes (Bookman, 2023; Ryan et al., 2021). In line with the maxim *ubi jus ibi remedium* (where there is a right, there is a remedy), the rights-of-nature doctrine posits a suite of implementation and enforcement mechanisms, varying according to jurisdictional norms, from standing to bring suit to the creation of legal guardianships and injunctive remedies in the case of violations (Ryan et al., 2021).

On their face, and even more evident upon review of the terms' histories and applications to date, the rights of nature and ecocide approaches are paradigm-shifting. Borrowing from the now-familiar concept of genocide, both as historical fact and as legal framework, ecocide forces us to confront the idea that, in some cases, illegal acts affecting the environment

DOI: 10.4324/9781003333906-15

occur at either such a devastating scale or wreak such devastating consequences that traditional environmental crime approaches are inadequate.

The rights-of-nature doctrine goes against the grain of orthodox environmental law—and even environmental thinking, period—by recasting nature as a holder of rights for its own benefit and for the benefit of wildlife (albeit with plenty of spillover benefits for humans) rather than simply as subject matter to be regulated by the law according to human interests.[2]

A historical review of wildlife law reveals a systemic anthropocentrism that prioritises the so-called sustainable use of wildlife that values it as a resource and human wants as paramount. It is against this backdrop that we find dire data points, such as a 20 billion USD wildlife black market (INTERPOL, 2023). Under the frameworks of most wildlife laws, wildlife is merely of instrumental value, and the object of the law is to ensure the viability of wildlife populations for ongoing exploitation. Habitat is rarely protected as a means of wildlife conservation beyond the fencing of national parks, the creation of game reserves, or the identification of other protected areas or critical habitat. The consequence of this is twofold: First, habitat destruction and degradation are the primary threats to most biodiversity. Second, direct human exploitation is the second most impactful threat, and the most significant in the context of marine biodiversity (Díaz et al., 2019).

Given these facts, the Intergovernmental Panel on Biodiversity and Ecosystem Services has repeatedly called for transformational change of our legal systems (Díaz et al., 2019). Rights of nature and ecocide are complementary concepts that have the potential to fundamentally transform how both international and domestic legal systems apply to, and address, harms suffered by nature, including wildlife. While the rights of nature concept is a critical tool for pushing back against rampant human exploitation and the role of the State actor in approving and even prioritising initiatives that damage or degrade nature or wildlife, ecocide offers the opportunity for criminal sanctions when natural or legal persons are responsible for significant destruction of the environment.

With the latter point in mind, this chapter explores the role of ecocide in addressing wildlife crimes. Beginning with a discussion of the emergence of the ecocide concept, this chapter canvasses the extent to which ecocide has been embraced in various legal contexts and explores the possibility of its application to wildlife crimes. Whether ecocide is a promising tool for combatting wildlife crime depends on the scale of anthropocentrism embedded in the concept, threshold questions that depend on definitional permutations of ecocide, and whether direct harm to wildlife falls within the ecocide ambit.

Ecocide as an international crime

In the wake of the devastating environmental and human costs of the United States' use of Agent Orange during the Vietnam War, Arthur Galston, a biology professor at Yale, coined the phrase "ecocide" (Kusnetz et al., 2021). Almost immediately, activists and lawyers pursued codification of the concept. The most well-known was a barrister from the United Kingdom, the late Polly Higgins, whose tireless advocacy helped build the case for ecocide in both domestic and international settings (Fischels, 2021). With the contributions of a growing cast of actors, the ecocide movement has since transitioned from a compelling concept to a tangible—and, in several jurisdictions, controlling—rule of law.

At the international level, the concept rose to prominence in the late 1990s as negotiations to establish the International Criminal Court (ICC) were underway. Early drafts of the founding treaty of the ICC, called the Rome Statute, contained provisions that would have allowed prosecution of ecocide as a criminal offence before the ICC (Kusnetz et al., 2021). However, the United States, the Netherlands, and the United Kingdom opposed the inclusion of environmental crimes, other than certain offences during wartime, severely limiting the ICC's jurisdiction over crimes that affect the environment in most contexts (Kusnetz et al., 2021).

As ultimately constituted, the Rome Statute articulates four core crimes over which the ICC has jurisdiction: genocide, crimes against humanity, war crimes, and the crime of aggression (Rome Statute, 1998). Of these four jurisdictional crimes, the language of "war crimes" acknowledges, however modestly, the possibility of military crimes that purposely destroy the natural environment. Specifically, under Article 8(2)(b)(iv), "war crimes" include

> [I]ntentionally launching an attack in the knowledge that such attack will cause incidental loss of life or injury to civilians or damage to civilian objects *or widespread, long-term and severe damage to the natural environment which would be clearly excessive in relation to the concrete and direct overall military advantage anticipated.*
>
> *(Rome Statute, 1998, emphasis added)*

As a measure of ecocide's penetration within international criminal law, the Rome Statute can be seen as a mixed bag. On the one hand, the mere fact that ecocide was proposed and considered for inclusion in the treaty, even if ultimately scrapped, shows the international community's willingness to take the concept seriously. On the other hand, the proposal's rejection is hard to classify as anything other than a setback, even if only temporary. Further, the war-crimes provision's coverage of environmental harm, though hardly meaningless, remains a far cry from the ambitious scope of ecocide. The "environmental destruction" iteration of a war crime is limited to wartime and, by way of further narrowing, only applies when the conduct fails a test of proportionality (where the damage to the environment is "clearly excessive" when compared to the associated "military advantage"). Finally, and perhaps not surprisingly given these limitations, the "environmental destruction" provision of the Rome Statute has never been applied (Kusnetz et al., 2021).

Although not fully successful in the context of the original Rome Statute, activists and civil-society leaders continue the push for recognition of ecocide as the "fifth crime" under ICC jurisdiction.[3] Higgins' work, in particular, has offered a definition of ecocide around which many proponents have rallied (Fischels, 2021). Specifically, Higgins has defined ecocide as

> the extensive damage to, destruction of or loss of ecosystem(s) of a given territory, whether by human agency or by other causes, to such an extent that peaceful enjoyment by the inhabitants of that territory has been or will be severely diminished.
>
> *(European Law Institute, 2023)*[4]

Not limited to times of war and not subject to a balancing test of proportionality, Higgins' iteration is clearly an order apart from the narrow "environmental destruction" provision of the Rome Statute.

Higgins' proposed definitions and many others in circulation pose an important question regarding the utility of ecocide as a means of addressing wildlife crime. The definitions appear to rely on deep-rooted anthropocentrism in their framing, leading to legal standards that focus on the impacts of the harm to humans, as opposed to the environment or wildlife specifically.

While some might argue Higgins' use of the word "inhabitants" in her definition of ecocide encompasses nonhuman animals, the most likely reading of the proposed definition appears to set as a marker that humans have been impacted by environmental distress (Sollund, 2020). Even if one were to read "inhabitants" as including wild animals, the harm remains as an indirect effect experienced by the wildlife from some amount of habitat destruction. Of course, although this is a more helpful interpretation than one that perceives "inhabitants" as including exclusively humans, it does not address the harm experienced by wildlife from direct exploitation.[5]

Ecocide as a wildlife crime: examples and additional considerations

While Higgins and other advocates' mission to add ecocide to the Rome Statute remains a work in progress, other jurisdictions have, in fact, adopted ecocide laws. In one form or another, several jurisdictions have codified ecocide, and several more are considering codification. In fact, the general concept of ecocide is already a crime in, among other nations, Armenia, Belarus, Belgium, France, Georgia, Kazakhstan, Kyrgyzstan, Moldova, Russia, Tajikistan, Ukraine, and Vietnam (Hillsdon, 2023; Khattabi, 2022; "Ecocide/serious environmental crimes," n.d.). And, at last count, authorities in the UK, Italy, Spain, Mexico, Brazil, the Netherlands, Belgium, Scotland, and more are considering similar legislation, while the EU moves to include a version of ecocide in a revised directive on crime (Kaminski, 2023).[6]

Although these legislative advances may be viewed as integration of the concept of ecocide into legal frameworks, a universal definition of ecocide does not exist, and some provisions that may be considered "ecocide-like" nonetheless do not refer explicitly to ecocide. Additionally, in considering whether or to what extent these national-level ecocide laws apply to wildlife crime, each of the provisions varies regarding requisite *mens rea*, the significance of the harm as a threshold element, and the application to individual animals versus impact to a species, for example. Consideration of the types of harms that may be experienced by wildlife is instructive as a means of discerning whether current or future iterations of ecocide may be helpful tools to address the nature and scale of wildlife crime.

In recent years, one of the most common references to potential ecocide is, in fact, a wartime situation, specifically, the explosion of the huge Nova Kakhovka dam in June 2023 during the Russia–Ukraine war (Yermak & Wallström, 2023). The explosion caused a catastrophic flood that spilt tons of oil and other toxins into the Dnipro River, killing countless animals (Yermak & Wallström, 2023). According to the Ukrainian Nature Conservation Group (UNCG), many species now face a much more dire conservation outlook, including the near-total loss of habitat or gross losses of the majority of individuals comprising an already endangered species. For example, according to UNCG, 70% of the world population of Nordmann's birch mouse (*Sicista loriger*) may have been lost in the flood, and the population of the sandy blind mole-rat (*Spalax arenarius*) and Falz-Fein's thick-tailed jerboa (*Stylodipus telum falzfeini*) may have been reduced by up to 50% (Moisiienko et al., 2023). The UNCG also reported the impacts on reptiles, bird-nesting colonies, wetlands, marine animals, and more (Moisiienko et al., 2023).

According to local experts and other international advocates, Russia's invasion of Ukraine also resulted in the occupation of national parks, some of which have been landmined (Platt, 2023). Landmines are a significant problem for people and wildlife, as they kill indiscriminately and can also cause toxic pollution and even fire. Many have described land-mining as ecocide (Platt, 2023). Landmines planted during wartime are the most typical concerns, but landmines are also sometimes used for hunting.

Several advocates and scholars have identified non-wartime scenarios that could also constitute ecocide. In some cases, the act described as ecocidal is government decision-making; other times it is an act perpetrated by an individual or group. Instances of mass destruction of wildlife that have been described as ecocide include wildfires, illegal logging, and oil spills. During the Australian bushfires of 2019, at least some of which were directly human-caused and widely noted as a particularly devastating fire season for Australia, conservative estimates conclude that over a billion reptiles, mammals, and birds lost their lives and that at least a hundred times as many insects suffered the same fate (Brew et al., 2020; "A statement about the 480 million animals," 2020). The fires brought many species closer to the brink of extinction; at least 73 species listed as threatened in Australia lost at least 50% of their known habitat to the fires.

The ongoing environmental crises in both the Amazon and the Niger Delta might be described as ecocide perpetrated by individuals to the extent that their acts are illegal under domestic law or as government-sanctioned ecocide that might be considered an international crime, given the nature and extent of the impacts. For example, the Amazon rainforest is home to more than 10% of all known animal and plant species globally, and each hectare of Amazon rainforest contains dozens of species of vertebrates and a billion invertebrates. Each hectare destroyed thus has massive impacts on the lives of many millions of individual animals and contributes significantly to climate change and species' extinctions (Costa, 2020).

The Niger Delta has been the home of an environmental crisis for years (Ratcliffe, 2019). Tens of millions of litres of oil are spilt every year across the Niger Delta, impacting the local air, land, and water (Ratcliffe, 2019). This oil spillage has had devastating consequences for the area's wildlife, including the Niger Delta red colobus and the Niger Delta pygmy hippopotamus (Burgess & Martin, n.d.). Large areas of the mangrove forests have been destroyed, decimating habitats and ecosystems essential to wildlife (Numbere, 2020).

Other harms that may qualify as ecocide might be less indicative of mass destruction but are nonetheless significant due to their impact on species' survival. Organised crime and trafficking, particularly when attributable to activities over a long period of time and involving significant quantities of wildlife or trafficking that involves species threatened with extinction, could be considered ecocide. For instance, the European Commission has reported that the world's tiger population has decreased from 100,000 a century ago to less than 3,500 today, and the number of elephants killed has doubled in the last decade (Neyret, 2017). It hardly seems hyperbolic to suggest that these trends may constitute ecocide; even if these kill-offs are often the result of numerous and disparate actors, especially large trafficking rings or complicit state actors may bear enough of the blame to warrant prosecution for ecocide. This would surely mark a new, more muscular approach, but it is one that seems justified on both ecological and broader security grounds. After all, according to the United Nations Secretary-General, organised crime against wildlife is a threat to peace and security (United Nations, 2015).

As the examples demonstrate, wildlife-related harms can range in severity, risk to species, and extent of animal or plant life lost, among other variables, and the fluctuation in these factors may determine whether a particular harm rises to a level constituting ecocide. The various iterations of ecocide in national legislation suggest that other than the existence of some threshold regarding the nature or scale of the harm, definitional permutations abound, making the examples of potential ecocide described above sometimes, but not always, ecocide under national law.

For instance, in the United Kingdom, proposed legislation defines "ecocide" as "unlawful or wanton acts or omissions" with "a substantial likelihood of severe and either widespread or long-term damage to the environment being caused by those acts" or omissions. While the definitions of the component terms suggest that harm to wildlife may fit the bill in some cases, those definitions simultaneously reveal a concern with impacts at the species level and related impacts to ecosystems rather than, for example, concerns over harm to individual wild animals, even in the aggregate, unless that aggregate number is such that humans, nonhuman species, or ecosystems are affected. Specifically, in the UK bill (House of Lords, 2023), the following definitions create thresholds that may be difficult for some wildlife crimes to meet:

- "'severe' means damage which involves very serious adverse changes, disruption or harm to any element of the environment, including serious impacts on human life or natural, cultural or economic resources"; and
- "'widespread' means damage which extends beyond a limited geographic area, to impact upon an ecosystem or species or a significant number of human beings."

Under this iteration, proving "ecocide" could be difficult, though not impossible, in a hypothetical case where many wild animals are killed but the broader ecosystem is not seriously impacted. As Table 12.1 demonstrates, most national legislation sets as a threshold either "severe" or "mass" harm. But legislation varies whether the subject of the crime is the environment, however defined, or enumerated elements of the environment. Illustrative examples of the latter include Russia, Georgia, and Armenia. Although the UK bill defines "environment" to include "any element," and thus would at least arguably include wildlife, other broad references to "environment" may present challenges to application to wildlife in cases where harm to an ecosystem may be difficult to prove. This means that while ecocide may be a useful tool for some wildlife crimes, it is unlikely to be relevant to all wildlife crimes.

In a more positive vein, the UK bill and many other legislative efforts contemplate liability for both natural and legal persons; and whereas the crime when committed by natural persons requires a *mens rea* of either knowledge or recklessness as to the likelihood of damage to the environment, the standard is strict liability for businesses and other legal entities (House of Lords, 2023). In other cases, strict liability may apply to natural persons as well.

Conclusions

There is no doubt that transformational change of our legal systems is fundamental to a future rich in biodiversity. These shifts must strike at our legal systems in multiple ways: the deep anthropocentrism of wildlife management and conservation laws must be complemented with approaches like the rights-of-nature movement in order to find a balance between what humans need, on the one hand, and what wildlife needs, on the other. Only true balance and an understanding of long-term consequences and gains will provide the path out of our current extinction crisis. Rights of nature as a regulatory model points us in a healthier direction, and ecocide, offering a criminal approach to gross violations of environmental and wildlife regulatory frameworks, adds muscle to the change in trajectory.

Ecocide offers much promise in light of the nature and scale of wildlife crimes and other crimes that affect wildlife, like illegal habitat destruction and degradation. As these laws are

Table 12.1 Examples of definitional permutations

Source (jurisdiction or author)	*Definition of "ecocide"*	*Notes*
EU (European Commission, 2021)	"Any conduct causing severe and either widespread or long-term or irreversible damage."	The disjunctive "or" in the EU definition means that while the damage must be "severe," it also needs to be only *one* of the following: "widespread," "long-term," or "irreversible."
UK (House of Lords, 2023)	"Unlawful or wanton acts or omissions" with "a substantial likelihood of severe and either widespread or long-term damage to the environment being caused by those acts" or omissions.	By recognising "unlawful or wanton acts or omissions," the UK definition encompasses not only otherwise illegal acts that produce ecocide effects but also potentially lawful acts that *become* unlawful as a result of a reckless disregard for the risks involved.
Scotland (Lennon, M., MSP, 2023)	"Unlawful or wanton acts committed with knowledge that there is a substantial likelihood of severe and either widespread or long-term damage to the environment being caused by those acts."	Unlike the UK bill, the Scottish proposal does not appear to contemplate ecocide arising from omissions. Yet, in some cases, omissions contribute to ecological crises. For example, if a country's government knows of an illegal mining operation but does nothing to stop it, the government's inaction is arguably just as problematic as the mining group's action.
Belgium (Khattabi, 2022)	"The deliberate commission of an unlawful act causing serious, widespread and long-term damage to the environment, in the knowledge that such acts cause such damage."	Compared to the EU and UK iterations, the Belgian approach would require the act to cause "serious, widespread *and* long-term damage." The EU and UK approaches require severe and *either* widespread *or* long-term damage (or irreversible damage, in the case of the EU) (Heller, 2022). Conceivably, some cases that would qualify under the latter definitions may not qualify under the Belgian definition.
Vietnam (Penal Code of Vietnam, 1999)	"Those who, in peace time or war time, commit acts of annihilating en-mass [*sic*] population in an area, destroying the source of their livelihood, undermining the cultural and spiritual life of a country, upsetting the foundation of a society with a view to undermining such society, as well as other acts of genocide or acts of ecocide or destroying the natural environment, shall be sentenced to between ten years and twenty years of imprisonment, life imprisonment or capital punishment."	The Vietnamese legislation does not explicitly define "ecocide," and it is unclear from the surrounding language what the term encompasses. Because the provision speaks of "ecocide *or* destroying the natural environment," this could mean that ecocide is something distinct from "destroying the natural environment." Still, that leaves unanswered the question of what that "something" is—and how it is distinct from the destruction of the environment.

(*Continued*)

Table 12.1 (Continued)

Source (jurisdiction or author)	*Definition of "ecocide"*	*Notes*
Uzbekistan (Criminal Code of the Republic of Uzbekistan, 1994; "Ecocide/ serious environmental crimes," n.d.)[7]	Article 196 states that "Pollution or damage of land, water, or atmospheric air, resulted in mass disease incidence of people, death of animals, birds, or fish, or other grave consequences – shall be punished with fine from 100 to 200 minimum monthly wages or deprivation of certain right up to five years, or correctional labour up to three years."	On the one hand, the language seems to be most concerned with pollution or damage leading to "mass" harms and "grave consequences" as opposed to direct harm to wildlife, such as from poaching or taking. On the other hand, the text appears potentially to apply to conduct resulting in the death of only a few animals, for example.
Russia (Criminal Code of the Russian Federation, 1996)	Article 358 defines "ecocide" as the "massive destruction of plant and animal life, the poisoning of the atmosphere or water resources, and the perpetration of any other actions that could cause an ecological disaster," with the crime "punishable by deprivation of liberty from 12 to 20 years."	Adding to the sense of gravitas flowing from the elevated custodial penalty (12–20 years), the Russian Criminal Code sets forth ecocide immediately following the crime of genocide.
Georgia (Criminal Code of Georgia, 1999)	Article 409 ("Ecocide") provides as follows: "1. Ecocide i.e. contamination of the atmosphere, soil, water resources, mass destruction of fauna or flora, or any other act that could have led to an ecological disaster, – shall be punished by imprisonment for a term of 12–20 years. 2. The same act committed during armed conflicts, – shall be punished by imprisonment for a term of 14–20 years or by life imprisonment."	Note that, on its face, the provision does not indicate what, if any, *mens rea* is required.
Armenia (Criminal Code of the Republic of Armenia, 2003)	Article 394 ("Ecocide") of Armenia's criminal code provides as follows: "Mass destruction of flora or fauna, poisoning the environment, the soils or water resources, as well as implementation of other actions causing an ecological catastrophe, is punished with imprisonment for the term of 10 to 15 years."	As with other iterations, Armenia's approach to ecocide uses words like "*mass* destruction" and "catastrophe" to emphasise ecocide's targeting of extreme ecological offences—crimes an order of magnitude greater than "ordinary" environmental crimes.

crafted, cognisance of their salience to the fight against the extinction crisis is key. Definitional aspects of ecocide laws should ensure that the law also applies to harm experienced by wildlife—not just to situations threatening a species' survival, or even more tangentially to destruction or severe degradation of ecosystems, which may be hard to prove in some cases involving wildlife. Further, definitions of terms like "severe" and "irreversible" should be calibrated to account for context, such as when the poaching of a few remaining animals hastens or culminates in a species' extinction, even if mass quantities of lives were not necessarily lost during the commission of that particular crime. Finally, ecocide laws should be based on the premise that harm to the environment, ecosystems, or wildlife is the fundamental threshold, as opposed to having to prove harm to humans flowing from the predicate harm to the environment, ecosystems, or wildlife.

Notes

1 Although our composite definition is not verbatim reflected in any existing text, it reflects a fair synthesis of leading definitions and discussions of the term "ecocide." Other definitions and key terms are discussed below.

2 Ryan and colleagues (2021) have traced two lines of thought underpinning the rights of nature movement:

> The rights of nature movement springs from two wells, one philosophical and the other pragmatic. For philosophical adherents, the movement offers a critical reframing of environmental rights and values, from an anthropocentric worldview arranged around conventional property and tort legal constructs to a biocentric system that vindicates legal rights in ecosystems and other natural features unrecognised at common law. For pragmatists, the movement provides an alternative mechanism for environmental protection to replace or complement conventional regulatory schemes that have proved inadequate to the task.

3 *See* Rome Statute of the International Criminal Court, July 17, 1998, 37 I.L.M. 999, *available at* https://www.icc-cpi.int/sites/default/files/RS-Eng.pdf; Kate Surma, *The International Criminal Court Turns 20 in Turbulent Times. Should 'Ecocide' Be Added to Its List of Crimes?*, Inside Climate News (July 10, 2022), https://insideclimatenews.org/news/10072022/international-criminal-court-20-years-ecocide/; Joanna Gill, *Ecocide: Should Destroying Nature Be an International Crime?*, Context (Mar. 31, 2023), https://www.context.news/nature/ecocide-should-destroying-nature-be-an-international-crime; Josie Fischels, *How 165 Words Could Make Mass Environmental Destruction an International Crime*, NPR (June 27, 2021), https://www.npr.org/2021/06/27/1010402568/ecocide-environment-destruction-international-crime-criminal-court.

4 Higgins continued her campaign until her death in 2019 and Stop Ecocide International has been working on the cause ever since (Fishels, 2021; "Who We Are," n.d.).

5 Concern also exists that the concept of ecocide is inherently, not merely definitionally, insufficient to address wildlife and other animal-related harms. Professor Eichler, for example, has expressed concern that constructions of ecocide must take care to avoid the traditional Western view that "frequently assumes a hierarchy between humans, animals, and other beings" (Eichler, 2020). Others argue that for the concept to be meaningful to animal advocates, the loss of an animal's sentient capacity should feature as a central, defining characteristic of the crime (Legge & Brooman, 2020). Thus, for example, Legge and Brooman (2020) propose that animal ecocide should be defined as

> the unnecessary killing or slaughter of a wild or wild-caught animal, by any human agency, or allowing such killing or slaughter to be so caused by any governmental organisation, to such an extent that an animal, or group of animals, lose their sentient capacity to live a natural life according to their species.

6 Jojo Mehta, one of the leaders of the international ecocide movement and close colleague of Polly Higgins, recently commented on the rise in domestic ecocide laws, including Mexico's possible adoption of such legislation. Mehta admitted that she used to be concerned that domestic

legislation would distract from the goal of making ecocide an international crime but has recently changed her perspective (Kaminski, 2023):

> Actually, it seems to be that they're mutually reinforcing. International criminal law is ultimately supposed to be about elevating the worst crimes to an international level, rather than inventing new crimes. And as more and more national governments start talking about ecocide, more states are starting to listen at the international level.

7 Note that our quote derives from the quoted text available on the Ecocide website, as opposed to the translation available through Sherlock. Nonetheless, the differences seem merely syntactical; they do not appear to differ materially.

Bibliography

A statement about the 480 million animals killed in NSW bushfires since September. (2020, January 3). The University of Sydney. https://www.sydney.edu.au/news-opinion/news/2020/01/03/a-statement-about-the-480-million-animals-killed-in-nsw-bushfire.html

Bookman, S. (2023). Rights of Nature in Comparative Perspective. *Natural Resources & Environmen't, 37*(4), 4–8. https://doi.org/10.2139/ssrn.4541848

Brew, N., Richards, L., and Smith, L. (2020, March 12). *2019–20 Australian bushfires—frequently asked questions: a quick guide*. Parliament of Australia. https://www.aph.gov.au/About_Parliament/Parliamentary_departments/Parliamentary_Library/pubs/rp/rp1920/Quick_Guides/AustralianBushfires

Burgess, N., and Martin, E. (n.d.). Niger Delta Swamp Forests. *One Earth*. https://www.oneearth.org/ecoregions/niger-delta-swamp-forests/

Costa, C. (2020, May 21). Amazon under threat: Fires, loggers and now virus. *BBC News*. https://www.bbc.com/news/science-environment-51300515

Criminal Code of Georgia. (1999). (as amended), Art. 409. Retrieved from https://matsne.gov.ge/en/document/download/16426/157/en/pdf

Criminal Code of the Republic of Armenia. (2003). (as amended), Art. 394. Retrieved from http://www.parliament.am/legislation.php?sel=show&ID=1349&lang=eng

Criminal Code of the Republic of Uzbekistan. (1994). (as amended), Art. 196. Retrieved from https://sherloc.unodc.org/cld/uploads/res/document/criminal-code-of-the-republic-of-uzbekistan_html/Uzbekistan_Criminal_Code_1994.pdf

Criminal Code of the Russian Federation. (1996). (as amended), Art. 358. Retrieved from https://wipolex-res.wipo.int/edocs/lexdocs/laws/en/ru/ru080en.html.

Destruction of Ukraine's Kakhovka Dam Has Decimated Wildlife. (2023, June 7). Yale Environment 360. https://e360.yale.edu/digest/kakhovka-dam-destruction-wildlife

Díaz, S. et al. (2019). *Summary for policymakers of the global assessment report on biodiversity and ecosystem services of the Intergovernmental Science-Policy Platform on Biodiversity and Ecosystem Services*. IPBES. https://doi.org/10.5281/zenodo.3553579

Ecocide/serious environmental crimes in national jurisdictions (n.d.). Ecocide Law. https://ecocidelaw.com/existing-ecocide-laws/

Ecocide/serious environmental crimes in national jurisdictions: Uzbekistan (n.d.). Ecocide Law. https://ecocidelaw.com/existing-ecocide-laws/

Eichler, L. J. (2020). Ecocide Is Genocide: Decolonizing the Definition of Genocide. *Genocide Studies and Prevention: An International Journal, 14*(2), 104–121. https://doi.org/10.5038/1911-9933.14.2.1720

European Commission. (2021, December 15). *Proposal for a Directive of the European Parliament and of the Council on the protection of the environment through criminal law and replacing Directive 2008/99/EC*. https://eur-lex.europa.eu/legal-content/EN/TXT/HTML/?uri=CELEX:52021PC0851

European Commission. (2023, November 16). *Commission welcomes political agreement on protecting the environment through criminal law* [Press release]. https://ec.europa.eu/commission/presscorner/detail/en/ip_23_5817

European Law Institute. (2023). *ELI Report on Ecocide: Model Rules for an EU Directive and a Council Decision* (quoting Higgins, P. (2010). *Eradicating Ecocide: Laws and Governance to Stop the*

Destruction of the Planet.) https://www.europeanlawinstitute.eu/fileadmin/user_upload/p_eli/Publications/ELI_Report_on_Ecocide.pdf
European Parliament. (2023, November 16). Environmental crimes: deal on new offences and reinforced sanctions [Press release]. https://www.europarl.europa.eu/news/de/press-room/20230929IPR06108/environmental-crimes-deal-on-new-offences-and-reinforced-sanctions
Fischels, J. (2021, June 27). How 165 Words Could Make Mass Environmental Destruction An International Crime. *National Public Radio.* https://www.npr.org/2021/06/27/1010402568/ecocide-environment-destruction-international-crime-criminal-court
Gill, J. (2023, March 31). Ecocide: Should destroying nature be an international crime? *Context.* https://www.context.news/nature/ecocide-should-destroying-nature-be-an-international-crime
Heller, K. J. (August 11, 2022). Belgium Set to Criminalise Ecocide (Kinda Sorta). *Opinio Juris.* https://opiniojuris.org/2022/11/08/belgium-set-to-criminalise-ecocide-kinda-sorta/
Hillsdon, M. (2023, February 20). Society Watch: Drive to make ecocide an international crime gains momentum. *Reuters.* https://www.reuters.com/business/sustainable-business/society-watch-drive-make-ecocide-an-international-crime-gains-momentum-2023-02-20/
House of Lords. (2023, November 30). House of Lords Bill 21: Ecocide Act 2024. https://bills.parliament.uk/publications/53197/documents/4084
ICC. (1998). Rome Statute of the International Criminal Court, UN Doc. 2187 UNTS 3. (Cited as Rome Statute). https://treaties.un.org/doc/Treaties/1998/07/19980717%2006-33%20PM/volume-2187-I-38544-English.pdf
INTERPOL. (2023, November 6). *Illegal wildlife trade has become one of the 'world's largest criminal activities'* [Press Release]. https://www.interpol.int/en/News-and-Events/News/2023/Illegal-wildlife-trade-has-become-one-of-the-world-s-largest-criminal-activities#:~:text=With%20the%20black%20market%20for,other%20forms%20of%20organized%20crime
Kaminski, I. (2023, August 26). Growing number of countries consider making ecocide a crime. *The Guardian.* https://www.theguardian.com/environment/2023/aug/26/growing-number-of-countries-consider-making-ecocide-crime?CMP=twt_a-environment_b-gdneco
Kaminski, I. (2023, November 17). EU criminalises environmental damage 'comparable to ecocide'. *The Guardian.* https://www.theguardian.com/environment/2023/nov/17/eu-criminalises-environmental-damage-comparable-to-ecocide
Khattabi, Z. (2022, November 5). Zakia Khattabi gets the crime of ecocide written into the criminal code [Press release].
Kusnetz, N. et al. (2021, April 7). As the Climate Crisis Grows, a Movement Gathers to Make 'Ecocide' an International Crime Against the Environment. *Inside Climate News.* https://insideclimatenews.org/news/07042021/climate-crisis-ecocide-vanuatu-the-fifth-crime/
Legge, D., and Brooman, S. (2020). Reflecting on 25 Years of Teaching Animal Law: Is it Time for an International Crime of Animal Ecocide?. *Liverpool Law Review, 41*, 201–218. https://doi.org/10.1007/s10991-020-09253-0
Lennon, M., MSP. (2023, November 8). Proposed Ecocide Prevention (Scotland) Bill. https://www.parliament.scot/-/media/files/legislation/proposed-members-bills/consultation-document-final-version--(1).pdf
Moisiienko, I. I. et al. (2023, June 7). *The consequences of the Russian terrorist attack on the Kakhovka Hydroelectric Power Plant (HPP) for wildlife.* Ukraine Nature Conservation Group. https://uncg.org.ua/en/the-consequences-of-the-russian-terrorist-attack-on-the-kakhovka-hydroelectric-power-station-hps-for-wildlife/
Neyret, L. (2017). From ecocrimes to ecocide: protecting the environment through criminal law. *C-EENRG Reports, 2017*(2), 1–198. https://www.ceenrg.landecon.cam.ac.uk/system/files/documents/report-002.pdf
Numbere, A. O. (2020). The Impact of Landscape Reclamation on Mangrove Forest and Coastal Areas in the Niger Delta, Nigeria. *IntechOpen.* https://www.intechopen.com/chapters/64407
Penal Code of Vietnam (No. 15/1999/QH10) (1999). (as amended), Art. 342. Retrieved from https://natlex.ilo.org/dyn/natlex2/r/natlex/fe/details?p3_isn=56207
Platt, J. R. (2023, February 21). In Ukraine, Saving Wildlife Harmed by War. *The Revelator.* https://therevelator.org/ecocide-ukraine/
Puleo, V. L. (2021). A New International Crime of Ecocide?. *University of Bologna Law Review,* 6(2), 163–170. https://doi.org/10.6092/issn.2531-6133/13921

Ratcliffe, R. (2019, December 6). 'This place used to be green': the brutal impact of oil in the Niger Delta. *The Guardian*. https://www.theguardian.com/global-development/2019/dec/06/this-place-used-to-be-green-the-brutal-impact-of-oil-in-the-niger-delta

Ryan, E., Curry, H., and Rule, H. (2021). Environmental Rights for the 21st Century: A Comprehensive Analysis of the Public Trust Doctrine and Rights of Nature Movement. *Cardozo Law Review, 2*(6), 2447–2576 (2021). https://ir.law.fsu.edu/articles/721

Sollund, R. (2020). Wildlife Management, Species Injustice and Ecocide in the Anthropocene. *Critical Criminology, 28*(3), 351–370. https://heinonline-org.library.lcproxy.org/HOL/Page?collection=journals&handle=hein.journals/ctlcrm28&id=351&men_tab=srchresults

Stop Ecocide International. (2019, November 18). *Pope Francis: Destroying the earth is a sin and should be a crime* [Press release]. https://www.stopecocide.earth/press-releases-summary/pope-francis-destroying-the-earth-is-a-sin-and-should-be-a-crime.

Surma, K. (2022, July 10). The International Criminal Court Turns 20 in Turbulent Times. Should 'Ecocide' Be Added to its List of Crimes? *Inside Climate News*. https://insideclimatenews.org/news/10072022/international-criminal-court-20-years-ecocide/

UK 'Ecocide Bill' Introduced in House of Lords. (2023, November 30). https://www.stopecocide.earth/breaking-news-2023/uk-ecocide-bill-introduced-in-house-of-lords

UK Parliament. (2023, February 13). House of Lords Business: Bills in Progress. https://lordsbusiness.parliament.uk/?sectionId=41&businessPaperDate=2024-02-13

United Nations. (2015, February 26). *Secretary-general urges consumers, governments to treat crimes against wildlife as 'threat to our sustainable future', in message for international day* [Press Release]. https://press.un.org/en/2015/sgsm16551.doc.htm

Van Brugen, I. (2023, June 8). Putin Killed Tens of Thousands of Animals in Kakhovka Dam 'Ecocide'—Rescuer. *Newsweek*. https://www.newsweek.com/putin-animals-flooding-kakhovka-dam-ecocide-kherson-evacuation-war-1805310

Who we are (n.d.). Stop Ecocide International. https://www.stopecocide.earth/who-we-are-

Yermak, A., and Wallström, M. (2023, August 16). Russia is committing grave acts of ecocide in Ukraine – and the results will harm the whole world. *The Guardian*. https://www.theguardian.com/commentisfree/2023/aug/16/russia-ecocide-ukraine-world-war-crimes

13

WILDLIFE HARMSCAPES AND WHOLE-OF-SOCIETY RESPONSES

Beyond criminalisation towards inclusive conservation

Annette Hübschle

Introduction: wildlife crime through a harmscapes lens

The concept of *harmscapes* offers a transformative framework for understanding wildlife crime as embedded within complex landscapes of harm shaped by historical, socioeconomic, and ecological forces. Originally developed by Berg and Shearing (2018), harmscapes (a contraction of "harm landscapes") were applied to urban security governance. The scholars examined how harm is produced and maintained through systems of inequality and socio-political structures that favour elite interests. Berg and Shearing (2018) used this framework to challenge conventional approaches to urban security that focus on crime control rather than addressing the structural conditions that generate persistent harm.

As Berg and Shearing (2018) explain, *harmscapes* are conceptual spaces that capture the layered and interconnected nature of harm—ranging from crime and ecological degradation to structural injustices. These spaces are not bounded or isolated but are shaped by the intersecting dynamics of global capitalism, governance failures, and socio-political inequalities. Rather than viewing harm as a discrete event, their framework positions it within broader institutional, historical, and structural contexts.

I extend this framework to wildlife crime contexts, where the notion of harmscapes reveals how harm against wildlife exists within broader systems of inequality, environmental degradation, and contested power dynamics. This application moves beyond traditional wildlife crime approaches that emphasise enforcement, criminal justice responses, and individual wrongdoing, instead highlighting how wildlife crimes are symptomatic of deeper structural injustices affecting both human and nonhuman communities.

Wildlife crime harmscapes capture the multidimensional spaces where various forms of harm—illegal hunting, snaring, poisoning, trafficking, habitat destruction, and exploitation—intersect with social, political, and economic structures. These spaces are not isolated in time and space, but are actively shaped by historical legacies, governance failures, and socioeconomic disparities that create environments where harm to wildlife becomes normalised rather than exceptional. This critical perspective aligns with an understanding of

DOI: 10.4324/9781003333906-16

wildlife crime as encompassing both illegal activities and broader notions of harm (Nurse & Wyatt, 2020, p.2).

My research on rhino poaching and illegal succulent harvesting illustrates how global market demand interacts with local socioeconomic vulnerability to create conditions where wildlife crime flourishes (Hübschle, 2017a; Hübschle & Shearing, forthcoming). These cases show how global trade systems and local livelihood needs intersect within regulatory frameworks that are often inconsistent, contested, or inadequately enforced. Central to this analysis are two concepts that recur throughout the chapter. First, contested illegality, which arises when laws on the books diverge from what is socially or culturally recognised as legitimate (Hübschle, 2016; Hübschle & Margulies, 2024). Second, fluid interfaces, which describe the ease with which actors move between legal, grey, and illegal markets, exploiting loopholes and ambiguities to launder wildlife contraband and contest its legal status (Hübschle, 2017b). These concepts are developed further in subsequent sections, with empirical examples from rhino horn and succulent markets.

Taken together, the harmscapes framework and these concepts provide a way to see wildlife crime not as discrete criminal acts but as embedded in wider systems of structural violence, global demand, and socioecological vulnerability. This chapter applies this lens to different wildlife contexts—conservation, rural, maritime, and global trafficking harmscapes—before turning to whole-of-society responses that move beyond criminalisation toward inclusive, justice-oriented approaches to conservation.

The structural foundations of wildlife harmscapes

Wildlife harmscapes exist within interconnected systems of power that determine which harms are recognised, prioritised, and addressed. These landscapes reflect Johan Galtung's (1969) concept of structural violence, where harm is embedded within institutional structures, policies, and laws that maintain societal inequalities and facilitate ongoing harm to wildlife and marginalised human communities.

The harmscapes framework highlights how harm is often embedded within institutional structures and legitimised by dominant ideologies, rather than resulting from isolated individual actions (Berg & Shearing, 2018, p.75). This systemic perspective allows us to conceptualise wildlife harmscapes as landscapes shaped by socioeconomic and environmental factors that continuously produce harm, particularly for vulnerable populations, including nonhuman populations.

In conservation contexts, wildlife harmscapes are shaped by historical processes of land dispossession, resource extraction, and the criminalisation of traditional practices. Colonial and post-colonial conservation models often prioritised wildlife protection through exclusionary approaches that marginalised indigenous peoples and local communities (IPLCs), creating persistent landscapes of socioecological harm. These historical legacies continue to influence contemporary conservation practices, where local populations face restricted access to ancestral lands and traditional resources, often forcing them into activities defined as "wildlife crime" for basic subsistence (Hübschle, 2016).

The concept of contested illegality becomes central here, as many activities defined as wildlife crime exist in legal grey areas shaped by competing claims to legitimacy. As highlighted in my research on rhino poaching, the contested nature of illegality in wildlife crime often arises from the grey areas in law where certain actions are legal in some jurisdictions but illegal in others, and where cultural practices may be deemed criminal in one context but legitimate in another (Hübschle, 2017a; 2025). These contested spaces reveal how wildlife crime is not simply about breaking laws but reflects deeper tensions between differing value systems, governance approaches, and understandings of human–wildlife relationships.

As Berg and Shearing (2024) argue, neoliberal governance logics—marked by privatisation, deregulation, and market-driven solutions—have reshaped responses to harmscapes, often privileging economic development over social equity. While such dynamics are well documented in urban contexts, similar global forces shape wildlife harmscapes by fuelling demand for natural resources while externalising socioecological costs. The commercialisation and commodification of wildlife—whether legally or illegally—often occurs through systems that prioritise profit over ecological sustainability and social equity. Corporate actors, state entities, and criminal networks all participate in these markets, blurring the boundaries between legal and illegal activities and creating complex landscapes of harm that transcend simplistic categorisations (see, e.g., Margulies, Massé & VandeBerg, 2023).

Historical legacies further contribute to these wildlife harmscapes, especially in post-colonial societies. For example, South African conservation practices established during colonial and apartheid eras displaced Indigenous communities, criminalised traditional practices, and privileged elite interests. While promoted as biodiversity sanctuaries, many such protected areas function in practice as restricted-access regimes for local populations, where legal and economic barriers to resource use contribute to ongoing social and environmental harms (Dore et al., 2022; Hübschle, 2017b).

The structural foundations of wildlife harmscapes are particularly evident in contexts where IPLCs engage in practices like traditional hunting, gathering, harvesting, capturing, or fishing, which, though criminalised by the state, are crucial for survival. These conflicts highlight the disconnect between state-imposed regulations and the lived realities of communities dependent on natural resources, illustrating the need for inclusive governance frameworks that account for local needs (Hübschle, 2017b; Mackenzie et al., 2024). While the threat to wildlife from commercial poaching is real, militarised, and securitised approaches to address wildlife poaching, further exacerbate these harms against vulnerable communities through the use of violence and dispossession (Annecke & Masubelele, 2016; Duffy et al., 2019; Witter & Satterfield, 2019).

Climate change has emerged as an additional structural factor shaping contemporary wildlife harmscapes, reframing harm as an outcome of shifting environmental realities. Climate-induced impacts highlight the cumulative effects of environmental changes—such as habitat loss, altered migration patterns, and resource scarcity—that disproportionately affect vulnerable wildlife populations and the communities that depend on them. This introduces a dimension of "slow violence" (Nixon, 2011), where impacts accumulate gradually and unevenly across regions and populations, creating new forms of harm that may not be immediately visible or criminalised.

Across these varied dimensions, the structural foundations of wildlife harmscapes emphasise how harm against wildlife exists within broader systems of power, inequality, and environmental change. By understanding these structural drivers, we can move beyond focusing solely on individual incidents of wildlife crime towards addressing the systemic conditions that create landscapes where harm to both human and nonhuman communities become normalised and persistent.

Types of wildlife harmscapes: from local to global

Wildlife harmscapes manifest across diverse contexts, reflecting different dimensions of harm against and involving wildlife.

Conservation harmscapes and the criminalisation of Indigenous peoples and local communities

Conservation areas, while designed to protect biodiversity, can become sites of socioecological harm when implemented through exclusionary, militarised approaches. In many parts of the Global South, protected areas established during colonial periods displaced IPLCs and criminalised traditional practices, creating landscapes where both wildlife and human populations remain vulnerable to exploitation. Contemporary "fortress conservation" models that prioritise security over community engagement further entrench these harms, increasing hostility between local populations and conservation authorities (Lunstrum, 2014; Ramutsindela et al., 2022).

The systematic criminalisation of IPLCs represents a profound injustice within conservation harmscapes. Activities that were once central to cultural identity and livelihood strategies—hunting, gathering, and resource harvesting—become reclassified as "poaching," "land invasions," or "resource theft" through conservation laws and policies that fail to recognise Indigenous rights and traditional knowledge systems. Traditional livelihood strategies are reframed as criminal, while local communities are treated as potential criminals rather than conservation partners (Hübschle & Shearing, 2018).

The disproportionate use of force against marginalised communities in anti-poaching operations represents a form of structural violence embedded within these conservation harmscapes. Such approaches can redirect resources away from wildlife conservation and care to militarised policing approaches, creating conditions where wildlife crime may actually increase as a form of resistance against perceived injustice (Witter, 2021). This contestation of illegality emerges when laws criminalising traditional resource use lack social legitimacy among local communities who see their relationships with wildlife as historical, legitimate, and necessary for their survival (Hübschle, 2022).

My fieldwork in communities living adjacent to conservation areas revealed numerous instances of human rights abuses in the name of wildlife protection (Hübschle & Jojo, 2021; Hübschle & Shearing, forthcoming). Community members reported physical violence, intimidation, and harassment by anti-poaching units, exclusion from decision-making about resources that affect their lives, and economic marginalisation through denial of access to natural resources.

Rural harmscapes

In rural areas, wildlife harmscapes are shaped by limited economic opportunities, environmental degradation, and resource exploitation. Extractive industries such as mining, logging, and industrial agriculture not only destroy wildlife habitat but also undermine traditional livelihoods dependent on healthy ecosystems. Communities facing economic hardship may turn to activities defined as wildlife crime—including hunting protected species or harvesting protected plants—as survival strategies in the absence of sustainable alternatives (Hübschle & Shearing, 2018; Hübschle et al., 2024).

These rural harmscapes highlight how wildlife crime often emerges from socioeconomic vulnerabilities rather than criminal intent. When traditional subsistence activities become criminalised through conservation regulations, rural communities face a double burden: the loss of access to natural resources and the risk of criminal penalties for engaging in practices essential to their survival. This creates landscapes where harm flows in multiple directions—to wildlife through unsustainable harvesting, to communities through criminalisation, and to ecosystems through the erosion of traditional stewardship practices (Iordăchescu & Vasile, 2022).

The contestation of illegality is particularly pronounced in these contexts, where communities may not accept the legitimacy of laws that criminalise their traditional practices.

Maritime harmscapes

Ocean environments constitute distinctive wildlife harmscapes shaped by transboundary challenges and governance gaps. Illegal, unreported, and unregulated (IUU) fishing, marine pollution, and the exploitation of marine species create complex patterns of harm that extend across national boundaries. Large commercial vessels from wealthy nations often deplete fish stocks in the waters of developing countries, pushing local fishers towards unsustainable practices that further damage marine ecosystems (Hübschle & Berg, 2024; Hübschle & Lindley, 2024).

These maritime harmscapes reflect global inequalities in resource access and control, where powerful economic interests exploit governance weaknesses to extract marine resources at the expense of both ecosystem health and local livelihoods. The transnational nature of these harms requires coordinated governance responses across scales, highlighting the limitations of traditional state-centred approaches to wildlife crime (Hübschle & Berg, 2024).

The fluidity of legal–illegal interfaces is particularly evident in maritime contexts, where fishing vessels may operate legally in one jurisdiction but illegally in another, or may engage in both legal and illegal fishing practices during the same voyage. This creates complex enforcement challenges and allows for the laundering of illegally caught fish through legal supply chains (Hübschle & Lindley, 2024).

Global trafficking harmscapes

Wildlife trafficking creates transnational harmscapes that connect source, transit, and destination countries through complex supply chains. These networks often involve both legal and illegal actors, from local harvesters to international criminal organisations and legitimate businesses that knowingly or unknowingly participate in the trade of protected and other species.

My research on rhino horn trafficking reveals the fluid interfaces that characterise these global harmscapes (Hübschle, 2019). Legal trophy hunting flows are appropriated and redirected to supply illegal markets through "pseudo-hunting" schemes. Wildlife products move between legal and illegal status as they cross jurisdictional boundaries, creating what I have termed "grey flows" that exploit ambiguities in regulation and enforcement. As I noted: "Actors capitalise on the room to manoeuvre between legal and illegal forms of rhino horn trade. Bolstered by sentiments of contested illegality and legality, criminal actors have no qualms to exploit or manipulate regulatory loopholes" (Hübschle, 2019, p.7).

The harms within these trafficking landscapes are unevenly distributed, with source communities in biodiverse regions of the Global South bearing the ecological costs of resource extraction while consumers in wealthy nations drive demand. This global dimension of wildlife crime harmscapes reflects broader patterns of resource exploitation and environmental injustice, where the environmental burdens fall disproportionately on marginalised communities while the benefits flow elsewhere (Massé & Margulies, 2020).

Wildlife harmscapes and fluid interfaces

Looking at the concept of harmscapes more broadly, one key characteristic that emerges across different contexts is the existence of fluid interfaces between legal, grey, and illegal activities. This fluidity creates environments where harm becomes embedded within both institutional structures and everyday practices.

The concept of fluid interfaces aligns with the broader harmscapes framework, which recognises that landscapes of harm are shaped by both spatial and temporal dimensions. Harm is unevenly distributed and produced through ongoing structural arrangements and historical legacies embedded in institutional systems (Berg and Shearing, 2018). Just as the original harmscapes concept identifies how harm is embedded within local and global processes that sustain inequality, wildlife crime similarly operates through networks that exploit spatial and temporal boundaries.

These fluid interfaces are facilitated by several factors:

1. *Jurisdictional complexity*: Wildlife products may have different legal status in different countries or regions within the same country. What is illegal to harvest in one jurisdiction may be legal to sell in another, creating opportunities for laundering.
2. *Regulatory loopholes*: Exemptions for certain forms of trade (e.g. pre-Convention specimens, hunting trophies, captive-bred specimens) create pathways that can be exploited to introduce illegal wildlife into legal markets.
3. *Document fraud and corruption*: Officials who validate documentation may be corrupted, allowing illegal wildlife to be "laundered" through document forgery or misrepresentation.
4. *Contested illegality*: When wildlife laws lack social legitimacy among key stakeholders, there is a greater willingness to circumvent or ignore them, creating spaces where illegal activities are normalised and justified.

This analysis connects to the original harmscapes framework's emphasis on how harmscapes encompass a variety of environments that are shaped by specific socio-political and economic histories. In wildlife contexts, these fluid interfaces create "contested spaces" where legality and illegality blur, reflecting how harm is embedded within institutional structures, policies, and laws that maintain societal inequalities and facilitate ongoing harm, particularly for marginalised groups.

The phenomenon of wildlife laundering exemplifies these fluid interfaces. Illegally harvested wildlife enters legal markets through document forgery, corruption, or exploitation of legal loopholes. For example, breeding facilities may augment their legal stock with wild-caught specimens, creating a "legal" origin story for illegally harvested wildlife (Dickinson, 2022).

Much like how the original harmscapes framework addresses "transboundary threats" through emphasising polycentric governance—a networked model where state and non-state actors collaborate across scales—addressing the fluid interfaces in wildlife crime requires similar collaborative approaches that can adapt to the complex, shifting nature of these harm landscapes.

From crime-centric to whole-of-society approaches: decriminalising communities

Addressing wildlife crime harmscapes requires moving beyond traditional enforcement-first approaches towards more holistic, justice-oriented models of governance that decriminalise IPLCs. A crime-centric approach to wildlife harmscapes may crowd out innovative alternatives and longer-term solutions. It often ignores harms which are not criminalised yet are devastating in impact and presumes that there exists a capable law enforcement body, state apparatus or "functional equivalent" to identify and prosecute wildlife crime, which may not be the case in some contexts (Berg & Shearing, 2018).

A crime-centric approach may also presume a dichotomy between "victim" and "offender" when the reality is more complex, with wildlife, ecosystems, and human communities all experiencing interrelated forms of harm. The main governance tactic of blaming and punishing individual actors may do little to address the structural drivers of wildlife crime or prevent future harms (Hübschle et al., 2021). In fact, over-policing and criminalisation of marginalised communities often exacerbate the very social and economic conditions that drive wildlife crime in the first place.

Criminalisation of traditional resource use and subsistence harvesting not only infringes on cultural rights and livelihood security but can also undermine conservation objectives by alienating communities who might otherwise serve as conservation allies (Hübschle, 2017b).

Shifting to a harm-centric sensibility allows for a broader range of solutions that recognise the complexity of wildlife harmscapes. This requires what Berg and Shearing (2011, 2018) call "whole-of-society" responses—governance approaches that engage multiple stakeholders across different sectors and scales. A whole-of-society approach means finding appropriate solutions and aligning resources to the specific context and nature of wildlife harmscapes, rather than starting with predetermined, usually state-centric solutions (like law enforcement) and attempting to apply them universally.

This approach recognises that the centre of authority and responsibility may shift depending on which entity is best positioned to address particular aspects of wildlife harm—whether government agencies, community organisations, private sector entities, or civil society groups (Berg & Shearing, 2020). Whole-of-society approaches are particularly important in contexts where state capacity may be limited, compromised, or illegitimate, which may be the case in wildlife conservation contexts in the Global South.

Crucially, whole-of-society approaches seek to decriminalise communities whose traditional relationships with wildlife have been disrupted by colonial and post-colonial conservation models. Rather than positioning IPLCs as potential criminals or threats to wildlife, these approaches recognise their rights, knowledge systems, and critical role as stewards of biodiversity. Moving from a criminal justice paradigm to a social justice paradigm requires a fundamental shift in how we conceptualise the relationship between people and wildlife—from one of prohibition and control to one of stewardship and mutual flourishing (Hübschle & Shearing, 2018).

Alternative approaches to addressing wildlife harmscapes

Community-based natural resource management

Community-based natural resource management (CBNRM) represents a transformative whole-of-society approach to wildlife harmscapes by centring the knowledge, rights, and needs of local populations. Rather than imposing external conservation models that criminalise traditional practices, CBNRM recognises IPLCs as key partners in wildlife protection. By integrating Indigenous knowledge and promoting sustainable livelihoods, these initiatives address both social and ecological dimensions of harm (Booker & Roe, 2017; Massé et al., 2017).

In southern Africa, community conservancies demonstrate how collaborative governance can transform landscapes of harm into spaces of care and well-being. When communities gain rights to manage wildlife resources and benefit from conservation, they develop stronger incentives to protect species and report illegal activities. This shift from exclusion to engagement addresses the structural drivers of wildlife crime by creating economic alternatives to illegal harvesting and building social capital around conservation goals (Hübschle & Shearing, 2018, forthcoming).

Sustainable use

The sustainable use framework offers another whole-of-society approach to transforming wildlife harmscapes by recognising that controlled, legal utilisation of wildlife can support both conservation and community well-being. Rather than criminalising all use of wildlife, this approach distinguishes between destructive exploitation and sustainable harvesting that maintains ecological balance while providing economic benefits to local communities (Roe & Booker, 2019).

Sustainable use approaches seek to understand how the use of wildlife by local communities for consumptive food security purposes and for livelihoods through the development of wildlife-based economies can yield socioeconomic and ecological benefits. By extending rights to local populations and recognising their role in conservation, these models address the structural inequalities that create wildlife harmscapes in the first place.

When communities derive tangible benefits from wildlife through sustainable use programmes, their relationship to wildlife shifts from one of alienation and potential conflict to one of stewardship and care. This transformation can reduce motivations for illegal wildlife harvesting by creating alternative, legal pathways for communities to benefit from wildlife. Moreover, sustainable use approaches recognise that not all use of wildlife constitutes harm—traditional, regulated harvesting practices can be compatible with conservation goals and may contribute to them by maintaining cultural connections between communities and wildlife (Hübschle & Shearing, 2018).

Restorative justice

Restorative justice offers a promising framework for addressing wildlife crime harmscapes by focusing on reparation and redress, thereby addressing root causes rather than punishment alone. As explored in the harmscapes framework, conventional criminological approaches often view crime as an isolated, event-driven outcome to be addressed through punitive measures. However, when viewed through the lens of harmscapes, wildlife crime emerges as embedded within larger social institutions, governance failures, and structural inequalities that create environments where harm is an expected rather than incidental outcome (Berg & Shearing, 2018).

Restorative approaches to wildlife crime acknowledge harm to multiple stakeholders—including wildlife, ecosystems, communities, and future generations—and seek to repair these relationships through collaborative, inclusive processes. These approaches align with the polycentric governance model advocated in our harmscapes research, where multiple stakeholders coordinate efforts to address harm across scales.

The harmscapes framework, with its emphasis on structural and historical drivers of harm, aligns naturally with restorative justice principles. Wildlife crime often emerges within landscapes shaped by colonial legacies, socioeconomic disparities, and contested access to natural resources. Many protected areas in South Africa were established through forced removals and land dispossession during the colonial and apartheid eras, creating persistent landscapes of socioecological harm where both wildlife and marginalised communities experience interconnected forms of suffering and exploitation.

Recent efforts in South Africa have explored how restorative justice principles might be applied to wildlife crime, particularly in contexts shaped by histories of dispossession, marginalisation, and contested illegality. These initiatives (compare with Hübschle et al., 2021; Dore et al., 2022) foreground the importance of addressing structural injustice, creating space for multi-stakeholder dialogue—including conservation authorities, communities, and offenders—and centring the experiences and knowledge of IPLCs. Such approaches reflect the broader harmscapes framework, which emphasises inclusivity, the relational dimensions of harm, and the need to engage with the socio-political contexts that shape environmental offences. Rather than focusing solely on individual wrongdoing, these efforts seek to surface the underlying systemic conditions that enable wildlife crime and envision more equitable, community-centred responses—aligning with calls for "whole-of-society" strategies to address complex harmscapes (Berg & Shearing, 2018).

The harmscapes framework and restorative justice share a commitment to addressing both individual cases of harm and their broader structural contexts. Both approaches recognise that environmental harms are not isolated events but emerge from complex systems of inequality, exclusion, and power. By integrating these frameworks, we can develop more effective, equitable responses to wildlife crime that promote healing and transformation across human–wildlife relationships. This integration represents a shift from crime-centric approaches towards whole-of-society responses that foster equity, justice, and sustainability in the face of complex, interconnected wildlife harmscapes.

Socioecological harm reduction

Building on restorative justice approaches, socioecological harm reduction offers a pragmatic alternative to conventional prohibition and enforcement-first strategies for addressing wildlife crime. Prohibitionist approaches to illegal wildlife trade may inadvertently increase harm to both human and nonhuman communities (Hübschle and Margulies, 2024). Drawing inspiration from Elinor Ostrom's work on managing social-ecological systems, this approach recognises that ecological issues cannot be divorced from their social dimensions.

Socioecological harm reduction directly engages with the reality that many wildlife crime harmscapes are characterised by complex trade-offs between human well-being and wildlife protection. Rather than pursuing idealised but often unachievable outcomes through strict prohibition and enforcement, this approach seeks pragmatic, context-specific interventions that minimise the most serious harms to both wildlife and human communities.

The harmscapes framework helps us understand these trade-offs by revealing how different forms of harm intersect within specific landscapes. For instance, in communities bordering protected areas, strict anti-poaching enforcement may reduce harm to wildlife but simultaneously exacerbate socioeconomic hardships and deepen historical generational trauma for human communities (e.g. forceful eviction or dispossession). These interrelated harms cannot be effectively addressed through conventional crime-centric approaches focused solely on deterring wildlife crime.

A socioecological harm reduction approach to wildlife trade

> provides a pathway to address wildlife trade that diverges from a one-size-fits-all prohibition approach. This approach acknowledges that wildlife conservation in conjunction with rural development, sustainable livelihood opportunities, and stakeholder approaches tailored to the realities of source sites will yield better conservation outcomes than traditional enforcement-heavy approaches.
>
> *(Hübschle and Margulies, 2024, p.e14335)*

Key elements of this approach include recognising and protecting local use rights, strengthening community governance structures, facilitating access to legitimate value chains, and implementing traceability systems that ensure transparency and accountability. By emphasising community engagement, sustainable use, and co-designed interventions, socioecological harm reduction seeks to balance ecological conservation and human dignity in responding to wildlife crime.

The harmscapes framework and socioecological harm reduction share a commitment to addressing the underlying structural drivers of harm rather than merely treating their symptoms. Both recognise that effective responses to wildlife crime must engage with the complex socioeconomic and political landscapes in which such activities occur. By integrating these approaches, conservation practitioners can develop more nuanced strategies that reduce harm across human and nonhuman communities while building more just and sustainable relationships between people and wildlife.

Conclusion: towards just and sustainable wildlife governance

This chapter has explored the application of the harmscapes framework to wildlife crime contexts, offering a transformative lens for understanding how illegal wildlife activities are embedded within complex landscapes of harm shaped by historical, socioeconomic, and ecological forces.

Wildlife crime harmscapes capture the multidimensional spaces where various forms of harm—illegal hunting, trafficking, habitat destruction, and exploitation—intersect with social, political, and economic structures. These spaces are not isolated in time and space but are actively shaped by historical legacies, governance failures, and socioeconomic disparities that create environments where harm to wildlife becomes normalised rather than exceptional.

The harmscapes framework, combined with the concepts of contested illegality and fluid interfaces between illegality and legality, reveals how wildlife crime exists within broader systems of inequality, environmental degradation, and contested power dynamics. By moving beyond enforcement-focused models towards more holistic, justice-oriented approaches, we can address the structural conditions that enable wildlife crime while promoting more just and sustainable relationships between human and nonhuman communities.

Rather than relying on predetermined, state-centric solutions like law enforcement, whole-of-society approaches engage multiple stakeholders across different sectors and scales to develop context-specific solutions to wildlife crime. This approach recognises that the centre of authority and responsibility may shift depending on which entity is best positioned to address particular aspects of wildlife harm—whether government agencies, community organisations, private sector entities, or civil society groups.

Importantly, whole-of-society approaches seek to decriminalise communities whose traditional relationships with wildlife have been disrupted by colonial and post-colonial conservation models. Rather than positioning IPLCs as potential criminals or threats to wildlife, these approaches recognise their rights, knowledge systems, and critical role as stewards of biodiversity. Moving from a criminal justice paradigm to a social justice paradigm requires a fundamental shift in how we conceptualise the relationship between people and wildlife—from one of prohibition and control to one of stewardship and mutual flourishing.

As we face escalating global challenges—including biodiversity loss, climate change, and resource scarcity—this integrated approach provides a critical lens for developing more effective, equitable approaches to wildlife protection. By understanding wildlife crime as embedded within complex landscapes of harm, characterised by contested illegality and fluid interfaces, we can move beyond simplistic enforcement paradigms towards transformative solutions that foster justice for both human and nonhuman communities.

Moving forward, whole-of-society responses to wildlife crime will require meaningful collaboration across traditional boundaries—between conservation authorities and local communities, between enforcement agencies and civil society organisations, and between scientists and traditional knowledge holders. This collaborative approach not only enhances the effectiveness of immediate interventions but also contributes to longer-term transformation of the systems and structures that shape wildlife harmscapes.

Ultimately, this chapter has shown that addressing wildlife crimes—and the broader harmscapes that shape many conservation contexts—requires more than enforcement. It calls for a fundamental reimagining of conservation as a collective endeavour rooted in social justice, historical recognition, and shared responsibility. The drivers of environmental harm are socially and economically embedded in human relationships, systems of inequality, and past and present injustices. Responding to the environmental crisis, then, also means confronting the social and political conditions that perpetuate harm.

References

Annecke, W., & Masubelele, M. (2016). A review of the impact of militarisation: The case of rhino poaching in Kruger National Park, South Africa. *Conservation and Society, 14*(3), 195–204.

Berg, J., & Shearing, C. (2011). The practice of crime prevention: Design principles for more effective security governance. *South African Crime Quarterly, 36*, 23–30.

Berg, J., & Shearing, C. (2018). Governing-through-harm and public goods policing. *The Annals of the American Academy of Political and Social Science, 679*(1), 72–85.
Berg, J., & Shearing, C. (2020). Polycentric security governance and sustainable development in the Global South. In J. Blaustein, K. Fitz-Gibbon, N. W. Pino & R. White (Eds.), *The Emerald Handbook of Crime, Justice and Sustainable Development* (pp. 153–173). Emerald Publishing Limited.
Booker, F., & Roe, D. (2017). *First Line of Defence? A Review of Evidence on the Effectiveness of Engaging Communities to Tackle Illegal Wildlife Trade*. IIED.
Dickinson, H. (2022). The material politics of the European caviar grey market. *Geoforum, 135*, 163–171.
Dore, A., Hübschle, A., & Batley, M. (2022). Towards environmental restorative justice in South Africa: How to understand and address wildlife offences. In B. Pali, M. Forsyth & F. Tepper (Eds.), *The Palgrave Handbook of Environmental Restorative Justice* (pp. 333–360). Palgrave Macmillan.
Duffy, R., Massé, F., Smidt, E., Marijnen, E., Büscher, B., Verweijen, J., … & Lunstrum, E. (2019). Why we must question the militarisation of conservation. *Biological Conservation, 232*, 66–73.
Galtung, J. (1969). Violence, peace, and peace research. *Journal of Peace Research, 6*(3), 167–191.
Hübschle, A. (2016). A game of horns: Transnational flows of rhino horn. *International Max Planck Research School on the Social and Political Constitution of the Economy.*
Hübschle, A. (2017a). Contested illegality: Processing the trade prohibition of rhino horn. In J. Beckert & M. Dewey (Eds.), *The Architecture of Illegal Markets* (pp. 177–197). Oxford University Press.
Hübschle, A. (2017b). The social economy of rhino poaching: Of economic freedom fighters, professional hunters and marginalized local people. *Current Sociology, 65*(3), 427–447.
Hübschle, A. (2019). Fluid interfaces between flows of rhino horn. *Global Crime, 20*(3), 198–217.
Hübschle, A. (2025). Contested illegality: Understanding legitimacy challenges in wildlife governance. *The Palgrave Handbook of Environmental Policy and Law*. London: Palgrave Publishers. https://doi.org/10.1007/978-3-031-30231-2_2-1
Hübschle, A., & Berg, J. (2024). Southern blue criminology: Rethinking ocean harmscapes in a global context. *Frontiers in Conservation Science, 5*, 1–6. https://doi.org/10.3389/fcosc.2024.1422829
Hübschle, A., Dore, A., & Batley, M. (2021). Focus on victims and the community: Applying restorative justice principles to wildlife crime offences in South Africa. *The International Journal of Restorative Justice, 4*(1), 140–150.
Hübschle, A., & Lindley, J. (2024). Blue crimes and ocean harmscapes: Strategies for tackling transnational maritime environmental crimes in the global south. *Frontiers in Conservation Science, 5*, 1–12. https://doi.org/10.3389/fcosc.2024.1448316
Hübschle, A., Mackenzie, S., Yates, D., & Bērziņa, D. (2024). Irregularly regulated collecting markets: Antiquities, fossils, and wildlife. *Crime, Law and Social Change, 82*, 1111–1130. https://doi.org/10.1007/s10611-024-10171-9
Hübschle, A., & Margulies, J. (2024). The need for a socioecological harm reduction approach to reduce illegal wildlife trade. *Conservation Biology, 38*(5), e14335.
Hübschle, A., & Shearing, C. (2018). Ending wildlife trafficking: Local communities as change agents. *Global Initiative against Transnational Organised Crime.*
Hübschle, A., & Shearing, C. (forthcoming). *Governing Wildlife Security: Towards Pragmatic Conservation*. Routledge.
Iordăchescu, G., & Vasile, M. (2022). Forest crisis narratives: Illegal logging, datafication and the conservation frontier in the Romanian Carpathian Mountains. *Political Geography, 96*, 102600.
Lunstrum, E. (2014). Green militarization: Anti-poaching efforts and the spatial contours of Kruger National Park. *Annals of the Association of American Geographers, 104*(4), 816–832., 101, 102818.
Massé, F., Gardiner, A., Lubilo, R., & Themba, M. N. (2017). Inclusive anti-poaching? Exploring the potential and challenges of community-based anti-poaching. *South African Crime Quarterly, 60*, 19–27.
Massé, F., & Margulies, J. D. (2020). The geopolitical ecology of conservation: The emergence of illegal wildlife trade as national security interest and the re-shaping of US foreign conservation assistance. *World Development, 132*, 104958.
Margulies, J. D., Masse,F., & VandeBerg, B. (2023). Illicit geographies and contested environments: An introduction. *Political Geography, 101*, 1–3.
Nixon, R. (2011). *Slow Violence and the Environmentalism of the Poor*. Harvard University Press.

Nurse, A., & Wyatt, T. (2020). *Wildlife Criminology*. Bristol University Press.
Ramutsindela, M., Matose, F., & Mushonga, T. (2022). Conservation and violence in Africa. In Ramutsindela, M., Matose, F., & Mushonga, T. (eds.), *The Violence of Conservation in Africa* (pp. 2–21). Edward Elgar Publishing.
Roe, D., & Booker, F. (2019). Engaging local communities in tackling illegal wildlife trade: A synthesis of approaches and lessons for best practice. *Conservation Science and Practice, 1*(5), e26.
Witter, R. (2021). Why militarized conservation may be counter-productive: Illegal wildlife hunting as defiance. *Journal of Political Ecology, 28*(1), 175–192.
Witter, R., & Satterfield, T. (2019). Rhino poaching and the "slow violence" of conservation-related resettlement in Mozambique's Limpopo National Park. *Geoforum, 101*, 275–284.

14

RESPONDING TO ILLEGAL WILDLIFE TRADE IN HONG KONG, SAR, CHINA

Amanda S. Whitfort

Introduction: wildlife trade in Hong Kong

A city of 7.5 million people, Hong Kong has been repeatedly identified as a hub for wildlife trade (both legal and illegal) (Sung and Fong, 2018; Wong, 2019; Andersson et al., 2021; Bager Olsen et al., 2021). Wildlife is imported into the city for use in food, medicine, as pets and for decoration. While certain species are imported for local consumption, a large quantity is transhipped, both legally and illegally, to Mainland China (Zhang et al., 2008). Seizures of species controlled under Convention on International Trade in Endangered Species (CITES) are regularly made at the border between Hong Kong and China (Cheng et al., 2017), and Hong Kong's ports are a major entry point for wildlife smuggled by sea from Southeast Asia and Africa (Jiao and Lee, 2022).

In 2018, Hong Kong raised its maximum penalty for serious wildlife crimes to ten years' imprisonment and banned ivory trading. These amendments to the *Protection of Endangered Species of Animals and Plants Ordinance* (Cap 586) raised Hong Kong's penalties for wildlife crimes to among the highest in the world. This chapter examines and evaluates the impact of the recent legislative changes in Hong Kong on the problem of illegal wildlife trade. It finds that while higher penalties have led to increased sentences for wildlife offenders in Hong Kong, a subsequent deterrent effect on wildlife offending is not clear.

Legal protection for endangered species in Hong Kong

Hong Kong has been a party to the CITES since 1976[1] and implements its obligations through Cap 586. Until 2018, all prosecutions brought under Cap 586 were necessarily tried in the lowest level of criminal court, the magistracy. In that jurisdiction, nominal fines were imposed for serious cases of trafficking wildlife, and the maximum penalty which could be imposed was two years' imprisonment. Even for the most serious cases, the penalties imposed by the magistrates generally involved only a few weeks' imprisonment. In 2017, the highest penalty that was imposed for an offence under Cap 586 was three months' imprisonment for smuggling 62 kg of ivory. Lenient sentences and a tendency to regard wildlife crime as a regulatory offence raised concerns that Hong Kong was not doing enough to curb illegal wildlife trade (Lau, 2021).

 DOI: 10.4324/9781003333906-17

In the years before Hong Kong increased its maximum penalty for wildlife trafficking the international pressure to curb illegal wildlife trade in the region increased. The United Nations resolution on Tackling Illicit Trafficking in Wildlife, adopted in July 2015, had urged member countries to consider the trafficking of protected species involving organised crime groups as 'serious crime'. The maximum penalty of a serious crime (as defined under the UN Convention Against Transnational Organised Crime) is at least four years' imprisonment. Research by the Hong Kong government's Agriculture Fisheries and Conservation Department (AFCD) in 2017, under the then Director for the Environment, had shown several jurisdictions were imposing a maximum of ten years or more for wildlife offences (Greece, Australia, Botswana, Mexico, Namibia, Zimbabwe, China, and Kenya). The Director proposed to increase Hong Kong's maximum penalty for trade in CITES Appendix I species to ten years' imprisonment on the basis that severe penalties were necessary to provide a strong deterrent against illegal wildlife trade and Hong Kong wanted to show that the government was very serious about deterrence.[2]

Other important precursors also led to the 2018 amendments. The extension of China's membership of the Convention on Biological Diversity (CBD) to Hong Kong, in 2011, had raised concerns as to the status of Hong Kong as a wildlife trading hub.[3] At that time, the Hong Kong government began the process of formulating the Territory's first Biodiversity Strategy and Action Plan under the CBD. CBD Aichi Biodiversity Target 12 required that 'by 2020 the extinction of known threatened species has been prevented and their conservation status, particularly of those most in decline, has been improved and sustained'. In compliance with Art 3 of the CBD, Hong Kong has an obligation to ensure that activities within its jurisdiction or control do not cause damage to the environment of other States or areas beyond the limits of the Territory's national jurisdiction. In addition, while the legal import of newly sourced ivory for trade in Hong Kong had ended with the CITES ban in 1990, Hong Kong had continued to be identified as one of the largest markets in the world for ivory carvings.[4] Increasing global criticism of Hong Kong's large ivory market was being fuelled by the impending extinction of Asian and African elephants.[5] In the decade preceding Hong Kong closing its ivory market, an increase in the poaching and trafficking of ivory had been observed in the scientific literature, raising an imminent threat of extinction (Underwood et al., 2013). In parallel, Hong Kong had recorded several large-scale ivory seizures Wong, 2018). A resolution at the CITES 17th Conference of the Parties in 2016 had recommended all parties with a domestic ivory market, which may be contributing to poaching, close their markets as a matter of urgency.[6]

In his January 2016 Policy Address, the then Chief Executive of Hong Kong, Leung Chun-Ying, signalled an intention to put an end to Hong Kong's ivory trade. In June 2016, the Environment Bureau presented a proposal to the Legislative Council's Panel on Environmental Affairs to both phase out local trade in elephant ivory and impose heavier penalties for smuggling and illegal trade of endangered species.[7] By December 2017, China had fully completed its own ban on the ivory trade on the Mainland and closing the market in Hong Kong was consistent with the national policy. The 2018 amendments to Cap 586 brought into effect a three-phase ban on the local ivory trade. In 2018, the first two phases banned the import and export of elephant trophies and pre-CITES ivory. The commercial sale of registered stocks of ivory was permitted, however, until the end of 2021, to allow all trading licences in place to expire. In May 2018, the increased maximum penalties for wildlife offences involving all CITES-listed species came into effect. At the discretion of the Hong Kong Department of Justice, offences could, for the first time, be tried as serious crimes in the District and High Court of Hong Kong. The following year, the Hong Kong judiciary, at all levels, received training highlighting the impact of wildlife crime.[8]

The impact of the increased penalties

Between May 2018 and December 2022, 74 cases involving the import, export and possession for trade of derivatives of endangered species were prosecuted in Hong Kong. These cases were collected and analysed by the author. Judgements for cases prosecuted in the District Court were published by the Hong Kong judiciary and made accessible at their website.[9] Outcomes from cases heard in the magistrates' cases were analysed from notes taken in court, as magistrates' decisions are not published unless appealed. The 74 cases comprised 12 pangolin prosecutions, 12 seahorse prosecutions, 10 American Ginseng prosecutions, 9 prosecutions for elephant ivory, 9 rhino horn prosecutions, 7 agarwood prosecutions, 4 hawksbill sea turtle prosecutions, 3 shark fin prosecutions, 1 prosecution for importing endangered orchids, 1 for importing Asiatic bear bile medicine, 1 helmeted hornbill prosecution, 1 prosecution for pink conch pearl, 1 for stony coral, 1 for costus root, 1 for importing tiger bone, and 1 prosecution for the import of saiga horn. Most of the prosecutions involving endangered species imported as derivatives were from species known to be used in traditional Chinese medicine. These included rhino horn, ginseng, pangolin scales, seahorse, bear bile, tiger bone, saiga horn (Cheung et al., 2021), and costus root (Nadda et al., 2020).

During the same period there were ten prosecutions for the import of live species, including one for 20 African spurred tortoises, one for 60 Japanese Leaf turtles, one for three star tortoises, one for 223 roofed, crown river and flat shell tortoises, one for 210 roofed tortoises, one for 55 radiated and 2 ploughshare tortoises, one for 6 African grey parrots, one for 27 yellow headed amazon parrot eggs, and two for European eel. All the live species seized are highly collectable in the pet trade (Cheung and Dudgeon, 2006; Nijman and Shephard, 2015; Petrozzi, 2018; Willows and Courtnee, 2020), other than eel, which is sourced as a luxury food (Stein et al., 2016).

Table 14.1 shows the eight most common endangered species encountered in *completed prosecutions* (charged and prosecuted) from May 2018 to December 2022. The tabled year is the year of charge, with most cases prosecuted within 12 months of charge.

It is apparent, from Table 14.1, that very few wildlife crime cases have been pursued to prosecution in Hong Kong since 2019. This could be construed as a sign that Hong Kong's stronger penalties for wildlife crime have had a significant deterrent effect. However, after COVID-19 was first detected in China in late 2019, Hong Kong introduced a two-week quarantine period for those crossing into Hong Kong from the Mainland and, by March 2020, a compulsory quarantine period was imposed on all international travellers (Hartley and Jarvis, 2020). From March 2020, trafficking wildlife in luggage through the Hong Kong airport and on to other cities in the region became impossible. Air cargo was also significantly delayed as fewer planes were flying into and out of Hong Kong airport due to limited passengers (Kan et al., 2021). Container shipments also became increasingly difficult to move internationally as maritime connectivity decreased (Guerrero et al., 2022). Hong Kong's quarantine requirements for inbound travellers were not lifted until September 2022. Three years of measures adopted by the Hong Kong government to combat COVID-19 rendered the usual methods used by traffickers to smuggle wildlife into Hong Kong useless.

Table 14.1 Endangered species identified in charged cases, May 2018–December 2022

	2018	*2019*	*2020*	*2021*	*2022*
Pangolin	5	6	1		
Seahorse	1	10	1		
Elephant ivory	3	6			
Rhino horn	4	5			
American Ginseng	5	5			
Shark fin		2		1	
Hawksbill sea turtle		4			
Agarwood	5	1	1		

Against this background, it is difficult to determine, by looking at the number of cases charged, whether the 2018 amendments have effectively deterred traffickers from Hong Kong. Accordingly, the next section of this chapter examines the judicial approaches taken to sentencing cases tried by the courts in 2018 and 2019 for three of the most traded derivatives: pangolin scales, rhino horn and elephant ivory.

Pangolin scales

As shown in Table 14.1, 12 cases involving the import and collection of pangolin scales were charged and prosecuted between May 2018 and December 2022. In three cases, large weights of pangolin scales had been brought to Hong Kong by Chinese nationals for shipment to China, where pangolin scales are used in traditional Chinese medicine. The first case was heard less than two weeks after pangolins were transferred from Appendix II to Appendix I of CITES and involved a Chinese man who flew from the Democratic Republic of Congo to Hong Kong with two suitcases containing scales, weighing a total of 48 kg. He bought a ferry ticket to Macau and checked his luggage through to the Chinese Territory. The scales were detected by Hong Kong Customs officers, and the defendant was arrested at the airport. The man claimed the scales had been given to him by his boss in Congo. The judge's sentence of three years' imprisonment was reduced by one-third for a guilty plea and another six months to reflect the recent change of CITES Appendix listing for pangolins.[10]

In a second case in 2018, two Chinese women carried four suitcases containing a total of 110 kg of pangolin scales on a plane from the Democratic Republic of Congo to Hong Kong for onward transport to China. They were intercepted before boarding the Hong Kong ferry for Macau. The court sentenced each of the offenders to two years' imprisonment, reduced by one-third for their guilty pleas.[11] In another case in 2018, a Chinese national living in Hong Kong was arrested at the airport when he collected six boxes of pangolin scales, weighing 56 kg, in air freight from Nigeria. He claimed to be collecting the boxes for a friend on the Mainland. While the name of the consignee was given to the prosecuting authorities, he was not prosecuted. The Hong Kong man was sentenced to 18 months' imprisonment.[12]

In a targeted delivery case in 2018, customs officers intercepted eight boxes containing 65 kg of pangolin scales posted from Malaysia and delivered them to an address under surveillance. The defendant, a Chinese national illegally staying in Hong Kong, received the box. Inside the room where he worked were found to be nine repackaged cartons containing 181.82 kg of pangolin scales, a cardboard box containing 8.08 kg of pangolin scales and an electronic balance. The defendant was prosecuted in the District Court and convicted of conspiracy to import unmanifested goods. In mitigation, he claimed that he was merely the transporter, responsible for repackaging the scales and posting them to China. He was sentenced to 48 months' imprisonment, reduced to 32 months for pleading guilty.[13] In other cases, heard in the lower magistrates' courts, defendants had picked up parcels containing 10–30 kg of pangolin scales sent from Malaysia by airmail for others. Most were charged with possession under Cap 586. Sentences of between four and six months were imposed by the magistrates, depending on the weight of the scales involved. In no cases were the persons who allegedly were to receive the scales prosecuted. One case involved a man attempting to export 3 kg of pangolin scales over the Hong Kong (HK) border to the Mainland in his backpack in early 2019. He was sentenced to community work.

In the last case involving pangolin scales imported in 2019 by air passenger luggage before COVID-19 effectively ended air travel to Hong Kong, a male Chinese national living in the Democratic Republic of Congo flew to Hong Kong with four suitcases containing 100 kg of pangolin scales in exchange for payment. The organiser of the shipment arranged for another male Chinese national to fly to Hong Kong, from China, to help carry the cases back to the Mainland. The two men were intercepted at the airport, prosecuted, and sentenced to three years' imprisonment, reduced by one-third for their guilty pleas. The organiser was not identified for prosecution.[14]

Rhino horn

Table 14.1 shows nine prosecutions for importing rhino horn tried in the District Court between 2018 and 2020: seven were heard in the District Court, and two in the magistracy. Each of these cases involved defendants who brought the horn to Hong Kong secreted in their luggage or on their person, by aeroplane. The largest seizure prosecuted was for 40 kg of horn, carried by two men transiting through Hong Kong to Ho Chi Minh City in February 2019. The South African men boarded the plane in Johannesburg. One worked as a customer service officer in the airport and the other as a security provider. Both men carried no other luggage than the horns in their checked baggage. Both men received a one-third discount from their sentence of 39 months' imprisonment for pleading guilty. Investigation of their offences by law enforcement authorities in Hong Kong led to no further prosecutions.[15]

The next large seizure occurred in October 2019, when a South African woman transiting through Hong Kong to Ho Chi Minh City was found to be carrying 15 kg of rhino horn in her luggage. She had boarded the plane in Johannesburg and was unemployed. In an interview with law enforcement authorities, the woman claimed she had been paid by a man (whose name she provided) to carry the horn. Her information to the authorities led to no further arrests. The woman received a one-third deduction from her sentence of 36 months' imprisonment for her timely guilty plea.[16] The other rhino horn cases involved significantly less horn and were all related to Chinese nationals carrying horn into Hong Kong for friends or relatives to use as medicine. In June 2018, a Chinese national who had lived with his family in South Africa for 12 years carried 6 kg of horn and 4 g of worked ivory on a flight from South Africa to Hong Kong for a friend. The defendant was sentenced in the magistracy to 12 months' imprisonment, reduced by a third for his plea and another four months because his case was the first heard by the courts after the penalty for trafficking in rhino had increased to a maximum of ten years.[17] Before the 2018 amendment, the standard tariff for smuggling two rhino horns had been just three months' imprisonment.[18]

In October 2018, a Chinese national arriving in Hong Kong from Nigeria via Doha was found in possession of 0.065 kg of ivory, 0.025 kg of rhino horn and 0.095 kg of pangolin scales. He claimed to be bringing them to China as gifts for his relatives. He was sentenced to three weeks' imprisonment in the magistracy.[19]

In addition to these cases between 2018 and 2019, five other Chinese nationals were prosecuted for bringing 3 kg or less of rhino horn to Hong Kong. Two boarded planes in South Africa, two in Mozambique and one in Thailand. All claimed to be carrying the horn to give as medicine to friends or relatives in China. District Court sentences ranged from 12 to 18 months' imprisonment.[20]

Elephant ivory

All nine cases involving the import and possession of elephant ivory have been tried in the magistrates' courts, except one. Four cases in 2019 involved Chinese nationals importing ivory bracelets and other small pieces into Hong Kong on their return from African countries. Each of these defendants was sentenced to three to nine weeks' imprisonment. Several cases involved stings on shops by law enforcement officers checking the age of the ivory on sale. In a 2018 case involving a trader who sold a set of post-Convention chopsticks to an undercover officer, the court imposed an order to be of good behaviour for 18 months. The trader concerned was a member of the Hong Kong government's advisory committee on endangered species.[21] In the only ivory-related case to have been heard in the District Court, a Hong Kong trader was prosecuted for possession of 20 kg of worked ivory and 35 kg of raw tusks without a licence. In late 2019, he had offered to sell some of the ivory to an undercover law enforcement officer. The trader was sentenced to 36 months' imprisonment, reduced by one month for his guilty plea.[22] The trader appealed against his sentence to Hong Kong's Court of Appeal. On appeal, the trader argued that possession of an endangered species should be treated more leniently than import; however, the Court ruled that the sentence was appropriate.[23]

Comparing the sentences for these cases with those imposed before the 2018 amendments, it is apparent that sentences for wildlife offences have been significantly increased under the Cap 586 amended higher penalties. Allowing cases to be tried in the District Court has raised the possibility of a higher sentence for cases transferred to that court. With the two-year limit on imprisonment in the magistracy no longer applied to very serious cases, far lengthier sentences have been imposed than were seen prior to 2018. Only two of these have been appealed as excessive, and both sentences were upheld by the High Court.[24]

The importance of judicial training

While raising the maximum penalty for wildlife crime has the potential to enhance sentences, particularly when the cases are heard in the District Court, increased legal protection of species also relies on ensuring the understanding of judges of the serious effects of wildlife crime. In one of the few wildlife crime cases to have reached the Hong Kong High Court before the 2018 amendment to Cap 586, *HKSAR v Xie Jin Bin* [2011] 2 HKLRD 631, the Honourable Mr Justice To, on hearing an appeal related to a conviction for theft of incense trees (CITES Appendix II) gave guidance to the courts below stating:

> In ordinary cases of theft from a supermarket or ordinary goods, even of trees which are not endangered species, value or quantity or weight may be the only consideration and the same is true for trafficking in dangerous drugs. But where the stolen property involved is a protected endangered species the considerations are wholly different. The end to which the law seeks to achieve is protection of our endangered flora. Thus, the evil which the law seeks to prevent is not theft as such but injury to the protected flora. Unlike theft of ordinary goods, restitution is impossible. The recovered wood could not be restored to the plant. The focus should therefore be on protection of the plant, rather than the value of the plant or part stolen. The factors to be taken into consideration are the injury done to the tree, profit-motive, the manner of commission and the gravity of the offence.

In passing sentence in the lower court, the trial judge had relied on IUCN (International Union for the Conservation of Nature) Red list data and the obligation of Hong Kong to give force to CITES. He also considered a statement produced by an expert from AFCD, detailing not only the likely harm to the tree that was damaged but also the ecological impact to the forest and the few healthy populations of the trees in China. The trial judge's use of expert scientific evidence to determine that there had been significant harm to the trees was approved as good practice by the appellate court. However, while *Xie Jinbin* has been followed in other cases involving incense trees, it had limited immediate effect on cases involving other endangered plants and animals, particularly as the case was prosecuted for theft of the tree by the police and not as a wildlife crime.

In the wake of the legislature's significant increase in maximum penalties for serious wildlife crime in 2018, judges began to seek more detailed information from prosecutors of the nature of the offending and its consequences. Hong Kong prosecutors immediately recognised that for effective sentences for wildlife crimes to be passed, judges needed to be presented with detailed admissible and reliable evidence illustrating the full impact of the alleged offending. To assist in providing this information to the court, species victim impact statements for Hong Kong's most commonly smuggled wildlife were developed in a collaboration between the University of Hong Kong's Law Faculty and Conservation Forensics Laboratory in the School of Biological Sciences. Along with legal and scientific data on the species concerned, current black-market values for the species in trade and smuggling trends were included in information provided by a local NGO, the Kadoorie Farm and Botanic Garden, where endangered species seized by government authorities are housed. The species victim impact statements allowed prosecutors and judges to easily ascertain the conservation status of the species concerned, including the CITES listing and current threat assessment for the species by the IUCN, as well as access relevant scientific literature to assess the ecosystem impacts of the illegal trade on the animals or plants concerned.

As noted in *HKSAR v Xie Jinbin*, the manner in which the crime is committed and the gravity of the offence should be considered by the court assessing the seriousness of the case. Harms caused in the commission of the offence will not always be obvious to the court unless expert evidence is provided. Where animals are traded live and not killed outright, they may be harmed by the use of inappropriate transport methods. Secreted to avoid detection, they may suffer crushing, asphyxiation, dehydration, starvation, temperature shock, and stress. Judges may not be made aware that illegal wildlife trade not only affects the animals targeted. Poaching may disrupt family and social groups and can severely undermine the stability of fragile populations, particularly where juveniles are taken before breeding. Poaching methods may also impact non-target species as collateral damage. Removal of top predators from ecosystems upsets the ecological balance, undermining the food chain and affecting other vulnerable species. Animals that are caught live for transport are also known to be more likely to suffer from disease or injuries than those which elude their captors. Disease in wild animals is not only a problem for the animals but also a potential public health risk. The risks from the import of animals that pose an invasive species threat may also be relevant for certain species.

Using species victim impact statements in wildlife crime prosecutions helps to ensure these types of harms are made known to the judge. With the assistance of species victim impact statements, judges since 2018 have increasingly underscored, in their published reasons for sentence, the serious nature of wildlife crime and its impacts, enabling longer sentences to be fortified against appeal. In 2021 and 2022, two lengthy sentences in which species victim impact statements had been referenced to provide trial judges with a greater understanding of the impacts of wildlife crime were upheld as appropriate on appeal. In both cases, the trial and appellate judges referred to scientific data on the loss of the species and the impact on biodiversity as an important factor supporting deterrent sentencing.[25]

The OSCO amendment

The 2018 amendments to Cap 586 and an increased understanding of the impact of wildlife crime have resulted in longer sentences than were recorded before the legislative change. However, an examination of the cases in 2018 and 2019 reveals that only certain types of cases have been pursued and significant seizures have gone unprosecuted. All but one prosecution concluded, since the 2018 amendments, has involved the defendant being caught red-handed carrying the contraband or collecting it in a shipment. None of the cases have involved the prosecution of those operating at any higher level in international wildlife trafficking networks.

In November 2020, a strategy paper was prepared by ADM Capital Foundation (a local NGO) and provided to Dr Elizabeth Quat, a local legislator. The paper noted that despite the amendments to the maximum penalties for serious wildlife crimes in 2018, Hong Kong was continuing to report record-breaking seizures for illegal wildlife. In the first four months of 2019, Hong Kong broke its own seizure records when 8.3 tonnes of pangolin scales were seized from a container shipped from Nigeria, along with 2 tonnes of ivory. Three further record-breaking seizures occurred in 2019 and 2020. In 2019, 82.5 kg of rhino horn was seized from cargo transshipment at the airport. In 2020, two seizures, which took place within weeks of one another, each comprised 13 tonnes of threatened shark fins. The strategy paper noted that failing to prosecute seizures of this volume was undermining the efforts of the government to address international perceptions of Hong Kong as a wildlife trading hub. Further, while Hong Kong has strong legislation to deter money laundering, its provisions had never been used to forfeit the financial benefits of wildlife crime.

Under criticism for prosecuting only those caught red-handed at the airport, senior officers of AFCD and Customs expressed difficulties in investigating the local and international criminal networks behind Hong Kong's large seizures. On the back of the strategy paper's recommendations, Dr Quat determined to introduce a private member's bill seeking to incorporate certain serious offences involving wildlife trafficking and illegal trade into Hong Kong's *Organised and Serious Crimes Ordinance*, Cap 455. By listing serious wildlife offences under Cap 586 in Schedule 1 to Cap 455, law enforcement agencies tasked with combatting wildlife trafficking and illegal trade would have access to important investigative powers applicable to other organised crimes. They could require offenders to answer questions under compulsory orders and produce documents for inspection. Powers under the Ordinance to freeze and confiscate assets related to wildlife crimes could also be used to deter offending. The Bill was adopted by the government as an effective means to strengthen authorities' investigation and enforcement powers in combatting wildlife crimes. The amendment to Cap 455 was passed by the legislature and came into effect in August 2021. No local or international kingpins involved in wildlife trafficking have yet been prosecuted using Hong Kong's OSCO powers.

Conclusion

In the absence of prosecutions against those controlling and benefitting from wildlife trafficking, it is difficult to claim Hong Kong has made serious inroads in deterring wildlife crime. While sentencing tariffs for wildlife trafficking have significantly increased since 2018 and judicial training on the impacts of wildlife crime has ensured longer sentences are upheld on appeal, law enforcement authorities have concentrated their efforts on prosecuting only mules and couriers, both of which can easily be replaced. With the resumption of international travel, it will become possible to more reliably determine the impact of Hong Kong's new approach to deterring wildlife offenders. However, until law enforcement efforts start to impact key players in the criminal syndicates behind the trade, Hong Kong's role as a wildlife trafficking hub will remain an international concern.

Notes

1 Hong Kong began to implement CITES in 1976 under the Animals and Plants (Protection of Endangered Species) Ordinance, Cap. 187. That Ordinance was replaced by the Protection of Endangered Species of Animals and Plants Ordinance, Cap. 586, in 2006.
2 Legislative Council Panel on Environmental Affairs, Background brief on the legislative proposal to phase out the local trade in elephant ivory prepared by the Legislative Council Secretariat, Special Panel meeting 6 June 2017. LC Paper No. CB (1)1018/16–17(01). Available at: https://www.legco.gov.hk/yr16–17/english/panels/ea/papers/ea20170606cb1–1018-1-e.pdf.
3 In 2014, the Terrestrial Biodiversity Working Group, a group of scientists, invited to advise the Hong Kong government on achieving CBD targets released an information note stating that Hong Kong was one of the major importers of wildlife internationally with not all legally sourced.
4 See Note 2.
5 Legislative Council Panel on Environmental Affairs, Updated Background Brief on Control of Legal Trade in Elephant Ivory prepared by the Legislative Council Secretariat, 27 March 2017, LC Paper No. CB (1)705/16–17(06). Available at: https://www.legco.gov.hk/yr16–17/english/panels/ea/papers/ea20170327cb1–705–6-e.pdf.
6 CoP17 Com. II. 6.
7 Environment Bureau, Hong Kong SAR Government, Proposed Plan for Phasing Out the Local Trade in Elephant Ivory, Submitted to the Legislative Council Panel on Environmental Affairs for discussion on 27 June 2016. CB (1) 1054/15–16(06). Available online at: ea20160627cb1–1054–6-e.pdf (legco.gov.hk).
8 Hong Kong Judiciary Annual Report 2019. Available at: https://www.judiciary.hk/en/publications/annu_rept_2019r/eng/training.html.
9 Available at: Hong Kong Judiciary - Judgments & Legal Reference.

10 *HKSAR v Lin Jinbao* [2019] HKDC 616.
11 *HKSAR v Wei Binxue and Wei Yanhong* [2019] HKDC 1162.
12 *HKSAR v Xie Cuo* [2021] HKDC 472.
13 *HKSAR v Huang Zengxuan* [2019] HKDC 1258.
14 *HKSAR v Zhang Guozhu and Another* [2020] HKDC 368.
15 *HKSAR v Tshirobo, Confidence and Matheblua Rhulani* [2020] HKDC 210.
16 *HKSAR v Muller Canlyn Yolande* [2020] HKDC 630.
17 *HKSAR v Liu Xin* (unreported, 13 July 2018) West Kowloon Magistrates Court Case No.2565/2018.
18 *HKSAR v Cheung Mo Tak*, High Court Magistracy Appeal 89/2012.
19 *HKSAR v Huang Yuan Wang* (unreported, 26 October 2018) West Kowloon Magistrates Court Case No 4845/2018.
20 *HKSAR v Wei Bin* [2018] HKDC 1311; *HKSAR v Zhao Zhengxin* [2019] HKDC 223; *HKASR v He Dongchen* [2019] HKDC 1420; *HKSAR v Yan* [2019] HKDC 1433; *HKSAR v Cao Xiaocheng* [2019] HKDC 1522.
21 *HKSAR v Yim Ka Shiu* (unreported, 16 April 2019) Kwun Tong Magistrates Court Case No. 22555/2018.
22 *HKSAR v Fung Tak Chueng* [2020] HKDC 778.
23 [2021] HKCA 1273.
24 *HKSAR v Xiao Rongqiang* [2021] HKCA 23 and HKSAR *v Fung Tak Cheung* [2021] HKCA 1273.
25 *HKSAR v Xiao Rongqiang* [2021] HKCA 23 and HKSAR *v Fung Tak Cheung* [2021] HKCA 1273.

References

ADM Capital Foundation. (2020). Protection of Endangered Species Enhanced Enforcement Strategy White Paper, Hong Kong.

Andersson, A. et al. (2021). CITES and beyond: Illuminating 20 years of global, legal wildlife trade. *Global Ecology and Conservation*, 26. https://doi.org/10.1016/j.gecco.2021.e01455

Bager Olsen, M. et al. (2021). Thirty-six years of legal and illegal wildlife trade entering the USA. *Oryx*, 55(3), 432–441.

Cheng, W., Xing, S. and Bonebrake, T.C. (2017). Recent pangolin seizures in China reveal priority areas for intervention. *Conservation Letters*, 10, 757–764.

Cheung, H. et al. (2021). Understanding traditional Chinese medicine to strengthen conservation outcomes. *People and Nature*, 3, 115–128. https://doi.org/10.1002/pan3.10166

Cheung, S.M. and Dudgeon, D. (2006). Quantifying the Asian turtle crisis: Market surveys in southern China, 2000–2003. *Aquatic Conservation: Marine and Freshwater Ecosystems*, 16, 751–770. https://doi-org.eproxy.lib.hku.hk/10.1002/aqc.803

Guerrero, D., Letrouit, L. and Pais-Montes, C. (2022). The container transport system during Covid-19: An analysis through the prism of complex networks. *Transport Policy,* 115, 113–125.

Hartley, K. and Jarvis, D. (2020). Policymaking in a low-trust state: Legitimacy, state capacity, and responses to COVID-19 in Hong Kong. *Policy and Society,* 39, 403–423.

Hong Kong Legislative Council Paper LC Paper No. CB (1)491/17–18.

Hong Kong Legislative Council Paper LC Paper No CB (1) 1393/16–17(152).

Hong Kong Legislative Council Paper LC Paper No. CB (1) 384/16–17(02).

Hong Kong Legislative Council Panel on Environmental Affairs Proposal Plan for Phasing out the Local Trade in Elephant Ivory CB (1) 1054/15–16(06).

Jiao, Y. and Lee, T. (2022). The global magnitude and implications of legal and illegal wildlife trade in China. *Oryx*, 56(3), 404–411.

Kan,W.H.T. et al. (2021). Analyzing Hong Kong's inbound tourism: The impact of the COVID-19 pandemic, 45 *IATSS Research*, 440–450.

Lau, W. (2021). *Ebbing Away: Hong Kong's Ivory Trade*. TRAFFIC International.

Nadda, R.K. et al. (2020). Aucklandia costus (Syn. Saussurea costus): Ethnopharmacology of an endangered medicinal plant of the himalayan region. *Journal of Ethnopharmacol*, 263, 113199. https://doi: 10.1016/j.jep.2020.113199

Nijman, V. and Shepherd, C.R. (2015). Analysis of a decade of trade of tortoises and freshwater turtles in bangkok, thailand. *Biodiversity & Conservation*, 24(2), 309–318. https://doi.org/10.1007/s10531-014-0809-0

Petrozzi, F. et al. (2018). Exploring the main threats to the threatened african spurred tortoise centrochelys sulcata in the west african sahel. *Oryx*, 52(3), 544–551. https://doi.org/10.1017/S0030605316001125

Stein, F.M. et al. (2016). First genetic evidence of illegal trade in endangered european eel (anguilla anguilla) from europe to asia. *Conservation Genetics Resources*, 8(4), 533–537. https://doi.org/10.1007/s12686-016-0576-1

Sung, Y.H. and Fong, J. (2018). Assessing consumer trends and illegal activity by monitoring the online wildlife trade. *Biological Conservation*, 227, 219–225.

Tsui, K. et al. (2021). Analysing Hong Kong's inbound tourism: The impact of the COVID-19 pandemic. *IATSS Research*, 45, 440–450.

Underwood, F. et al. (2013). Dissecting the illegal ivory trade: An analysis of ivory seizures data. *PloS One*, 8, e76539.

UN Resolution A/RES/69/314. Available online at www.un.org/en/ga/69/resolutions.shtml.

Willows-Munro, S. and Kleinhans, C. (2020). Testing microsatellite loci for individual identification of captive african grey parrots (psittacus erithacus): A molecular tool for parentage analysis that will aid in monitoring legal trade. *Conservation Genetics Resources*, 12(3), 489–495. https://doi.org/10.1007/s12686-019-01127-6

Wong, R.W. (2018). 'Do you know where I can buy ivory?': The illegal sale of worked ivory products in Hong Kong. *Australian & New Zealand Journal of Criminology*, 51, 204–220.

Wong, R.W. (2019). *The Illegal Wildlife Trade in China*. Palgrave Macmillan.

Zhang, L., Hua, N. and Sun, S. (2008). Wildlife trade, consumption and conservation awareness in southwest China. *Biological Conservation*, 17, 1493–1516.

15

SOUTHERN PERSPECTIVES ON WILDLIFE CRIME

Non-punitive responses to IUU in Argentina[1]

Valeria Vegh Weis

Introduction

Argentina 2021. The media continue to broadcast satellite images of South Korean, Chinese, Taiwanese, and Spanish vessels fishing illegally in Argentine waters and on the high seas. The damage to the ecosystem, the millions of dollars in losses to the economy and the risks to sovereignty are being highlighted. Criminalisation is called for as a response (El Cronista, 2021; la Nación, 2021; Perfil, 2021; Tais Gadea, 2021; Economía Sustentable, 2022; Prefectura Nacional Argentina, 2020). However, the Prefectura Naval Argentina (PNA), the security agency responsible for Argentine waters, has decided otherwise. This chapter examines the case of Huali 8, in which the PNA did not pursue the problematic strategy of physical pursuit and subsequent criminalisation of the offending vessel, nor did it call for the creation of new criminal offences. Instead, the PNA promoted an innovative manoeuvre by requesting the international apprehension of the vessel through Interpol and the boarding of the conflict outside the criminal channels. The procedure was successful and culminated in the resolution of the case through administrative channels, followed by the seizure of the vessel and the payment of the economic cost of the fishing carried out, as well as the payment of the corresponding fine for the cost of the seizure. After this case, boats detected by the PNA began to voluntarily surrender to the authorities and pay the corresponding costs instead of fleeing. In this sense, the Huali 8 case is perceived by the PNA as a "paradigm shift" (Prefectura Nacional Argentina, 2020a, 2020b).

Overall, this chapter seeks to enrich green and blue criminology with a case study that brings a Southern perspective and the experience of those who, acting on the ground, propose critical non-punitive solutions despite media and political pressure. The research is based on the case study method (Yin, 1984), based on extensive data collection of public and private documents, including regulations, jurisprudence and relevant documentation in the databases of the PNA (Petintseva & Eski, 2020), as well as six open-ended in-depth interviews with the main actors involved in the case (van Meter, 1990). Part 2 briefly introduces Southern perspectives within green and blue criminology. Part 3 presents the IUU situation

DOI: 10.4324/9781003333906-18

in Argentina and the role of the PNA. Part 4 develops the non-punitive, cross-sectoral, and interdisciplinary solutions developed by the PNA. Part 5 provides brief conclusions.

Southern perspectives within green and blue criminology

Green criminology studies focused on Southern cases, written from a Southern perspective and by Southern authors, have emerged only over the last few years. However, scholars have emphasised that encounters between criminological analysis focused on the North–South divide and green and blue criminology had been happening long before these studies emerged. As Brisman et al. put it:

Southern criminology and green criminology complement each other in various ways … to some extent, both these 'new' criminologies are old. They are statements of the accumulation of thought, and they necessarily draw upon earlier thinking that has questioned, resisted and called for re-balancing of knowledge-power and knowledge-claims (2018, p. 4).

This entanglement raises the need to position this new literature as one dimension within a wider scholarship: green criminology, critical criminology, Southern criminology, and even interdisciplinary research on environmental hazards in the Global South. In more detail, green criminology

> has always recognised global inequalities, the north-south divide and the weight of exploitation bearing down on Indigenous and colonised peoples, non-human species and resource-rich territories … [and] in so doing … it has connected fruitfully and powerfully to the important new area of Southern criminology and investigated the environmental contexts of migration, injustice against Indigenous Peoples and forms of resistance.
>
> *(Brisman & South, 2020, p. 44)*

In turn, critical criminologists have turned to the South

> as a context in which exploitative global forces may exercise power causing biodiversity loss due to patterns and practices of resource extraction … and where the theft of nature and the pollution of air, land and water follow from ever more intensive agribusiness and the food/profit chain.
>
> *(Brisman & South, 2020, p. 625)*

but also where resistance can be found (Vegh Weis, 2021).

Then, Southern criminology (Carrington et al., 2016; Carrington et al., 2018) was born to contest the North–South divide, including the environmental hazards that result from it. Also, researchers from the field of peace ecology, such as Amster (2015, p. 159), have noted the relevance of geopolitical conflicts when exploring harms against nature, as "resource wars and patterns of economic colonisation are often initiated by the nations of the Global North vis-à-vis those of the Global South, yielding a two-tiered world of privileged consumers at the top and vulnerable producers on the bottom". Likewise, from a social science and legal perspective, one relevant work on environmental destruction in the South is that of Viale and Swampa (2020). From an activist-scholar perspective, they have intervened in different cases of state-corporate environmental harms and made them accessible to the general public. Bohm, following Swampa and Viale, discussed the study of the crimes and harms of "maldevelopment" (2019, 2022), which they linked to structurally invisible forms of violence committed in Latin America for the sake of Northern economic interests. Arroyo-Quiroz and Wyatt (2018) referred to green criminology with a focus on "environmental degradation". Budó et al. (2022) also argue that the field of green criminology can be enriched with knowledge from the periphery of capitalism.

Concerning terminology, Brisman and South referred to "Southern green cultural criminology" as an example of green cultural criminology, and they have refrained from delineating any clear boundaries. They perceive green criminology as a *perspective*; green cultural criminology as a *perspective within a perspective* and, consequently, the theoretical framework of green cultural criminology "enriched by engaging with examples of literary cultural productions of various kinds from Latin America … as a *perspective within a perspective within a perspective*" (2020, pp. 625–626). This third-level perspective considers literature of and about nature in Latin America – specifically, eco-critical scholarship encompassing both fiction and non-fiction (2020, p. 625). Nevertheless, Brisman and South have acknowledged their Western, Global Northern (urban) perspective and have argued that they therefore cannot guide or prescribe "a genuinely Southern criminology" (2020, p. 625), which should instead be developed by authors from the South.

Within this realm of studies, specific analysis started to look at the particularities of the aquatic environments under the label of "blue crimes" (Bueger & Edmunds, 2020). Going beyond the letter of the law, "blue criminology" also emerged as a concept that studies harmful behaviours in rivers, seas, and oceans, regardless of whether they constitute criminal offences under criminal law (García Ruiz et al., 2022). In turn, the notion of "Southern blue criminology" zooms in to the particularities of these harmful behaviours in the Global South (Hübschle & Berg, 2024; Vegh Weis, 2023). In this line of reasoning, a critical perspective exposes the more drastic levels of selectivity of the criminal justice systems when dealing with harmful behaviours in the aquatic environment in the Global South, where what prevails is the over-criminalisation of marginalised subsets of society (even when defending nature from human destruction) and the under-criminalisation of the powerful (regardless of whether their actions endanger the aquatic system) (Vegh Weis, 2019).

Upon this background, this chapter inquires whether there are specific examples that can be explored and that can shed light on innovations developed in the Global South to confront blue harms and crimes, in an effort towards the development of "Southern blue criminology". More specifically, are there Southern innovations compatible with the critical criminology's rejection of increased crime control as a solution to crime? The following sections provide some indications of what is available for the development of a Southern blue criminology that can shed light on non-punitive responses to blue harms and crimes.

IUU in Argentina

(Limited) primary criminalisation – more criminal law?

At the national level, Law No. 24,922 establishes the Federal Fisheries Regime, which regulates the surveillance and control of fishing vessels and the exploitation of living marine resources in the maritime areas under Argentine jurisdiction. Meanwhile, three "Protocols" regulate illegal fishing, establishing uniform criteria and actions for the surveillance of the different maritime areas under national jurisdiction, as well as for the collection and exchange of information. Vessels operating within the 200-mile exclusive economic zone (EEZ) may be boarded by the PNA for violation of the federal fisheries regulations. This is an administrative offence, and the criminal law only intervenes when the PNA attempts to detain the offending vessel and a situation of disobedience is created, thus constituting the crime of assault and resistance to authority, in accordance with Article 239 of the Penal Code. The Prefect of the PNA, Mariana Baggini (2022), clarifies that when the PNA intervenes in these situations, it is not fishing and the marine environment that is protected, but the exercise of public administration.

Given the limited primary criminalisation, many point to the need to create specific offences for the protection of the aquatic environment, either in national or international legislation. In this sense, Magistrate Eva Parcio, Federal Judge of the Federal Court of Comodoro Rivadavia, explains that at present, "crimes *with* environmental *repercussions*" (e.g. resistance to authority) are prosecuted, but that we lack environmental crimes per se (Parcio, 2020). However, the authorities of the PNA state that:

It is not necessary to add criminal sanctions to the administrative ones. The fines provided for in the Federal Fisheries Regulation are effective and sufficiently high to deter activity. Effective control of areas under national jurisdiction, together with the current regulatory framework, should be sufficient (Almada, 2022).

Indeed, there are risks in the punitive solution, whether through the application of the existing Article 239 or through the creation of new offences. First, the criminal justice system tends to target the most vulnerable individuals (Zaffaroni et al., 2002; Vegh Weis, 2017). On this basis, and in line with what is happening in the criminal justice system as a whole, the PNA authorities warn that if fishing is treated as a crime, criminal control runs the risk of being used to over-criminalise the last link in the fishing chain:

If illegal fishing is to be included as a criminal offence in the Penal Code, the legislator should be very careful when drafting the law, because behind the fishing vessels that plunder the oceans there are companies and even states that would penalise captains or crew members, but those who actually promote and finance the commission of the illegal act could remain outside the scope of the law (Interdisciplinary Team for the Control of Maritime Spaces and Resources, 2022).

Second, the criminal law comes into play when the damage has already been done, in this case when the vessel has already initiated the harmful fishing activity. In other words, even if fishing in Argentinean waters were criminalised, this would not solve the damage to the aquatic environment. Again, in the words of the Interdisciplinary Team (2022), "going through the courts and criminal proceedings implies a legal process that takes years, that is very long, that is very cumbersome … all the time that elapses undermine the effectiveness of the process". Third, there is the broader critique of the effectiveness of criminal law. If the concern is not retribution but the damage done to sovereignty, the economy, and nature, would criminalising fishing serve to prevent further acts by the crew, as theories of special prevention suggest? Would it serve to deter other vessels from engaging in such harmful behaviour, as established by theories of general prevention? Adaro (2022) replies: "The penalty that would be applied is a prison sentence, which does not solve the fishing problem". Finally, a criminal approach could be disproportionate to the damage caused. As Baggini (2022) explains: "In criminal proceedings, let's say, the maximum penalty that can be sought is imprisonment, and to respond to illegal fishing by having a person end up in prison is, to say the least, too much of a punishment".

Secondary (limited) criminalisation: more weapons, personnel, and technology?

The PNA has both law enforcement and investigative powers over more than 4,700 km of coastline, in addition to 11,325 km of coastline in Antarctica and the Southern islands, which contain commercial fisheries, hydrocarbon basins and mineral deposits of great economic importance (Prefectura Naval Argentina, no date; UNODC, 2020). To control these vast waters, the PNA uses aircraft and ships, as well as surveillance by the Coast Guard, although this is not enough as no state has enough capacity to patrol the entire coastline (Almada, 2022). However, even with these tools, the prosecution of vessels is complex. As mentioned above, prosecution can only take place if the PNA's order to stop is disobeyed, and it is very difficult to prosecute, with statistics showing that the PNA only manages to arrest the offending vessels two or three times a year.

Is it therefore necessary to have more technology, personnel, or weapons to increase the effectiveness of secondary criminalisation? Again, the PNA authorities are reluctant to expand the punitive response. The first scenario is that even with more staff, weapons, or technology, they could more easily catch vessels on the run. However, this is likely to involve the use of force, which raises the issue of proportionality in relation to a minor offence (the offence of disobedience). Second, the pursuit may cause very serious damage to the vessels and risk loss of life at sea. This is all the more serious when we consider that on many vessels the captain is the only person who can decide to stop the vessel and make themselves available to stop the escalation of violence, while the rest of the crew are often in a situation of extreme vulnerability (EICEMAR Legal Team, 2021). Third, pursuing the vessel at all costs confronts the coastguard with the question of how far to escalate the situation: is it legitimate, ethical, and moral to go so far as to shoot at the boat and sink it, causing harm to people? As Almada (2022) points out: "You join the coastguard to save lives, not to take them, so hot pursuit was a big burden and a complex problem for our captains". Ultimately, these actions could lead to the criminalisation of coastguards for allegedly using excessive force to injure or harm crew members.

A second scenario is not to pursue or to stop pursuing the fleeing vessel and to refer the case to a competent court. In this case, even if there were more personnel, technology or weapons, there would still be delays and difficulties in identifying the crew of the offending vessel: "We are talking about foreign ships that do not necessarily return to the country, in addition to the impossibility of our legislation to try in absentia" (Adaro, 2022). Moreover, even if the perpetrators were identified and extradited, the punitive solution remained short-sighted:

> If you think about it, well, OK, fine, fine, we take it to a criminal court, the criminal court is favourable. What is the sentence? What is the penalty? Prison? In other words, we have to ask for the extradition of a whole fishing fleet or a captain to China, Spain or wherever to come here. That is not very effective.
>
> *(Adaro, 2022)*

The search for approaches outside of criminal punishment: illegal fishing and the Hua Li 8 case as a leading case or paradigm shift

In light of these criticisms, the question remains: how can IUU cases be addressed without resorting to harmful and ineffective punitive responses? The key case for understanding the PNA's actions from this perspective is that of the vessel Hua Li 8. This was a Chinese-flagged vessel that was detected in 2016 by the Coast Guard Cutter GC-26 Thompson while conducting illegal fishing within the EEZ. Despite being ordered to cease its activities and allowed to be inspected, the traditional sequence of events took place: the captain fled and was pursued for several hours by Coast Guard personnel.

The innovation in the case of the Hua Li 8 is that the PNA did not follow either of the two usual paths: neither did it resort to force, with the risk of collapse that this entails, nor did it abandon the pursuit without further procedural progress. Instead, the case was referred to the Federal Court, where Dr Parcio issued a request for international arrest under case 1142/16 for the offence of assaulting and resisting authority. The novelty was that, since the identity of the captain and crew members was unknown, as usual, the PNA, in agreement with the judge, asked Interpol to issue the arrest warrant for the vessel, thus setting an innovative precedent. Only 40 days after the warrant was issued, the Indonesian maritime authorities contacted the court to inform it that they had complied with the international arrest warrant and that the vessel was at the disposal of the Argentine state. On this basis, the PNA identified the crew, and the Secretariat of Fisheries calculated and enforced the corresponding fines. The same procedure was repeated in subsequent cases (Parcio, 2020).

Thus, from the Hua Li case onwards, attempts were made to identify the captain of the ship so that the judiciary could act by requesting his international arrest from Interpol, but if this was not possible, the court could request the international arrest of the ship. This change was radical and by no means spontaneous: "It was a very thorough work carried out by the PNA for Interpol to agree to order the capture of ships" (Almada, 2022). In this sense, the PNA, in line with international recognition, refers to the Hua Li case as a "paradigm shift in the capture of foreign vessels operating illegally in the waters of Argentina's Exclusive Economic Zone (EEZ)" (Prefectura Nacional Argentina, 2022). Similarly, the ICEMAR Legal Team clarifies that "it was a reference case at the international level that generated a paradigm shift in the capture of illegal foreign fishing vessels" (EICEMAR Legal Team, 2021).

This innovation facilitates

> the police action of the institution and its action in full compliance with the Law of the Sea, preventing an excessive escalation of the use of force by the coastguard to proceed with the capture from creating risks of loss of life at sea.
>
> *(EICEMAR Legal Team, 2021)*

Thus, "the Hua Li 8 case marks the end of the excessive use of force and protects the integrity of the vessel and crew" (Almada, 2022). It also avoids the potential criminalisation of the coastguard for excessive use of force. Thus, the creativity of the PNA in conjunction with the federal judiciary and Interpol in international cooperation "really brought us relief" (Almada, 2022). In this vein, the innovation of the PNA focuses on the instrument deployed to perpetrate the crime (the vessel) rather than on the perpetrator, which allows to remove the instrument used to perpetrate the crime and prioritise the recovery of the assets over punishment.

It is added that "it is no longer essential to capture the offending vessel, as the necessary legal and international cooperation instruments have been created to hold the vessel responsible for the infringement and the crime committed, if physical capture is not possible" (Prefectura Naval Argentina, 2022). As the vessel is usually located in third countries when the international arrest warrant is successful, Huali8 was also innovative in that "it is the first time that the capture of an infringing fishing vessel has been requested and achieved in waters other than those of the jurisdiction in which the infringement was committed or in which it was sought" (Adaro, 2022).

The impact of the case was even greater than intended, because for fishing fleets, the request for international arrest of a vessel is extremely serious, both economically and diplomatically. In economic terms, and in the case of China, the government deprives the vessels of subsidies under the international catch and thus ends their ability to operate. Faced with this dilemma, the shipowners or owners of the vessels spontaneously began to present themselves to the PNA through the relevant embassy, before the intervening Federal Court enforced the international fishing order (Almada, 2022). In diplomatic terms, and 'given the fact that it is possible to request international capture and that this means a much greater hassle and problem for the shipowners [owners], what has happened is that in some cases the embassy itself has ordered the offending vessels to return to Argentina and submit to the national authority' (Adaro, 2022).

This strategy has also brought procedural and financial benefits: procedural time is saved, illegally acquired goods are recovered, fines are collected, and even the operational costs of the coastguard or air pursuit are paid by the offending vessel. It should be added that, in some cases, the owners of the vessel present themselves spontaneously after the international alert has been issued. This makes it possible to speed up the procedure and to even establish whether the same vessel may have committed other infringements of the EEZ before, which can then be claimed. This happened, for example, in the case of the Portuguese fishing vessel Calvao, where the documents found on board (logbooks, engine logbooks, fishing logbooks) and the Automatic Identification Satellite System (AIS-S) of the offending vessel showed that it had registered other entries and fishing positions within the EEZ. In view of the above:

> The Fisheries Enforcement Authority charged the relevant offence for all unauthorised entries and imposed a fine.... In addition, the vessel owner company came forward and acknowledged the violations committed...
>
> *(Prefectura Nacional Argentina, 2022)*

Finally, the Huali 8 case had an international impact and raised the PNA's profile in the world. Almada comments that 'the case attracted the attention of many countries ... they considered it a good practice to be used at the international level to make progress against illegal fishing' (Almada, 2022). He adds that at the time of the case:

> I was working as an exchange officer with the United States and a lawyer from the US Coast Guard called me and asked me to explain to her and lawyers from all the armed forces what had happened with the Hua Li 8 and how the PNA had intervened; they had seen the impact of the case and adapted some of their protocols following this case, i.e. they accepted that it was not necessary to sink the ship, knowing that there was another way to bring those who had committed a crime to justice.
>
> *(Almada, 2022)*

In spite of all these advantages, the procedure followed in the case of Hua Li 8 was repeated in the following years with various ships. In order to standardise the procedure followed in this case, the Ministry of Security drew up a resolution setting out the action protocol for surface units. The protocol establishes the point at which the pursuit is to be stopped, the possibility of firing a warning shot, a knockout shot in safe places and, if the vessel does not stop and it is decided to stop the pursuit, to ensure that there is sufficient evidence to request international arrest, in line with the precedent set by the Huali 8: the case that brought about a "paradigm shift".

Conclusions

This chapter has built on the theoretical contributions of green criminology and its intersection with Southern criminology to open the door to Southern perspectives within blue criminology and the specificities of blue crimes and harms in the Global South. Furthermore, this chapter has shed light on how a Southern blue criminology could inform innovative non-punitive responses to blue harms that can inform other cases worldwide.

Particularly, this chapter allowed us to explore an Argentinean case study and, in particular, the role of the Argentinean Naval Prefecture (PNA) and its search for non-punitive solutions to IUU. In correlation with the lessons of blue criminology, the work of the PNA shows that the most harmful behaviours are not necessarily those that are criminalised and that greater criminalisation of fishing activities may not only fail to be the best solution but may even become an iatrogenic solution that reinforces the most negative aspects of the functioning of the criminal justice system. Specifically, this article has shown how the PNA has confronted the problem of fishing vessels operating within the 200-mile limit without resorting to greater punitive control.

For a long time, the vessels generally escaped (with the consequent accusation that the PNA had allowed foreigners to damage the national fishing reserve) or sank (with the scandal of the escalation of the use of force). Faced with this, and in line with the legacy of critical criminology on the futility of punitive power, the PNA developed an innovative solution based on the "leading case" of Huali 8. In this case, based on the cooperation of the judiciary and Interpol, it avoided prosecution and instead asked this international agency – normally limited to the arrest of persons – to focus on the arrest of the boat. In contrast to the sinking cases, this action made it possible (1) to avoid damage to the life and health of the offending crew (most of whom come from vulnerable sectors) and (2) to avoid criminalising both these offenders and the PNA personnel, who often had to be brought before the criminal justice system due to the escalation of violence. In addition, unlike cases where the PNA has abandoned prosecution, this action (1) repaired the damage and even managed to cover the costs of prosecution, fines and confiscation, and (2) even avoided prosecution by notifying the owner of the offending vessel through the Ministry of Foreign Affairs before proceeding with the execution of the international arrest warrant.

Overall, this chapter has attempted to focus on the Global South and how one of the agencies charged with exercising crime control has sought to explore alternative courses of action that prioritise the care of the aquatic environment without the need for further punishment. Further research should be done to explore whether this experience might bring lessons for terrestrial contexts and to the wide range of environmental harms. While different trends within green and blue criminology claim that we need to expand our criminological imagination to seek approaches to environmental harm that do not rely on more punishment, this pioneering experience in the Global South shows that the most innovative approaches are out there, waiting to be explored.

Note

1 This study was conducted as part of a consultancy with the Instituto Universitario de Seguridad Marítima (IUSM) under the direction of Enrique Font. A similar version in Spanish was published in Delito & Sociedad (2023).

Bibliography

Adaro, R. (2022). *'Entrevista - Secretario de Extensión de la Universidad de la Prefectura Nacional Argentina'*. Buenos Aires: Universidad Nacional del Litoral.

Almada, S. (2022). *'Interview - ICEMAR Coordinator'*. Buenos Aires, February 3.

Amster, R. (2015). *Peace Ecology*. London: Routledge.

Arroyo-Quiroz, I., & Wyatt, T. (2018). *Criminología verde en México*. Mexico DF: Centro Regional de Investigaciones Multidisciplinarias.

Baird, A., & Walters, R. (2020). 'Water theft through the ages: Insights for green criminology', *Critical Criminology*, 28(3), 371–388.

Belhabib, D., & le Billon, P. (2020). 'Illegal fishing as a trans-national crime', *Frontiers in Marine Science*, 7, 1–3.

Brisman, A., Nigel, S., & Reece, W. (2018). Southernising green criminology. Human dislocation, environmental injustice and climate apartheid | Criminología verde. Criminología Verde. https://www.criminologiaverde.com/publication/j16/

Brisman, A., & South, N. (2000). *Routledge International Handbook of Green Criminology;* Second Edition: Vol. Second Edition (A. Brisman & N. South, Eds.), London: Routledge.

Brisman, A., & South, N. (2020). Toward a green cultural criminology of the South. In Nigel South and Avi Brisman (Eds.). *Routledge International Handbook of Green Criminology*. London: Routledge. 529–564.

Budó, M., Goyes, D., Natali, L., Sollund, R., & Brisman, A. (2022). *Introdução à criminologia verde: Perspectivas críticas, descoloniais e do Sul*. Sao Paulo: Tirant Lo Blanch.

Bueger, C., & Edmunds, T. (2020a). 'Blue crime: Conceptualising transnational organised crime at sea', *Marine Policy*, 1–8 Available at: https://doi.org/10.1016/J.MARPOL.2020.104067.

Bueger, C., & Edmunds, T. (2020b). 'Blue crime: Conceptualising transnational organised crime at sea', *Marine Policy*, 119.

Carrington, K., Hogg, R., Scott, J., & Sozzo, M. (2018). Criminology, Southern Theory and Cognitive Justice. In Kerry Carrington, Russell Hogg, John Scott, and Máximo Sozzo (Eds.). *The Palgrave Handbook of Criminology and the Global South*. 3–18. London: Springer International Publishing.

Carrington, K., Hogg, R., & Sozzo, M. (2016). Southern criminology, *The British Journal of Criminology*, 56(1), 1–20.

Carrington, K., Hogg, R., & Sozzo, M. (2018). Criminología del Sur, *Delito y Sociedad*, 27, 9–33.

Economía Sustentable (2022). 'Argentina loses US$2 billion a year to illegal fishing in the Argentinean Sea', 14 January. Available at: https://economiasustentable.com/noticias/la-millonaria-perdida-por-la-pesca-ilegal-en-el-mar-argentino-en-la-milla-201 (Accessed: 11 July 2022).

EICEMAR (2021). *'Interview'*. Buenos Aires, 2 December.

El Cronista (2021). 'Pesca ilegal: prevén la llegada de más de 300 pesqueros chinos a la Argentina', 7 December. Available at: https://www.cronista.com/economia-politica/pesca-ilegal-preven-la-llegada-de-mas-de-300-pesqueros-chinos-a-la-argentina/ (Accessed: 11 July 2022).

García Ruiz, A., South, N., & Brisman, A. (2022). Eco-crimes and ecocide at sea: Toward a new blue criminology, *International Journal of Offender Therapy and Comparative Criminology*, 66(4), 407–429.

Hübschle, A., & Berg, J. (2024). Ocean as harmscapes: navigating the waters of southern blue criminology, Frontiers in Conservation Science.

ICEMAR Legal Team (2021). *'Interview'*. Buenos Aires, 2 December.

Interdisciplinary Team for the Control of Maritime Spaces and their Resources (2022). *'Interview'*. Buenos Aires, 2 December.

Lasslett, K. (2010). 'Crime or social harm? A dialectical perspective', *Crime, Law and Social Change*, 54(1), 1–19. Available at: https://doi.org/10.1007/s10611-010-9241-x.

Lynch, M.J. (1990). 'The greening of criminology: A perspective for the 1990s', *Critical Criminology*, 2(3), 3–12.

la Nación (2021). 'Pesca ilegal: una ONG denuncia cómo buques chinos "se ocultan" para "saquear" en aguas argentinas', 2 June. Available at: https://www.lanacion.com.ar/seguridad/pesca-ilegal-una-ong-denuncia-como-buques-chinos-se-ocultan-para-saquear-en-aguas-argentinas-nid02062021/ (Accessed: 11 July 2022).

Natali, L. (2014). 'Green criminology, environmental victimization and social harm. The case of Huelva (Spain)', *Crítica Penal y Poder*, 0(7), 5–34.

Página/12 (2022). '¿Qué es el Agujero Azul que está en Argentina?', 8 July. Available at: https://www.pagina12.com.ar/435013-que-es-el-agujero-azul-que-esta-en-argentina (Accessed: 11 July 2022).

Paulson, N. (2018). 'An emergent "blue criminology"? Review of a new critical criminology book on water', *Critical Criminology*, 28(3), 549–551.

Pauly, D., & Zeller, D. (2016). 'Catch reconstructions reveal that global marine fisheries catches are higher than reported and declining', *Nature Communications*, 7(1), 1–9.

Perfil (2021). 'Pesca ilegal: la Armada Argentina y Prefectura Naval dispusieron un patrullaje por 300 buques chinos', 6 December. Available at: https://www.perfil.com/noticias/politica/pesca-ilegal-armada-argentina-prefectura-naval-dispusieron-patrullaje-300-buques-chinos.phtml (Accessed: 11 July 2022).

Petintseva, O., Faria, R., & Eski, Y. (2020). *Interviewing Elites, Experts and the Powerful in Criminology*. London: Palgrave.

Prefectura Nacional Argentina (2020). Captured vessels. Available at: https://www.argentina.gob.ar/prefecturanaval/resumen-operativo/buques-capturados (Accessed: 8 April 2022).

Prefectura Nacional Argentina (no date). Cambio de paradigma. Available at: https://www.argentina.gob.ar/prefecturanaval/resumen-operativo/cambio-de-paradigma (Accessed: 8 April 2022).

Prefectura Naval Argentina (no date). Resumen operativo | Argentina.gob.ar. Available at: https://www.argentina.gob.ar/prefecturanaval/resumen-operativo (Accessed: 8 April 2022).

Ruggiero, V., & South, N. (2013). 'Green criminology and crimes of the economy: Theory, research and praxis', *Critical Criminology*, 21(3), 359–373. Available at: https://doi.org/10.1007/s10612-013-9191-6.

Segato, L. et al. (2017). 'Environmental crimes in the water sector', *European Water*, 60, 373–379.

Snyder, H. (2019). 'Literature review as a research methodology: An overview and guidelines', *Journal of Business Research*, 104, 333–339.

Tais Gadea, L. (2021). 'Pesca ilegal: qué ocurre en el Mar Argentino y cuáles son las medidas necesarias', *Chequeado*, 14 April. Available at: https://chequeado.com/el-explicador/pesca-ilegal-que-ocurre-en-el-mar-argentino-y-cuales-son-las-medidas-necesarias/ (Accessed: 11 July 2022).

UNEP/INTERPOL (2016). *The Rise of Environmental Crime: A Growing Threat to Natural Resources Peace, Development and Security*, Nairobi: UNEP.

UNODC (2020). Global Programme against Maritime Crime maritime crime.

van Meter, K. (1990). 'Methodological and design issues: Techniques for assessing the representatives of snowball samples'. In E. Lambert (ed.), *Collection and Interpretation of Data From Hidden Populations*. Rockville: National Institute on Drug Abuse, 69–84.

Vegh Weis, V. (2017). *Marxism and criminology: A history of criminal selectivity*, Leiden: Brill.

Vegh Weis, V. (2019). Green activism and indigenous claims under the language of criminality. The persecution of argentine indigenous peoples confronting state-corporate environmental harms, *International Journal for Crime Justice, and Social Democracy*, 8(3), 38–55.

Vegh Weis, V. (2021). *Criminalization of Activism. Historical, Present and Future Perspectives*. London: Routledge.

Vegh Weis, V. (2023). La Prefectura Naval Argentina y el "Cambio de Paradigma" en la Persecución de la Pesca Ilegal. Una mirada desde las Criminologías Verde y Azul, Delito y Sociedad, 55.

Viale, E., & Svampa, M. (2020). *El colapso ecológico ya llego. Bs.* As: Siglo XXI.

Wyatt, T., & Brisman, A. (2017). 'The role of denial in the "theft of nature": Comparing biopiracy and climate change', *Critical Criminology*, 25(3), 325–341. Available at: https://doi.org/10.1007/s10612-016-9344-5.

Yin, R. (1984). *Case Study Research: Design and Methods*. Beverly Hills: Sage.
Zaffaroni, R., Alagia, A., & Slokar, A. (2002). *Derecho Penal. Parte general*. Buenos Aires: Ediar.

Interviews and Talks

EICEMAR Team, CABA, 2021, Decemeber 2
EICEMAR Legal Team, CABA, 2021, Decemeber 2
Mag. Eva Parcio, Federal Judge of the Federal Court of Comodoro Rivadavia, 2020, https://www.youtube.com/watch?v=zUmyCDOt5es
Mariana Baggini, CABA, 2022
Roberto Adaro, CABA, 2022, February 3.
Sergio Almada, CABA, 2022, February 5.

16

REGULATORY COMPLIANCE AND THE ILLEGAL WILDLIFE TRADE

A case study on the factors affecting rhino horn prescription by traditional Chinese medicine practitioners in Guangdong province, China

Hubert Cheung, Lorraine Mazerolle, Hugh P. Possingham, and Duan Biggs

Introduction

Wildlife crime can cause serious ecological, social and economic harm (Cardoso et al., 2021; Fukushima et al., 2021). International, national and local regulations determine what behaviours constitute wildlife crime. Regulatory compliance is, therefore, critical to maintaining or restoring the health of socioecological systems. Understanding why people comply with regulations can help improve biodiversity outcomes. Regulatory measures must involve strong compliance and enforcement programmes that penalise and deter violations to effectively control environmental behaviours (Paddock, 2012).

The illegal wildlife trade (IWT) involves "all unlawful activities associated with the commercial exploitation and trade of wildlife specimens" ('t Sas-Rolfes et al., 2019). It overlaps with other serious transnational and organised crimes, including fraud, corruption, money laundering, criminal violence, and the trafficking of humans, narco tics, arms, and counterfeit goods (Anagnostou & Doberstein, 2022). From an ecological perspective, IWT can threaten the species' survival, both through direct exploitation and through indirect consequences like zoonotic disease spread and biological invasion (Cardoso et al., 2021).

Wildlife trade governance is centred on international and domestic trade restrictions. Success hinges on compliance by actors along the IWT chain ('t Sas-Rolfes et al., 2019; Phelps et al., 2016). However, trade controls alone are inadequate. Challender and MacMillan (2014) argued that although regulatory enforcement "remains a necessary instrument for governing trade ... reliance on this approach alone is doomed to fail because it cannot cope with the complexity of trade". The harvesters, intermediaries and consumers in IWT are diverse, each with distinct motivations (Phelps et al., 2016). Values-based drivers of compliance are needed to complement a strong deterrence programme (Challender & MacMillan, 2014; Paddock, 2012). In this chapter, we introduce the three motivations for compliance – calculated, social, and normative (Burby & Paterson, 1993; Winter & May, 2001) – and situate

 DOI: 10.4324/9781003333906-19

them in a case study on traditional Chinese medicine (TCM) practitioner compliance with rhino horn regulations.

Motivations for compliance

Calculated motivations (or instrumental motivations) revolve around the utilitarian idea that regulated entities act in accordance with the costs and benefits of compliance (Oyanedel et al., 2020; Paddock, 2012; Winter & May, 2001). Sanctions for violations are a formal deterrence measure (Grasmick & Bursik, 1990). Would-be offenders must perceive that: (1) the risk of being apprehended is high (certainty), (2) the punishment is sufficiently severe (severity), and (3) the punishment will be imposed without delay (celerity). Manipulating these three levers can increase deterrence and compliance (Nagin, 2013).

Social motivations stem from the desire to earn the respect or approval of significant others (family, friends, colleagues) (Winter & May, 2001). Embarrassment, loss of respect and feelings of shame associated with offending are informal deterrents that diminish the expected utility of crime (Grasmick & Bursik, 1990). These informal social controls discourage illegal behaviour because individuals value the judgement of significant others (Allen et al., 2017).

Normative motivations are driven by an individual's moral duty and agreement with the importance of a rule. These are based on internalised values and lend legitimacy to regulations – greater compliance is expected when people have a stronger sense of duty to comply (Burby & Paterson, 1993; Winter & May, 2001). Having pro-environmental attitudes is more likely to motivate compliance with environmental regulations (Paddock, 2012). Measures like connectedness-to-nature (CTN) scale (Mayer & Frantz, 2004) were developed around the idea that reconnecting people to nature can foster ecological behaviour (or reduce anti-ecological behaviour) (Tam, 2013).

Rhino conservation, poaching, trade, and TCM demand

Rhinoceroses are targeted by poachers because their horns can sell for USD $30,000–65,000/kg on the black market (Eikelboom et al., 2020). In Asia, rhino horn is used in cultural, social and medicinal applications (Cheung et al., 2021b; Dang & Nielsen, 2018; Gao et al., 2016). TCM records of rhino horn use date back 2,000 years (Liu et al., 2010). Rising demand has driven poaching to alarming levels in recent years; China and Vietnam are major demand sources (Di Minin et al., 2015).

Globally, the current approach to rhino conservation revolves around trade restrictions, anti-poaching actions (e.g. armed patrols, education, alternative livelihoods) and demand reduction (Eikelboom et al., 2020; Hanley et al., 2018). The Convention on International Trade in Endangered Species of Wild Fauna and Flora (CITES) lists of all five extant rhino species in Appendix I (the only exception being white rhino in Eswatini, Namibia, and South Africa in Appendix II; South Africa permits domestic trade in rhino horn, live animals and hunting trophies) (CITES, 2019a). This ban on all international trade has been in place since 1977, and all Parties are expected to comply with and implement it through domestic regulations and enforcement.

In China, the constitution, national laws and official circulars (normative documents functioning as law) provide for the protection of endangered wildlife (Wong, 2016).

Chief among these is the 1989 Law of the People's Republic of China on the Protection of Wildlife (most recently amended in 2022), the stated purpose of which is "protecting and saving the species of wildlife which are rare or near extinction; protecting, developing and rationally utilising wildlife resources and maintaining ecological balances" (People's Republic of China, 1989). The species covered under this legislation are listed and classed in the Directory on Special State Protection of Wildlife (People's Republic of China, 2006), which covers all CITES-listed species (People's Republic of China, 1993b). The State Council issued a further official circular in 1993 pertaining specifically to rhino horn (and tiger bone) to explicitly:

- Prohibit rhino horn import, export, sale, purchase, transport, carrying, and mailing.
- Abolish rhino horn's medicinal standards and prohibit further pharmaceutical use.
- Promote medicinal substitute development and use.
- Require all existing stock to be declared and registered with the authorities (People's Republic of China, 1993a).

In short, all activities related to the trade and medicinal use of rhino horn are banned in China. Wildlife crime penalties vary depending on circumstance and severity (per 1979 Criminal Law article 341), ranging from fines to 10-plus years of imprisonment (People's Republic of China, 1979). In this chapter, we refer to these domestic restrictions collectively as the 1993 ban on domestic rhino horn trade and medicinal use.

Rising demand has fuelled a sharp increase in poaching despite the decades-long implementation of international and domestic measures (Di Minin et al., 2015), calling into question the effectiveness of existing trade controls. Calls for trade legalisation have sparked heated debate among conservationists and remain hotly contested (Biggs et al., 2013; Cheung et al., 2018a; Dang Vu et al., 2022; Haas & Ferreira, 2016). Polarised, entrenched values have deadlocked policymaking (Biggs et al., 2017). These differences came to a head at the 2019 CITES conference when ten Southern African Development Community countries threatened to withdraw from CITES in response to the rejection of multiple proposals to legalise trade, criticising anti-trade and anti-sustainable use ideologies for restricting their use of natural resources (CITES, 2019b). Separately, China has also made clear its desire to reinstate rhino horn trade and medicinal use domestically. The State Council issued a circular in 2018 to revoke its 1993 ban to permit limited and conditional trade (People's Republic of China, 2018b), though implementation has been delayed (Cheung et al., 2018b; People's Republic of China, 2018a).

While conservationists remain polarised over trade legalisation and struggle to improve the implementation of existing measures, there is agreement that further research can improve conservation outcomes. In particular, more insight is needed into the dynamics of rhino horn demand (Hanley et al., 2018), the key actors along the IWT chain (Phelps et al., 2016; Rubino & Pienaar, 2020), and the factors affecting regulatory compliance (Barichievy et al., 2017). These will enable an assessment of how to improve the implementation of existing measures and a constructive, evidence-based evaluation of the available policy options, including trade legalisation proposals. Here, we investigate the factors influencing compliance with rhino horn trade controls and rhino horn prescription by TCM practitioners in China.

TCM is an inextricable part of China's culture, identity and healthcare landscape (Chen & Qian, 2019; Cheung et al., 2021a). For many Chinese, elements of TCM are ingrained into their dietary habits and daily lives (Kong & Hsieh, 2012; Koo, 1984). TCM concepts shape social representations of health and illness, most notably that of 'balance' and 'harmony' between *yin* and *yang* (陰陽, *yīnyáng*; duality of opposite yet complementary forces) within the body and between individuals and their surroundings (Jovchelovitch & Gervais, 1999). Rhino horn is used as a medicinal ingredient, or Chinese *materia medica*, for pharmacotherapy (But et al., 1990), which involves the preparation and consumption of herbal decoctions to treat illnesses or promote health by restoring holistic balance (Cheung et al., 2021a; Kaptchuk, 2000). TCM practitioners (physicians) are key stakeholders in the rhino horn trade (Cheung et al., 2018a, 2021b), whose actions affect both supply and demand. Examining their role more closely will facilitate the development of more nuanced conservation strategies.

In China, TCM is formally integrated with and practised alongside biomedicine (i.e. Western medicine) (Griffiths et al., 2010). A convoluted history of development and reform has produced a structurally complicated healthcare system: healthcare is delivered through a three-tiered system of hospitals, community health centres and smaller clinics, and is divided into public and private sectors (Chen & Qian, 2019; Daemmrich, 2013; Li et al., 2017; Lim et al., 2004; Xu & Yang, 2009; Zhang et al., 2015). TCM practitioners operate at all levels of the healthcare system, both publicly and privately. Physicians in China can both prescribe and dispense medication to patients (Fang et al., 2013; Sun et al., 2008; Zhang et al., 2015). As such, they have a unique intermediary role as 'middle people' who bridge the gap between harvesters and consumers (Phelps et al., 2016). They straddle both supply and demand: prescribing rhino horn triggers demand, dispensing provides supply. While hospitals typically dispense medications through in-house pharmacies after a physician has written a prescription, physicians operating at other tiers of the healthcare system (e.g. clinics, village doctors) prescribe and dispense drugs to patients directly with little or no oversight, particularly in the private sector (Fang et al., 2013; Sun et al., 2008; Xu et al., 2014; Zhang et al., 2015). TCM products and services can also be obtained from herbalist shops, independent pharmacies and unlicensed or untrained individuals who practice TCM illegally (Li et al., 2017).

Case study: TCM practitioners and compliance with rhino horn trade ban

The medicinal use of rhino horn – including prescribing and dispensing – is illegal in Mainland China. However, TCM practitioner compliance in Mainland China has not been studied; whether the domestic ban has deterred rhino horn prescription is unknown. We interviewed practitioners to investigate their perceptions of rhino horn's medicinal value, clinical applications, and substitutability to provide insight for policymaking and TCM stakeholder engagement. We designed our study around the calculated, social, and normative motivations for compliance to examine both the prevalence of and factors affecting rhino horn prescription. To address the knowledge gap with regards to calculated motivations, we measured the formal deterrence of legal and professional sanctions on rhino horn prescription (Allen et al., 2017; Grasmick & Bursik, 1990). We hypothesised higher compliance among practitioners who perceive greater deterrence. To investigate feelings of shame in relation to the social motivations of compliance, we measured shame-related emotions using a composite scale adapted to our context (Allen et al., 2017). For TCM practitioners, significant others include medical colleagues, and we hypothesised that practitioners who feel more shame from prescribing rhino horn would be less likely to do so. With regards to normative motivations, we administered a Chinese version of the CTN scale with TCM practitioners, hypothesising that practitioners who feel more closeness with nature are less likely to prescribe rhino horn.

Methods

This research complied with the Australian National Health and Medical Research Council's National Statement on Ethical Conduct in Human Research; the University of Queensland granted human research ethics approval (#2017002130).

We conducted semi-structured interviews face to face with TCM practitioners in Guangdong province in southeastern China. Previous exploratory research showed that TCM practitioners in Hong Kong prescribe rhino horn primarily to dispel heat, detoxify blood and treat *wenbing* (温病) (Cheung et al., 2018a). In TCM, geography is a key determinant of the illnesses that commonly manifest in a given place, and *wenbing* is most associated with acute infections and epidemics in southern China (Hanson, 2011). COVID-19 is an example of *wenyi* (瘟疫; epidemic of *wenbing*) (Ren et al., 2020). Guangdong is the province with the largest population (108,490,000 in 2015), the highest household per capita disposable income (~USD $4,170 in 2015) and one of the highest per capita GDP (~USD $10,900 in 2015) (National Bureau of Statistics of China, 2016). Rates of wildlife consumption as food and medicine tend to be higher in Guangzhou, the provincial capital, than in most other parts of the country (Kennaugh, 2016; Zhang & Yin, 2014). These indicate that the data collected here would likely represent the upper end of consumption.

We defined TCM practitioner broadly: any self-identified physician providing TCM services to the public (inpatient or outpatient; prescribes treatment). This included unlicensed individuals and herbalist shopkeepers who offer medical consultations and prescribe herbal decoctions (Li et al., 2017) in addition to licensed practitioners working in facilities like hospitals (ruling out random sampling). We used snowball sampling, which was most likely to generate a reasonable sample size given anticipated recruitment and issue sensitivity challenges. Snowball sampling is an established means for studying sensitive topics, whereby participants assist in the recruitment of further participants by referring personal acquaintances (Heckathorn, 2011; Newing, 2011). Similar sampling approaches have been used to study illegal tiger and ivory trading in China (Wong, 2016, 2017). We built a core group of interviewees through personal contacts, acquaintances from previous research, and self-introductory approaches.

Interviews were conducted in Chinese (Cantonese, Mandarin) for an average of 45 minutes (longest 2:15, shortest 0:25). Interviews were conducted at a convenient location for each participant, almost all of which were their primary workplaces (two personal residences, one secondary workplace). We asked interviewees about the clinical uses and substitutability of rhino horn before inquiring about past prescriptions. We were particularly interested in when interviewees last prescribed it (i.e. before or after the 1993 domestic ban). We adapted the CTN scale into Chinese to measure connection to nature (Cheung et al., 2020) and composite scales from a study of traffic law compliance to measure shame-related emotions and formal deterrence of legal and professional sanctions (Allen et al., 2017) in order to examine these variables' impact on compliance. To avoid exhausting potential participants in Guangdong, we piloted questions with 30 TCM practitioners in neighbouring Hong Kong.

We first used descriptive statistics to identify the clinical contexts for which TCM practitioners deem rhino horn suitable for patient treatment. We compiled a list of rhino horn substitutes recognised by our interviewees, and compared substitutes identified by multiple individuals with regard to perceived treatment efficacy. We used R (version 3.6.1) to conduct a binary logistic regression to assess the relationships between compliance and deterrence, shame, and connectedness to nature. Summated scores were calculated for scales measuring formal deterrence of legal and professional sanctions, shame-related emotion, and connectedness to nature. These scores are considered continuous and can be used in logistic regression (Tabachnick & Fidell, 2013). Demographic and professional independent variables were included in the regression as covariates. Input variables were entered into the model simultaneously using the 'enter' method (Ranganathan et al., 2017).

Results

We interviewed 84 TCM practitioners between July and November 2018. Table 16.1 summarises their demographic and professional characteristics. Two interviewees withdrew participatory consent mid-interview and were excluded.

The view that rhino horn imparts valuable medicinal properties for patient treatment was almost unanimous, the exception being one interviewee with no familiarity with it. More than half of our respondents (54.8%, $n = 46$) stated that rhino horn is only clinically appropriate for treating severe illnesses or urgent care in life-threatening cases (Figure 16.1); 19 (22.6%) stated that rhino horn's utility is limited, in TCM terms, to conditions at *ying* and

Table 16.1 Demographic characteristics of our sample of TCM practitioners

Variable	*B*	*SE*	*Wald*	*p*	*Odds ratio*
Constant	−3.158	5.518	−0.572	0.567	0.043
Demographics:					
Age	−0.054	0.091	−0.591	0.555	0.948
Gender (M/F)	1.481	1.010	1.466	0.143	4.399
Education	0.444	0.523	0.849	0.396	1.558
Experience	0.113	0.082	1.388	0.165	1.120
Workplace sector (public/private)	1.945	0.936	2.078	**0.038**	6.994
Test predictors:					
Connectedness to nature	−0.437	1.132	−0.386	0.700	0.646
Shame-related emotion	−0.200	0.670	−0.299	0.765	0.819
Formal deterrence (legal sanctions)	−0.053	0.861	−0.061	0.951	0.949
Formal deterrence (professional sanctions)	−0.660	0.883	−0.748	0.454	0.517

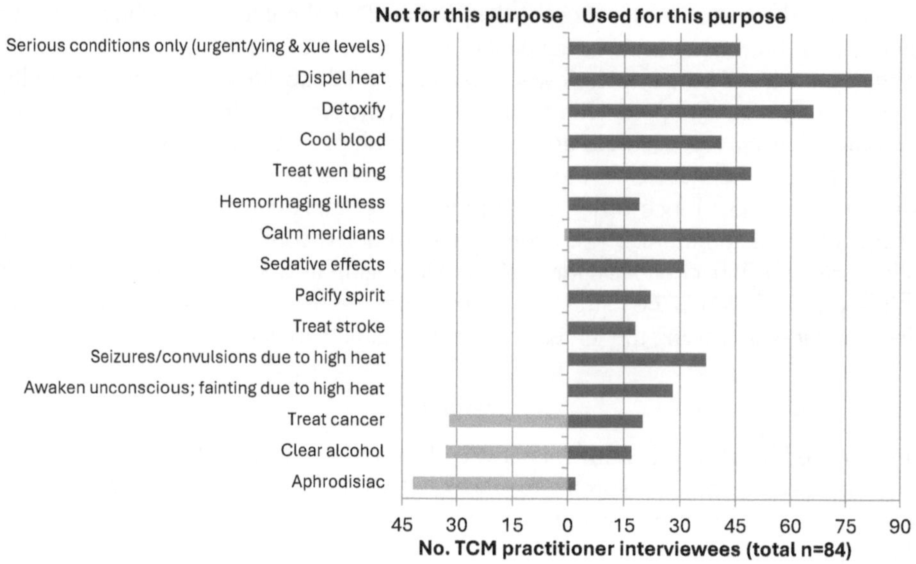

Figure 16.1 Summary of the conditions for which rhino horn can be prescribed as identified by our TCM practitioner interviewees.

xue levels of illness severity, while nine limited its utility to conditions at the deepest *xue* level only (衛氣營血; *wei-qi-ying-xue* describe progressive levels of illness severity – outer defences, vital energy, nutritive level, and blood) (Kaptchuk, 2000). Dispelling heat (97.6%, *n* = 82; 清熱 / 祛熱), detoxification (78.6%, *n* = 66; 解毒), cooling blood (48.8%, *n* = 41; 涼血), and treating *wenbing* (58.3%, *n* = 49; 治療溫病) were commonly identified functions of rhino horn. Calming the meridians (59.5%, *n* = 50; 定經), sedative effects (36.9%, *n* = 31; 鎮靜) and pacifying the spirit (26.2%, *n* = 22; 安神) (Kaptchuk, 2000) were also frequently described. Some described the use of rhino horn to treat seizures or convulsions (44.0%, *n* = 37; 高熱抽搐 / 驚厥 / 止痙) and fainting (33.3%, *n* = 28; 開竅 / 高熱昏迷 / 神志不清), though only when high heat is the underlying imbalance. Disagreement (38.1%, *n* = 32) outnumbered support (23.8%, *n* = 20) for treating cancer (治療癌症) with rhino horn. Disagreement (39.3%, *n* = 33) also outnumbered support (20.2%, *n* = 17) for clearing alcohol from the body (解酒) using rhino horn. Its use as an aphrodisiac (壯陽) was largely refuted (50.0%, *n* = 42); only two interviewees recognised such use (2.4%).

A majority of interviewees estimated that rhino horn was a medically appropriate treatment option for none (41.7%, *n* = 35) or up to 1% (22.6%, *n* = 19) of their patients seen in the last week (disregarding all other considerations, including legality, price, and procurement difficulties) (Figure 16.2). Thirty interviewees estimated that rhino horn could have been suitable for more than 1% of patients seen in the past week. Only 11 estimated that more than 10% of their patients could have been treated with rhino horn.

Our interviewees identified 16 rhino horn substitutes (Table 16.2). The most commonly identified substitute was water buffalo horn (90.5%, *n* = 76), followed by saiga antelope horn (27.4%, *n* = 23). Only three others were identified more than twice: Rehmannia root (4.8%, *n* = 4), gypsum (*n* = 3), and tree peony bark (3.6%, *n* = 3). These were all described as less effective than rhino horn (less potent, more complications, shorter-lasting ability to

Table 16.2 Summary of substitutes, where mean scores for effectiveness relative to rhino horn were only calculated for substitutes that were identified by at least three interviewees

Interviewee demographics – TCM practitioners (total n = 84)					
Variable		*N*	*Variable*		*Result*
Sex (*n*)	Male	64	Age (years)	Mean	46.4
	Female	20		Median	45.5
TCM education (*n*)	No formal TCM training	11		Maximum	80
	Vocational training	8		Minimum	22
	Undergraduate degree	44	Experience (years)	Mean	21.7
	Graduate degree	21		Median	20
TCM specialisation (*n*)	General/internal medicine	46		Maximum	55
	Orthopaedics	10		Minimum	0
	Acupuncture	6	Primary workplace (*n*)	Public sector	51
	Cardiology	4		- WM hospital	20
	Paediatrics	4		- TCM hospital	29
	Diabetes	3		- Community health centre	2
	Oncology	3		Private sector	33
	Neurology	2		- Private practice clinic	26
	Obstetrics and gynaecology	2		- Herbal medicine shop	3
	Arthritis	1		- Home clinic	3
	Gastroenterology	1		- WM-style pharmacy	1
	Hepatology	1	Geographical administrative division in Guangdong province (*n*)	Qingyuan	28
	Neonatology	1		Shenzhen	19
	Pain medicine	1		Shaoguan	14
	Pulmonary medicine	1		Shantou	10
	Recovery and rehabilitation	1		Foshan	6
	Stroke	1		Guangzhou	4
	Urology	1		Zhuhai	3

* 5-point Likert scale of effectiveness relative to rhino horn (compared to rhino horn, this substitute is: 1 – much less effective, 2 – somewhat less effective, 3 – about the same, 4 – somewhat more effective, 5 – much more effective).

suppress high heat). Dosages can be raised (53.6%, n = 45) and substitutes can be used in combination with other treatments (e.g. acupuncture) and other Chinese *materia medica* (8.3%, n = 7) to compensate for poorer treatment efficacy. However, very few felt that these strategies could produce equivalent treatment outcomes (8.3%, n = 7). Some felt that substitutes cannot fully replace rhino horn (20.2%, n = 17), or that even prescribing substitutes at higher dosages would be less effective (e.g. longer recovery times, more complications) (29.8%, n = 25). One interviewee explained that where rhino horn can reach illnesses that have penetrated the deeper *ying* and *xue* levels of the body, water buffalo horn and saiga antelope horn can only tackle illnesses at the shallower *qi* level (Interviewee GD-R09). Another suggested that water buffalo horn is "more consolation than substitute" (Interviewee GD-B21).

Overall, 22 interviewees (26.2%) reported having been prescribed rhino horn at any time in the past. Nine (10.7%) last did so before 1993. Eight (9.5%) last prescribed rhino horn over ten years ago, though recalled that such instances were after the domestic ban came into effect. One (1.2%) last prescribed rhino horn between five and ten years ago, two (2.4%) between one and five years ago, and two (2.4%) during the last year. Altogether, 15.5% of our sample (n = 13) self-reported non-compliance since the 1993 ban. They averaged 52.9 years of age with 30.9 years of clinical experience, both higher than those of the overall sample. Most had received some type of formal TCM training (3.6%, $n_{none} = 3$; 2.4%, $n_{vocational} = 2$; 7.1%, $n_{undergraduate} = 6$; 2.4%, $n_{graduate} = 2$). Nine of 33 private sector practitioners (27.3%) reported non-compliance, compared with four of 51 public sector practitioners (7.8%).

Table 16.3 summarises the means, standard deviations, and Cronbach's reliability coefficients for the CTN, shame-related emotion, and formal deterrence scales. The mean scores for the formal deterrence scales indicate that interviewees tended to have concerns about being caught and punished by the authorities with regard to rhino horn use, though they were less concerned about professional repercussions. The moderate mean score for shame-related emotions suggests that our TCM practitioners would feel some degree of shame if they prescribed rhino horn illegally. The CTN scale's mean score indicated that our interviewees feel somewhat connected to nature.

Table 16.4 summarises our binary logistic regression of variables predicting regulatory non-compliance. Non-compliance was significantly higher among private sector practitioners than their public sector colleagues (odds ratio = 6.994, p = 0.038). No other variables were significant non-compliance predictors, including CTN (p = 0.700), shame-related

Table 16.3 Means, standard deviations, and Cronbach's reliability coefficients of scales administered with TCM practitioners

Chinese materia medica	*Frequency of identification*	*Mean score of effectiveness relative to rhino horn* *
Water buffalo horn 水牛角	76	1.589
Saiga antelope horn 羚羊角	23	1.174
Rehmannia root 地黃	4	1.500
Gypsum 石膏	3	1.333
Tree peony bark 牡丹皮	3	1.666
Cow bezoar (牛黃); dwarf lilyturf tuber (麥冬)	2	-
Lacca (紫草茸); hawksbill carapace (玳瑁); gambir plant (鉤藤); earthworm (地龍); danshen root (丹參); red peony root (赤芍); figwort root (玄參); common anemarrhena rhizome (知母); cape jasmine fruit (梔子)	1	-

* 5-point Likert scales (1 – strongly disagree, 2 – disagree, 3 – neither agree nor disagree, 4 – agree, 5 – strongly agree).

Table 16.4 Binary logistic regression of variables predicting of TCM practitioner non-compliance with China's domestic ban on trade and medicinal use

Variable *	*No. of items in scale*	*M*	*SD*	*Cronbach's α*
Connectedness to nature	14	3.866	0.855	0.739
Shame-related emotion	3	3.552	0.950	0.828
Formal deterrence (legal sanctions)	3	3.738	0.815	0.899
Formal deterrence (professional sanctions)	3	3.540	0.881	0.899

Bold = significant at 95% confidence interval; R^2 = 0.202 (Hosmer–Lemeshow), 0.159 (Cox–Snell), 0.276 (Nagelkerke); model X^2 = 14.595, p = 0.103.

emotions (p = 0.765), and formal deterrence of legal (p = 0.951) and professional sanctions (p = 0.454). In reporting these results, we acknowledge the limitations of our model given our sample size of 84, noting that our model is significant at a 90% confidence level (X^2 = 14.595, p = 0.103).

Conservation implications

Rhino conservation relies heavily on trade controls, anti-poaching, and demand reduction measures. In China, the domestic trade and medicinal use of rhino horn has been banned since 1993; rhino horn has also been removed from the official TCM pharmacopoeia (People's Republic of China, 1993a). However, our study revealed considerable regulatory non-compliance by TCM practitioners: 15.5% of interviewees described having prescribed rhino horn since 1993 in response to direct questioning.

We sought to understand the context behind such non-compliance to help optimise interventions and enforcement (Arias, 2015). The perception that rhino horn imparts valuable medicinal qualities for patient treatment is near unanimous. Practitioners mostly see rhino horn as a valuable treatment option for certain severe or critical conditions (Figure 16.1), and do not find substitutes to be equally effective (Table 16.2). Practitioners do not expect to use rhino horn frequently, as patient cases for which it is 'medically appropriate' are uncommon, though the impact of such use could still be considerable given China's population (Figure 16.2). Rhino horn is seen as a potent 'cool' or 'cold' drug. Dispelling heat, detoxifying and cooling the blood, and treating *wenbing* were commonly cited uses by our interviewees, while its use for other conditions – seizures, convulsions, unconsciousness – is only suitable if high heat is the patient's underlying pattern of disharmony (Kaptchuk, 2000) (Figure 16.1). Interviewees largely disagree with rhino horn being used for treating cancer, removing alcohol from the body, and as an aphrodisiac – all of which have been reported in Vietnam (Milliken & Shaw, 2012). These findings are largely consistent with the results of our previous exploratory work in Hong Kong (Cheung et al., 2018a).

Neither CTN nor shame-related emotion was significant compliance predictors (Table 16.4). This may be because practitioners are primarily concerned with patient care; when rhino horn use is 'medically appropriate', practitioners may not view its prescription

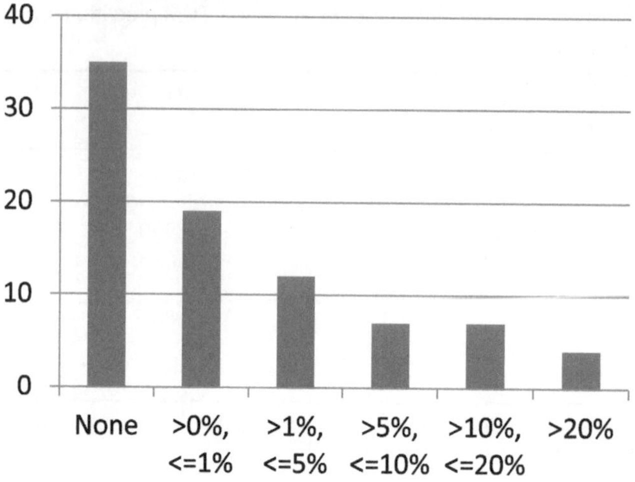

Figure 16.2 Estimated proportion of patient caseload for whom a prescription containing rhino horn is suitable for treatment, without considering any other factors like legality, price, ease of procurement, or patient affordability.

as morally wrong, even if banned. This reduces the utility of social motivations for compliance, as TCM colleagues may not judge offenders negatively. Likewise, this may diminish normative motivations of compliance, which stem from moral duty and agreement with regulations (Burby & Paterson, 1993; Winter & May, 2001). Even for individuals who otherwise hold pro-environmental attitudes, compliance would be weighed against patient care needs and treatment efficacy.

These findings are consistent with the characterisation of rhino horn trade controls as contested illegality, where legal rules are delegitimised due to their divergence from social, moral, and cultural norms (Hübschle, 2017). Absent values-based drivers of compliance, trade controls must rely on fear of apprehension and severe penalties (Nagin, 2013; Paddock, 2012). However, our analysis also shows that formal deterrence posed by legal and professional sanctions is not a significant predictor of compliance (Table 16.4). Existing trade controls have failed to deter many practitioners from prescribing rhino horn. China's intention to reinstate rhino horn trade and medicinal use domestically is likely to further erode the legitimacy of trade controls and add to the contested nature of rhino horn as a commodity (Cheung et al., 2018b; Hübschle, 2017).

The only significant compliance predictor was the type of workplace in which a practitioner primarily operates (Table 16.4). Nearly a third of private sector practitioners reported non-compliance, compared with less than 10% among public sector practitioners. We acknowledge that non-compliance may have been underreported due to its illegality and social desirability bias, the effects of which may have been more pronounced among public sector practitioners who are more likely to face consequences if past violations are uncovered. However, we posit that differences between the public and private sectors in how Chinese *materia medica* are accessed, procured, and distributed more likely explain this divergence. Public sector physicians are not typically involved in procuring medical supplies, including medicinal ingredients and pharmaceuticals, as these tasks are done by administrative personnel. Furthermore, physicians working at public hospitals are unlikely to prescribe ingredients which cannot be filled in-house. One interviewee stated that rhino horn was "very difficult to source [even before the ban] because hospitals did not stock it" (Interviewee GD-R17). We observed during hospital-based interviews that prescribing medications digitally is the norm. Several interviewees demonstrated how this works: physicians access inventory information on all Chinese *materia medica* and Western medicines stocked at the in-house dispensary, and issue prescriptions through the hospital's computer system by selecting and specifying the dosages of different Chinese *materia medica*. Even at hospitals where prescriptions have yet to be digitised, Chinese *materia medica* that are unavailable or out of stock cannot be filled in-house, and patients would have to source them externally.

Compliance can be influenced by the ease with which it is to comply or break the rules (Arias, 2015). Our findings indicate that institutional procedures, oversight, and separation of prescribing and dispensing present barriers for TCM practitioners to prescribe rhino horn. Three of the four public sector interviewees who prescribed rhino horn since the 1993 ban stated that they only gave usage instructions and medical advice to patients, who either needed to source it themselves or already had some in possession. In contrast, two of the nine non-compliant private sector interviewees provided directions for its use only; the other seven had procured, prescribed, and dispensed rhino horn. Private sector practitioners typically work in smaller (often independent) clinical settings, where they are subject to little or no oversight. They are more likely to be directly involved in procuring, prescribing, and dispensing Chinese *materia medica* (Daemmrich, 2013; Wang et al., 2011).

Limitations and conclusion

Our findings should not be interpreted as nationally representative, particularly given regional differences in socioeconomics, healthcare access, and TCM practices and norms (Li et al., 2018; Ling et al., 2011). Our inclusion criteria ruled out random sampling, and a larger sample would strengthen our analyses and results. Stressing these limitations, our results have significant conservation implications.

TCM practitioners influence both demand and supply – prescription stimulates rhino horn demand while dispensing provides retail supply. Similarities can be drawn with certain public health issues. For instance, 'supply side' solutions like restricting prescriptions, physician drug monitoring programmes, and regulatory actions that typify the US opioid epidemic response must be complemented with 'demand side' measures like educating physicians and requiring continuing education for opioid prescribing licensure (Srivastava & Gold, 2018). Likewise, conservation strategies targeting TCM practitioners should include components addressing both supply and demand.

Effective enforcement is essential for trade controls to work as intended (Challender & MacMillan, 2014), but numerous researchers have raised concerns over the effectiveness of wildlife crime enforcement in China (Shepherd et al., 2017; Wong, 2019; Zhang et al., 2008). Within the existing regulatory framework, optimising law enforcement would benefit conservation (Haas & Ferreira, 2018). This can be achieved without overhauling the legislative tools for combatting wildlife crimes in China that some researchers have advocated (Feng et al., 2019; Li, 2007). Criminologists have increasingly argued for 'focused deterrence' in law enforcement. Focused deterrence enhances the core tenets of deterrence (certainty, celerity, severity) by deploying traditional and nontraditional law enforcement tools innovatively (Braga & Weisburd, 2012), with "direct communications of increased enforcement risks ... to target groups and individuals' [being] defining characteristics of focused deterrence strategies" (Braga et al., 2018). Insight into compliers and non-compliers is required (Arias, 2015). On the basis of our results, we recommend selective targeting in the enforcement of China's domestic rhino horn regulations among TCM practitioners to better leverage calculated motivations of compliance. Specifically, enforcement efforts should be focused on the private sector, particularly practitioners who simultaneously prescribe and dispense.

Restructuring the way Chinese *materia medica* are distributed and procured, particularly by private sector practitioners, can reduce the financial incentives for prescribing and dispensing high-value endangered wildlife products and increase barriers. This is consistent with broader healthcare reforms that have been discussed both within China's public health community (Wang et al., 2011; Yip et al., 2014; Yip et al., 2010) and by the Chinese government (People's Republic of China, 2011). A state-funded study on the financial incentives behind antibiotic, injection and corticosteroid over-prescription highlighted the "need for more vigorous supervision of the pharmaceutical market and prescribing behaviour" (Li et al., 2012). Li et al. (2012) emphasised that "China must reform its health care financing and payment system to remove the profit incentives in prescribing". We echo these recommendations from the standpoint of addressing the financial incentives for prescribing and dispensing threatened species products.

Finally, we found that a considerable proportion of TCM practitioners have prescribed rhino horn in violation of longstanding trade controls. Improving enforcement may increase formal deterrence and compliance (Haas & Ferreira, 2018), but our findings add credence to the notion that existing trade controls are misaligned with the values held by the people they intend to regulate. This limits the legitimacy of policy measures and the utility of values-based drivers of compliance (Challender & MacMillan, 2014; Paddock, 2012). Attempting to orchestrate major value shifts for conservation is impractical (Manfredo et al., 2017), and it is clear that TCM will continue to be integral to the culture and everyday lives of millions of people. Conservationists must ensure that the values and interests of key stakeholders are carefully considered (Rubino & Pienaar, 2020) in weighing alternative policy options for rhino conservation.

Acknowledgements

We are grateful to the Lee Shau Kee Foundation for supporting this work. We thank those who helped us over the course of this project, in particular LKY, XJK, our translation team, and the individuals who connected us with stakeholders in the TCM community.

References

Allen S., Murphy K., Bates L. (2017). What drives compliance? The effect of deterrence and shame emotions on young drivers' compliance with road laws. *Policing and Society 27*, 884–898.

Anagnostou M., Doberstein B. (2022). Illegal wildlife trade and other organised crime: A scoping review. *Ambio 51*, 1615–1631.

Arias A. (2015). Understanding and managing compliance in the nature conservation context. *Journal of Environmental Management 153*, 134–143.

Barichievy C., Munro L., Clinning G., Whittington-Jones B., Masterson G. (2017). Do armed field-rangers deter rhino poachers? An empirical analysis. *Biological Conservation 209*, 554–560.

Biggs D., Courchamp F., Martin R., Possingham H. (2013). Legal trade of Africa's rhino horns. *Science 339*, 1038–1039.

Biggs D., Holden M.H., Braczkowski A. et al. (2017). Breaking the deadlock on ivory. *Science 358*, 1378–1381.

Braga A.A., Weisburd D.L. (2012). The effects of focused deterrence strategies on crime: A systematic review and meta-analysis of the empirical evidence. *Journal of Research in Crime and Delinquency 49*, 323–358.

Braga A.A., Weisburd D., Turchan B. (2018). Focused deterrence strategies and crime control: An updated systematic review and meta-analysis of the empirical evidence. *Criminology & Public Policy 17*, 205–250.

Burby R.J., Paterson R.G. (1993). Improving compliance with state environmental regulations. *Journal of Policy Analysis and Management 12*, 753–772.

But P.P.-H., Lung L.-C., Tam Y.-K. (1990). Ethnopharmacology of rhinoceros horn. I: Antipyretic effects of rhinoceros horn and other animal horns. *Journal of Ethnopharmacology 30*, 157–168.

Cardoso P., Amponsah-Mensah K., Barreiros J.P. et al. (2021). Scientists' warning to humanity on illegal or unsustainable wildlife trade. *Biological Conservation 263*, 109341.

Challender D.W.S., MacMillan D.C. (2014). Poaching is more than an enforcement problem. *Conservation Letters 7*, 484–494.

Chen X., Qian X. (2019). Overview of healthcare system in china. pp. 237–254 in J. Ratliff, T. J. Albert, J. Cheng, J. Knightly editors. *Quality Spine Care: Healthcare Systems, Quality Reporting, and Risk Adjustment.* Springer International Publishing, Cham.

Cheung H., Doughty H., Hinsley A. et al. (2021a). Understanding traditional Chinese medicine to strengthen conservation outcomes. *People and Nature 3*, 115–128.

Cheung H., Mazerolle L., Possingham H.P., Biggs D. (2018a). Medicinal use and legalised trade of rhinoceros horn from the perspective of traditional Chinese medicine practitioners in Hong Kong. *Tropical Conservation Science 11*, 1–8.

Cheung H., Mazerolle L., Possingham H.P., Biggs D. (2021b). China's legalisation of domestic rhino horn trade: Traditional Chinese Medicine practitioner perspectives and the likelihood of prescription. *Frontiers in Ecology and Evolution 9*, 1–11.

Cheung H., Mazerolle L., Possingham H.P., Tam K.-P., Biggs D. (2020). A methodological guide for translating study instruments in cross-cultural research: Adapting the 'connectedness to nature' scale into Chinese. *Methods in Ecology and EVolution 11*, 1379–1387.

Cheung H., Wang Y., Biggs D. (2018b). China's reopened rhino horn trade. *Science 362*, 1369–1369.

CITES. (2019a). Appendices I, II and III valid from 26 November 2019. in Convention on International Trade in Endangered Species of Wild Fauna and Flora editor.

CITES. (2019b). Summary record of the fourth plenary session - Eighteenth meeting of the Conference of the Parties, Convention on International Trade in Endangered Species of Wild Fauna and Flora. in *Convention on International Trade in Endangered Species of Wild Fauna and Flora editor.*, Geneva, Switzerland.

Daemmrich A. (2013). The political economy of healthcare reform in China: Negotiating public and private. *Springer Plus 2*, 448.

Dang V.H.N., Nielsen M.R. (2018). Understanding utilitarian and hedonic values determining the demand for rhino horn in Vietnam. *Human Dimensions Wildlife 23*, 417–432.

Dang Vu H.N., Nielsen M.R., Jacobsen J.B. (2022). Conserving rhinos by legal trade: Insights from a choice experiment with rhino horn consumers. *Ecological Economics 193*, 107287.

Di Minin E., Laitila J., Montesino-Pouzols F. et al. (2015). Identification of policies for a sustainable legal trade in rhinoceros horn based on population projection and socioeconomic models. *Conservation Biology 29*, 545–555.

Eikelboom J.A.J., Nuijten R.J.M., Wang Y.X.G. et al. (2020). Will legal international rhino horn trade save wild rhino populations? *Global Ecology and Conservation 23*, e01145.

Fang Y., Yang S., Zhou S., Jiang M., Liu J. (2013). Community pharmacy practice in China: Past, present and future. *International Journal of Clinical Pharmacy 35*, 520–528.

Feng L., Liao W., Hu J. (2019). Towards a more sustainable human-animal relationship: The legal protection of wildlife in China. *Sustainability 11*, 3112.

Fukushima C.S., Tricorache P., Toomes A. et al. (2021). Challenges and perspectives on tackling illegal or unsustainable wildlife trade. *Biological Conservation 263*, 109342.
Gao Y., Stoner K.J., Lee A.T.L., Clark S.G. (2016). Rhino horn trade in China: An analysis of the art and antiques market. *Biological Conservation 201*, 343–347.
Grasmick H.G., Bursik R.J. (1990). Conscience, significant others, and rational choice: Extending the deterrence model. *Law & Society Review 24*, 837–861.
Griffiths S.M., Chung V.C.H., Tang J.L. (2010). Integrating traditional Chinese medicine: Experiences from China. *Australasian Medical Journal 3*, 385–396.
Haas T.C., Ferreira S.M. (2016). Combating rhino horn trafficking: The need to disrupt criminal networks. *PLoS One 11*, e0167040.
Haas T.C., Ferreira S.M. (2018). Optimal patrol routes: Interdicting and pursuing rhino poachers. *Police Practice Research 19*, 61–82.
Hanley N., Sheremet O., Bozzola M., MacMillan D.C. (2018). The allure of the illegal: Choice modelling of rhino horn demand in Vietnam. *Conservation Letters 11*, e12417.
Hanson M.E. (2011). *Speaking of Epidemics in Chinese Medicine: Disease and the Geographic Imagination in Late Imperial China*. Routledge, Oxford.
Heckathorn D.D. (2011). Snowball versus respondent-driven sampling. *Sociological Methodology 41*, 355–366.
Hübschle A. (2017). Contested illegality: Processing the trade prohibition of rhino horn. pp. 177–197 in J. Beckert, M. Dewey editors. *The Architecture of Illegal Markets: Towards an Economic Sociology of Illegality in the Economy.* Oxford University Press, Oxford.
Jovchelovitch S., Gervais M.-C. (1999). Social representations of health and illness: The case of the Chinese community in England. *Journal of Community and Applied Social Psychology 9*, 247–260.
Kaptchuk T.J. (2000). *Chinese Medicine: The Web that Has No Weaver*, Revised ed. Rider, London.
Kennaugh A. (2016). *Rhino Rage: What is Driving Illegal Consumer Demand for Rhino Horn*. pp. 1–23. New York: Natural Resources Défense Council.
Kong H., Hsieh E. (2012). The social meanings of traditional Chinese medicine: Elderly Chinese immigrants' health practice in the United States. *Journal of Immigrant and Minority Health 14*, 841–849.
Koo L.C. (1984). The use of food to treat and prevent disease in Chinese culture. *Social Science & Medicine 18*, 757–766.
Li J., Shi L., Liang H., Ding G., Xu L. (2018). Urban-rural disparities in health care utilization among Chinese adults from 1993 to 2011. *BMC Health Services Research 18*, 102.
Li P.J. (2007). Enforcing wildlife protection in China: The legislative and political solutions. *China Information 21*, 71–107.
Li X., Lu J., Hu S. et al. (2017). The primary health-care system in China. *The Lancet 390*, 2584–2594.
Li Y., Xu J., Wang F. et al. (2012). Overprescribing in China, driven by financial incentives, results in very high use of antibiotics, injections, and corticosteroids. *Health Affairs 31*, 1075–1082.
Lim M.-K., Yang H., Zhang T., Feng W., Zhou Z. (2004). Public perceptions of private health care in socialist China. *Health Affairs 23*, 222–234.
Ling R.E., Liu F., Lu X.Q., Wang W. (2011). Emerging issues in public health: A perspective on China's healthcare system. *Public Health 125*, 9–14.
Liu R., Wang M., Duan J.-A., Guo J.-M., Tang Y.-P. (2010). Purification and identification of three novel antioxidant peptides from Cornu Bubali (water buffalo horn). *Peptides 31*, 786–793.
Manfredo M.J., Bruskotter J.T., Teel T.L. et al. (2017). Why social values cannot be changed for the sake of conservation. *Conservation Biology 31*, 772–780.
Mayer F.S., Frantz C.M. (2004). The connectedness to nature scale: A measure of individuals' feeling in community with nature. *Journal of Environmental Psychology Research 24*, 503–515.
Milliken T., Shaw J. (2012). The South Africa-Viet Nam Rhino Horn Trade Nexus: A deadly combination of institutional lapses, corrupt wildlife industry professionals and Asian crime syndicates. Species Reports. TRAFFIC, Johannesburg, South Africa.
Nagin D.S. (2013). Deterrence in the twenty-first century. *Crime and Justice 42*, 199–263.
National Bureau of Statistics of China. (2016). China statistical yearbook 2016. National Bureau of Statistics of China. China Statistics Press and Beijing Info Press. https://www.stats.gov.cn/sj/ndsj/2016/indexeh.htm
Newing H. (2011). *Conducting Research in Conservation: Social Science Methods and Practice*. Routledge, Oxford.

Oyanedel R., Gelcich S., Milner-Gulland E.J. (2020). Motivations for (non-)compliance with conservation rules by small-scale resource users. *Conservation Letters 13*, e12725.

Paddock L.C. (2012). Beyond deterrence: Compliance and enforcement in the context of sustainable development. *Environmental Law Reporter 42*, 10622–10638.

People's Republic of China. (1979). 中华人民共和国刑法（修订）[Criminal Law of the People's Republic of China (amended in 1997)]. National People's Congress.

People's Republic of China. (1989). 中华人民共和国野生动物保护法 [Law of the People's Republic of China on the Protection of Wildlife (amended in 2004)]. National People's Congress.

People's Republic of China. (1993a). 国务院关于禁止犀牛角和虎骨贸易的通知 [Circular of the State Council banning trade in rhinoceros horn and tiger bone]. State Council.

People's Republic of China. (1993b). 林业部关于核准部分濒危野生动物为国家重点保护野生动物的通知 [Circular of the Department of Forestry approving the listing of additional endangered species as special stated protected wildlife]. Department of Forestry (defunct; now State Forestry Administration).

People's Republic of China. (2006). 国家重点保护野生动物名录 [Directory of the People's Republic of China on Special State Protection of Wildlife]. State Forestry Administration, Beijing.

People's Republic of China. (2011). 人力资源社会保障部关于进一步推进医疗保险付费方式改革的意见 [Guidance on promoting provider payment reform]. Ministry of Human Resources and Social Security.

People's Republic of China. (2018a). Full transcript: State Council Executive Deputy Secretary-General Ding Xuedong Answers Media Questions. China Internet Information Center. State Council Information Office and China International Publishing Group, Beijing.

People's Republic of China. (2018b). 国务院关于严格管制犀牛和虎及其制品经营利用活动的通知 [Circular of the State Council on the trade ban regarding rhinoceros horn and tiger bone]. State Council.

Phelps J., Biggs D., Webb E.L. (2016). Tools and terms for understanding illegal wildlife trade. *Frontiers in Ecology and the Environment 14*, 479–489.

Ranganathan P., Pramesh C., Aggarwal R. (2017). Common pitfalls in statistical analysis: Logistic regression. *Perspectives in Clinical Research 8*, 148–151.

Ren X., Shao X.-X., Li X.-X. et al. (2020). Identifying potential treatments of COVID-19 from Traditional Chinese Medicine (TCM) by using a data-driven approach. *Journal of Ethnopharmacology 258*, 112932.

Rubino E.C., Pienaar E.F. (2020). Rhinoceros ownership and attitudes towards legalization of global horn trade within South Africa's private wildlife sector. *Oryx 54*, 244–251.

Shepherd C.R., Gray T.N.E., Nijman V. (2017). Rhinoceros horns in trade on the Myanmar–China border. *Oryx 52*, 1–3.

Srivastava A.B., Gold M.S. (2018). Beyond supply: How we must tackle the opioid epidemic. *Mayo Clinic Proceedings 93*, 269–272.

Sun Q., Santoro M.A., Meng Q., Liu C., Eggleston K. (2008). Pharmaceutical policy in China. *Health Affairs 27*, 1042–1050.

Tabachnick B.G., Fidell L.S. (2013). *Using Multivariate Statistics*, 6th ed., New International Edition. ed. Pearson Education Limited, Harlow.

Tam K.-P. (2013). Concepts and measures related to connection to nature: Similarities and differences. *Journal of Environmental Psychology 34*, 64–78.

't Sas-Rolfes M., Challender D.W.S., Hinsley A., Veríssimo D., Milner-Gulland E.J. (2019). Illegal wildlife trade: Scale, processes, and governance. *Annual Review Environment and Resources 44*, 201–228.

Wang H., Zhang L., Yip W., Hsiao W. (2011). An experiment in payment reform for doctors in rural China reduced some unnecessary care but did not lower total costs. *Health Affairs 30*, 2427–2436.

Winter S.C., May P.J. (2001). Motivation for compliance with environmental regulations. *Journal of Policy Analysis and Management 20*, 675–698.

Wong R.W.Y. (2016). The organisation of the illegal tiger parts trade in China. *The British Journal of Criminology 56*, 995–1013.

Wong R.W.Y. (2017). 'Do you know where I can buy ivory?': The illegal sale of worked ivory products in Hong Kong. *Australian & New Zealand Journal of Criminology 51*, 204–220.

Wong R.W.Y. (2019). *The Illegal Wildlife Trade in China: Understanding the Distribution Networks*. Palgrave Studies in Green Criminology, Cham.

Xu H., Zhang W., Gu L. et al. (2014). Aging village doctors in five counties in rural China: Situation and implications. *Human Resources for Health 12*, 36.
Xu J., Yang Y. (2009). Traditional Chinese medicine in the Chinese health care system. *Health Policy 90*, 133–139.
Yip W., Powell-Jackson T., Chen W. et al. (2014). Capitation combined with pay-for-performance improves antibiotic prescribing practices in rural China. *Health Affairs 33*, 502–510.
Yip W.C.-M., Hsiao W., Meng Q., Chen W., Sun X. (2010). Realignment of incentives for health-care providers in China. *The Lancet 375*, 1120–1130.
Zhang L., Hua N., Sun S. (2008). Wildlife trade, consumption and conservation awareness in southwest China. *Biodiversity and Conservation 17*, 1493–1516.
Zhang L., Yin F. (2014). Wildlife consumption and conservation awareness in China: A long way to go. *Biodiversity and Conservation 23*, 2371–2381.
Zhang S., Zhang W., Zhou H. et al. (2015). How China's new health reform influences village doctors' income structure: Evidence from a qualitative study in six counties in China. *Human Resources for Health 13*, 26.

17
BEYOND CITES
The rationale for a global agreement against wildlife trafficking

Alice Pasqualato and Aubrey Collins

Introduction

As the illegal exploitation of wild flora and fauna continues to escalate, its impact extends well beyond the immediate loss of iconic species. In addition to driving biodiversity decline, wildlife trafficking – including fish and timber – has grown into a multi-billion-dollar enterprise, ranking among the world's most profitable criminal activities. For those perpetrators, it remains low-risk and highly lucrative, with criminals exploiting loopholes and taking advantage of limited oversight or enforcement (ABA, 2023). Wildlife trafficking is enabled by corruption and threatens human and animal health, hinders security and sustainable development, and undermines the rule of law. By depleting wildlife populations, wildlife trafficking also significantly affects ecosystem services – including carbon sequestration – exacerbating climate change (UNODC, 2022).

The complex nature of wildlife trafficking requires a wide variety of solutions, adequate resources, and a robust but adaptable legal framework. Yet, despite publicly expressed support by governments, commendable attempts to raise concerns from civil society, and high-profile conferences, little has been done to improve the existing international legal framework. Not only is there no single international instrument addressing this issue, but there is also still no agreed-upon definition of the terms 'wildlife trafficking' or 'wildlife crime'.

The fight to prevent and combat wildlife trafficking has recently found new momentum in the form of a potential additional Protocol under the United Nations Convention Against Transnational Organised Crime (UNTOC). This newly proposed solution is drawing global attention as it would signify an ambitious rethink of existing international wildlife law, one of the most significant since the adoption of the Convention on International Trade in Endangered Species of Wild Fauna and Flora (CITES).

When the UNTOC and its protocols were negotiated and adopted, over 20 years ago, the international community had yet to grasp the nature, scale, and consequences of wildlife trafficking. The level of awareness has changed significantly since then – or at least the authors like to think so – and governments have started to look at this growing problem through a different lens. More countries now address wildlife trafficking as a form of transnational organised crime, rather than a biodiversity issue. In the European Union, for

DOI: 10.4324/9781003333906-20

instance, wildlife trafficking features in the EU Strategy to tackle Organised Crime 2021–2025 as one of the crime forms needing a "tailor-made response". The increased level of attention is also testified to by a growing body of UN resolutions (Economic and Social Council, 2001, 2003, 2008, 2011, 2013; United Nations General Assembly, 2015, 2016, 2017, 2019, 2021, 2023).

As the enforcement officers are faced with ever-changing challenges from criminals illegally trading in wildlife, the effectiveness of existing national and international frameworks is being called into question. While CITES is largely regarded as successful in regulating legal and sustainable trade, it does not have the mandate to address the illegal trade, smuggling, poaching, capture, and collection of wildlife. The status quo is no longer working; implementation and resources fail to keep up with the criminal industry. It raises questions as to the current regime's ability to adequately respond to the growing threat of wildlife trafficking and whether the barriers to prevention can be overcome. Next steps will require increased resources, training, dedicated national units, and courts, as well as enhanced political will to drive effective action. But it will also require countries to drastically improve existing laws and harmonise their approaches based on best practices to encourage international coordination and cooperation. One way to do that is through a new Protocol under the UNTOC.

The current international legal framework

Despite the growing body of Multilateral Environmental Agreements (MEAs) that has emerged since the 1970s, none of them contain specific measures to prevent and combat wildlife trafficking. There is no international legal instrument that defines the term, prohibits or criminalises specific international actions or practices, or that establishes criminal sanctions against offenders (Elliot, 2017). As a result, the current international legal framework against wildlife trafficking appears as a mosaic of different provisions, contained in different agreements – none of which was designed to deal with this very complex, multifaceted issue. These key MEAs include CITES, the Convention on Biological Diversity (CBD), the Convention on the Conservation of Migratory Species of Wild Animals (CMS), and the Convention Concerning the Protection of the World Cultural and Natural Heritage (World Heritage Convention). Outside of the MEA realm, international criminal law instruments – most notably the UNTOC – can also be employed to tackle transnational wildlife crime. For the sake of brevity, this chapter will only analyse two of these agreements, the most relevant: CITES and the UNTOC.

CITES

Signed in 1973, CITES is one of the most well-known MEAs, aimed to ensure that international trade in wild animals and plants does not threaten the survival of species. The Convention works by subjecting international trade in specimens of selected species, listed in its three Appendices, to certain controls. More specifically, all import, export, re-export, and introduction from the sea of species covered by the Convention has to be authorised through a licensing system.

While providing a thorough explanation of the functioning of CITES and assessing its effectiveness exceeds the scope of this chapter, a few aspects should be made clear. First and foremost, CITES is an international *trade* and conservation Convention. It is neither a transnational crime nor a law-enforcement convention. CITES Parties set up Management and Scientific Authorities tasked with overseeing the permit system for the legal trade, but they are not required to have an Enforcement Authority. The Convention, in fact, does not have any law-enforcement capacity (Wyatt, 2022), and even the former CITES Chief of Enforcement stated that CITES is a questionable forum to effectively discuss crime-related matters (Rosen, 2020).

Moreover, CITES does not require Parties to criminalise breaches of the Convention, but rather to "take appropriate measures" to "prohibit trade in specimens" in violation of its provisions, including through the penalisation of trade and possession of such specimens (Art. VIII). How violations are penalised is at the full discretion of Parties, who may opt for administrative or civil sanctions, rather than criminal ones.

As a result, the implementation of CITES in domestic legal systems is all but homogenous, with penalties ranging from minor administrative sanctions and fines to life imprisonment (Schloenhardt and Pitman, 2022). Finally, at the time of writing, CITES protects about 40,900 species (ca. 29,000 of which are orchid species), meaning only 0.5% of the 8.7 million species of wild animals, plants, and fungi in existence currently fall under its regulatory umbrella. These are excruciatingly low numbers when considering that over 1 million species are thought to be at risk of extinction, with overexploitation being one of the main drivers (IPBES, 2019). An argument could be made that more effort should be put into protecting species *before* they reach the point of being threatened with extinction and can therefore be listed under CITES. Besides, if wildlife is traded illegally, it should not matter whether the species is endangered or not. The lack of international protection for non-CITES-listed species, including when they are nationally protected in their country of origin, is a significant gap that needs to be addressed.

In conclusion, while CITES serves an important role in conservation and sustainable development, it should not be used as a "jack of all trades" to make up for the absence of other appropriate instruments, and it cannot address the growing criminal industry that is illegal wildlife trade.

UNTOC

The UNTOC is the first comprehensive and legally binding instrument designed to fight transnational organised crime, signed in Palermo, Italy, in 2000. It is complemented by three protocols, addressing trafficking in persons, smuggling of migrants, and illicit manufacturing of and trafficking in firearms. While the Convention established four offences – participation in an organised criminal group, money laundering, corruption, and obstruction of justice – it is not subject-specific. It was designed to "promote cooperation" (Art. 1) against all forms of criminal activity, as long as they meet the definition of "serious crime" provided by Art. 2: "conduct constituting an offense punishable by a maximum deprivation of liberty of at least four years or a more serious penalty". The offence also needs to be "transnational in nature" and involve an "organised criminal group" (Art. 3).

Because of this feature, the 'Palermo Convention' can adapt to new and emerging crime forms, including those affecting the environment, such as wildlife trafficking. The 2000 UN General Assembly (UNGA) resolution under which the Convention was adopted even explicitly mentions it, claiming that this instrument "will constitute an effective tool and the necessary legal framework for international cooperation in combating, *inter alia*, ... illicit trafficking in endangered species of wild flora and fauna".

Despite this potential, the UNTOC appears to be underutilised in cases of wildlife trafficking and other forms of environmental crime. In the United Nations Office on Drugs and Crime's (UNODC) SHERLOC (Sharing Electronic Resources and Laws on Crime) Case Law Database, containing over 3,000 cases dealing with transnational organised crime cases, we could only identify two 'crimes that affect the environment' cases in which the UNTOC was referred to, both dealing with fisheries crime. Similarly, no wildlife-related case could be identified in the 2021 UNODC Digest of Cases on 'International Cooperation in Criminal Matters Involving the United Nations Convention Against Transnational Organised Crime as a Legal Basis'.

This is likely because in many national jurisdictions, wildlife trafficking is not treated as a 'serious crime' under the definition of the UNTOC. As a result, even transnational and organised forms of illegal trade do not fall within the scope of the Convention. According to a recent UNODC report (UNODC, 2024), in fact, "fewer than half of Member States have criminal sanctions that meet the UNTOC threshold of a serious crime (86 Member States)". This is something the UNGA has been trying to rectify since 2015, with several resolutions encouraging Member States to "make illicit trafficking in protected species of wild fauna and flora involving organised criminal groups a serious crime" in accordance with the UNTOC.

The numbers are even lower when it comes to fishing offences and illegal deforestation and logging, with respectively only 34 and 49 (out of 193) Member States meeting the "serious crime" threshold.

Regardless of how it is used in the fight against wildlife trafficking, a number of authors and non-governmental organisations also argued that the UNTOC has a more general implementation problem. Some observed that the ratification status and coverage are highly uneven throughout the different world regions (RIOC, 2021), limiting its effectiveness, while others blamed the poor implementation of the Convention on the "apathy and indifference" of "many signatory states" (GI-TOC, 2018). The scope of this chapter does not allow for an in-depth analysis of these issues, but they are worth mentioning as a part of the reason why this Convention is not widely used to prevent and tackle wildlife crimes.

An additional Protocol to the UNTOC

Despite its limited application to cases of wildlife trafficking, the UNTOC has gained newfound notoriety within the international wildlife conservation community. In recent years, in fact, calls for an additional Protocol to the UNTOC, dealing with the growing issue of wildlife trafficking, have gained more and more traction.

Since the adoption of the mother Convention and its three protocols, wildlife trafficking has grown to become one of the largest forms of international crime. With a scale that rivals drugs or weapons, wildlife and their parts are highly desirable. Whether it is for medicine, jewellery, clothing, or fixtures for the home. Leading experts in the enforcement and customs sector have made it clear that those who engage in criminal actions have quickly adapted to available technology in a now global world (UNODC, 2020). Utilising the dark web (Thomaz, 2018) and social media, criminal networks are seeing their profits grow at the expense of wildlife and humans alike (Coalition to End Wildlife Trafficking Online, 2021). With greater awareness of the environmental, economic, social, security and, indeed, health consequences of wildlife trafficking, it has become harder to justify the lack of an international legal framework covering this issue.

Supporters of an additional UNTOC Protocol claim a global agreement would help harmonise national legal frameworks, foster cooperation, and raise the overall profile of wildlife trafficking worldwide (Slobodian, 2014). Indeed, while no single instrument can be considered the silver bullet for tackling such complex issues, embedding wildlife trafficking in the international criminal law framework represents a crucial step towards a more comprehensive and effective strategy.

Expressions of support

After being at the centre of international discussions for a number of years, the proposal for an additional Protocol to the UNTOC addressing wildlife trafficking started gaining political momentum in 2021. On 17 May 2021, the presidents of Costa Rica and Gabon released a joint statement calling for a new global agreement against wildlife crime, taking the form of an additional Protocol under the UNTOC (Global Initiative to End Wildlife Crime, 2021). "Given the scale of the risks to people and the planet, we simply cannot stand by and watch wildlife continue to disappear without strengthening our collective response, including international laws for combating and preventing wildlife crime", said the (now former) presidents in their joint statement. Within a matter of months, the presidents of Angola and Malawi also released similar statements (Global Initiative to End Wildlife Crime, 2023).

Further, a public expression of support came from the European Commission, through its April 2021 'EU Strategy to tackle Organised Crime 2021–2022', stating: "Finally, since the international dimension of wildlife trafficking is crucial, the Commission will promote the adoption of an additional Protocol under the UNTOC". The support of the European Union was later confirmed in the November 2022 'Revision of the EU Action Plan against wildlife trafficking', which includes support for a Protocol to the UNTOC covering wildlife trafficking.

Expressions of support have also come from civil society and academia. In June 2020, a dedicated international alliance, called Global Initiative to End Wildlife Crime (EWC), was created to promote and offer technical support towards the adoption of an additional Protocol to the UNTOC. At the time of writing, the objectives of the EWC, chaired by former CITES Secretary-General John Scanlon, are supported by 37 'international Champions'. These include well-known conservation non-governmental organisations, such as the Wildlife Conservation Society (WCS), the International Fund on Animal Welfare (IFAW), Conservation International and the Jane Goodall Institute (world-renowned primatologist Dr Jane Goodall is a vocal supporter of an additional Protocol herself), among others.

Finally, in February 2023, the American Bar Association adopted a resolution which urges all UNTOC Parties to

> adopt a wildlife crime protocol that would: (1) define the term wildlife crime; (2) identify the measures that Parties would be called upon to adopt in their domestic laws to prevent and combat wildlife crime; and (3) identify measures that would enhance cooperative global enforcement efforts to prevent and combat wildlife crime.
>
> *(ABA, 2023)*

Possible content

Over the past decade, the possible form and content of a global agreement focusing on wildlife crime have been debated by prominent scholars and technical experts. The most well-known draft for a potential UNTOC Protocol was created by the EWC in 2020, in an effort to "offer some tangible ideas on the possible form and content of a potential Protocol and thereby help advance the discussion on this necessary reform". The text takes inspiration from the existing UNTOC protocols, national laws, and UNODC's Guide on Drafting Legislation to Combat Wildlife Crime (2018). As highlighted on the EWC website, said draft "has no formal status and it will be for States alone to negotiate the content of a new Protocol" (Global Initiative to End Wildlife Crime, 2020). The draft itself presents insight into how the potential Protocol could be structured, with the focus to criminalise the intentional trafficking of specimens of wild fauna and flora.

Defining wildlife trafficking

The first and most frequently cited benefit of an additional UNTOC Protocol on preventing and combatting wildlife trafficking is the adoption of a universally agreed definition of the term itself. As mentioned above, no definition of the terms "wildlife trafficking" or "wildlife crime" currently exists at the international level, with different jurisdictions still employing different terminologies. This is significant because even when countries do criminalise wildlife trafficking, one may criminalise the mere possession of an illegally sourced wildlife product while another may criminalise only import or export (Roux, 2015). Such a divergence in understanding can pose challenges to international cooperation, for example, when dual criminality is made a requirement for extradition or mutual legal assistance.

A Protocol would provide an opportunity to draft such a definition, one that "takes into account modern forms of committing crimes, like cybercrime, and including the diversity of wildlife (including plants and fish)" (Wyatt, 2021). As highlighted by Wyatt and Hutchinson (2023) and Wyatt et al. (2021), in fact, certain national laws specifically exclude plants and fish from the legal definition of 'wildlife'.

Under the EWC draft Protocol, trafficking is defined as "the import, export, transport, sale (including by electronic means), receipt, introduction from the sea, acquisition, possession, purchase, delivery, movement or transfer of specimens of wild fauna and flora". A definition of 'Specimen' is also provided:

> ...not limited to: (a) any wild fauna or flora or parts thereof, whether alive or dead; (b) any seed, egg, gamete or propagule or part of any wild fauna or flora; (c) any readily recognisable part or derivative of any wild fauna or flora; or (d) any goods which contain a part or derivative of any wild fauna or flora.
>
> *(Global Initiative to End Wildlife Crime, 2020)*

The potential impact of a universally agreed definition can be estimated by looking at one of the existing UNTOC protocols, the Protocol to Prevent, Suppress, and Punish Trafficking in Persons. Those who have studied this instrument have described the incorporation of a common definition as the "single achievement that made all this possible" and a "genuine breakthrough because it provided the necessary prerequisite for the elaboration of a meaningful normative framework" (Gallagher, 2015). According to Gallagher (2015):

> The absence of a definition also reflected the marginal place of trafficking within the international human rights system and the associated reluctance of States to tie themselves to specific and detailed rules. As long as the concept of trafficking remained unclear, it was virtually impossible to formulate substantive obligations and to hold States to account for violations.

It was clear that this step was necessary to "facilitate convergence in national approaches with regard to the establishment of domestic criminal offenses that would support efficient international cooperation in investigating and prosecuting trafficking in persons cases" (UNODC, n.d.).

As recognised by the American Bar Association (ABA, 2023), an additional UNTOC Protocol focusing on wildlife trafficking would provide an explicit, uniform definition for the crime and ensure that its scope covers all relevant flora and fauna, including fish and timber species.

Criminalisation and harmonisation

As previously mentioned, CITES does not specify how violations of the Convention need to be penalised and indeed no international instrument currently mandates the criminalisation of wildlife trafficking offences. A new Protocol could be the first to do so, ensuring the adoption of a global criminal law approach and facilitating the harmonisation of national criminal laws. Much like the mother Convention and its existing protocols (Obokata, 2017), an additional UNTOC Protocol would encourage a degree of harmonisation by clearly defining the conduct to be criminalised, as well as promoting common definitions to be used by the relevant law-enforcement authorities. The EWC draft Protocol defines the conduct to be criminalised in its Art. 5, which mandates Parties to "adopt such legislative and other measures as may be necessary to establish as a criminal offence, *when committed intentionally*, the trafficking in any specimen, knowing that the specimen was taken, possessed, distributed, transported, purchased or sold" in contravention with relevant international and national law. The provision also criminalises document falsification or alteration, as well as attempt, aiding, and abetting.

Harmonisation of heterogeneous national approaches is desirable for a number of reasons, especially when it comes to transnational crimes such as wildlife trafficking. Firstly, it can prevent the well-known issue of 'forum shopping' on the part of criminals (i.e. choosing to operate in jurisdictions where the conduct is not criminalised or the punishment regimes are weak) (Obokata, 2017). Further, a shared understanding of wildlife crime offences and their serious nature would lead to smoother international cooperation in the phases of investigation, prosecution, and punishment (i.e. global intelligence-sharing, mutual assistance in police investigations, cooperation in confiscations and seizures, technical assistance and capacity building, extradition agreements, and better judicial cooperation) (Roux, 2015).

Criminalising trade in illegally sourced wildlife

As discussed, CITES only covers a fraction of wild animal and plant species in existence. Consequently, "millions of species that are not listed by CITES may be illegally harvested and traded internationally", including nationally protected ones (UNODC, 2016). One way to prevent illegally sourced nationally protected species from entering destination markets is to adopt legislation similar to the United States 'Lacey Act'. Following its 2008 amendment, the Lacey Act makes it illegal to import, export, transport, sell, receive, acquire, or purchase any animal or plant, with certain exceptions, taken or traded in violation of the laws of the country where it was harvested. Similar provisions exist in Australia and Canada while the European Union Timber Regulation only addresses illegally sourced timber, not other wild animals or plants.

If language similar to that of the Lacey Act was included in a potential additional UNTOC Protocol, it would fill a significant gap in the international legal framework. An example of such a provision is included in the EWC draft Protocol (Global Initiative to End Wildlife Crime, 2020), which states:

1 Each State Party shall adopt such legislative and other measures as may be necessary to establish as a criminal offence, when committed intentionally, the trafficking in any specimen, knowing that the specimen was taken, possessed, distributed, transported, purchased or sold in contravention of:
[...]

a Any applicable domestic or foreign law concerning the protection, conservation management, trade or use of wild fauna or flora.

Prevention

Reducing demand for illegal wildlife products is widely recognised as a key strategy to tackle wildlife trafficking, as it removes the economic incentive driving the trade (Ayling, 2015). Prevention and demand reduction strategies may include finding alternative replacements for items or ensuring local communities have an alternative source of income. Prevention is also key for reducing the risks of zoonotic disease, an issue that has grown in international importance since the COVID-19 pandemic (Doody et al., 2021).

International tools, such as a potential additional UNTOC Protocol, could play a critical role in encouraging and coordinating national demand reduction strategies in both source and destination countries. The draft Protocol created by EWC dedicates an entire article (Art. 7) to this issue, mandating, *inter alia*, that Parties "establish comprehensive policies, programmes and other measures" and "endeavour to undertake measures such as research, information and mass media campaigns and social and economic initiative" to prevent and combat wildlife trafficking (Global Initiative to End Wildlife Crime, 2020).

Repatriation of trafficked live animals

As the enforcement against wildlife trafficking increases, countries are experiencing a growing number of live animal seizures and confiscations. Although the current draft of the EWC Protocol does not address this particular issue, a potential new instrument could include arrangements on the repatriation of live wild animals that originated within one State's territory but entered into custodial management of another State as a result of a trafficking enforcement operation. In partnership with IFAW and other organisations, Legal Atlas (2023) recently published a model international legislation focusing on the mechanisms to return wildlife to authorities of the country of origin, which could serve as inspiration for one or more provisions within a potential additional Protocol.

Addressing other environmental crimes

Prominent voices from civil society and academia, as well as a number of governments (CCPCJ, 2023), have suggested that a potential additional Protocol to the UNTOC could address other environmental crimes, in addition to wildlife trafficking. A Protocol with a broader scope could include crimes involving pollution, waste trafficking, and illegal mining, among others. Interestingly, at the time of writing, a similar regional instrument is being negotiated at the Council of Europe, under the name of 'Convention on the Protection of the Environment through Criminal Law'.

A Protocol that addresses all environmental crimes is both ambitious and challenging. Political will, or lack thereof, represents the main obstacle, as "Often, there is a perception that environmental rules will slow down or impede development" (UNEP, 2019). Although the impact of a comprehensive environmental crime instrument would be paramount, it is likely to take significantly more time to negotiate and find common ground on the international level than a Protocol with a narrow scope.

Criticism

Like most proposed legal reforms, the proposal for an additional UNTOC Protocol has attracted both support and a level of criticism. Some of the criticism is addressed at the effectiveness of international law more generally and how painfully slow it is to negotiate and implement, while other observations are more targeted.

One argument is that the UNTOC was designed to cover all emerging crime areas, including wildlife trafficking; therefore, the expensive and time-consuming process of negotiating a new instrument can be avoided. Indeed, if States were determined to make wildlife trafficking (and related offences) a 'serious crime', they could act unilaterally and ensure it is not only criminalised but also punishable by a maximum deprivation of liberty of at least four years. A few observations can be made on this front. Firstly, waiting for over 190 individual States to spontaneously amend their national laws and define (or re-define) the crime might turn into a long, possibly disappointing wait.

Secondly, a separate Protocol would go significantly further and address key legal gaps. As explained above and summarised by Schloenhardt and Pitman (2022), it "would bring clarity, nomenclature and standards to an area of law that is presently marked by differences, discrepancies and a lack of understanding". It would also bring wildlife trafficking within the overall scope of the UNTOC regardless of whether it is a 'serious crime' in national legislation (ABA, 2023). This is desirable, as there are other penalties that a State may wish to consider, such as fines, restitution, restoration, or even community service. Especially at a time when some see carceral strategies as limited and inequitable (ABA, 2023). Some authors have also highlighted that in order for the UNTOC to be used, wildlife cases must be classified as 'organised crime'; without such thresholds met, there is a serious barrier to use (Vries, 2022).

Finally, if the UNTOC truly is a "catch-all" instrument, why did the international community go through the incredible effort of creating the existing three protocols? Why were time and resources spent to negotiate and implement them? To the co-authors, the answer is clear: the three protocols cover what were perceived, at the time of negotiations, as three priority crime areas, deserving of their own instruments. As one of the largest forms of criminal activity, wildlife trafficking now warrants a similar response.

Alternatively, others may argue that new international law is not the answer to the growing illegal trade in wildlife and that time and resources should be invested in bolstering national-level mechanisms. In particular, efforts should be directed towards combatting corruption and enhancing governmental capacity.

Indeed, these are valid concerns, which should be kept in mind if the international community decides to move forward with an additional Protocol. No single instrument can serve as a 'panacea' for the intricate problem of wildlife trafficking. However, we feel it is important to also highlight that the adoption of an additional Protocol does not and should not prevent efforts at the national level. Increased effort will be required at both levels. International regulation ideally serves as a tool to increase pressure for action, and a dedicated instrument would ensure that wildlife trafficking remains a part of the intergovernmental agenda for years, if not decades, to come. Each UNTOC Protocol, in fact, is supported by its own dedicated working group (UNODC, n.d.) and is included in the UNTOC Review Mechanism, which is meant to assess its implementation. It is reasonable to assume that an additional Protocol would be treated similarly, ensuring that Member States remain actively engaged in combatting wildlife trafficking and that ongoing efforts at the national level are regularly evaluated and reinforced.

Intergovernmental process within the United Nations

So, what does the future of an additional Protocol to the UNTOC, focusing on wildlife trafficking, look like? Will this proposal ever see the light of day? It might be too early to tell, but following the expressions of support mentioned above, some interesting developments have taken place.

In May 2022, the United Nations Commission on Crime Prevention and Criminal Justice (CCPCJ)[1] adopted by consensus resolution 31/1 titled "Strengthening the international legal framework for international cooperation to prevent and combat illicit trafficking in wildlife". The resolution invited Member States to provide the UNODC with their "views on possible responses, including on an additional Protocol under the United Nations Convention against Transnational Organised Crime, to address any gaps that may exist in the current international legal framework to prevent and combat illicit trafficking in wildlife". By doing so, the resolution initiated, for the first time, an intergovernmental discussion on strengthening the current international legal framework against wildlife trafficking, including the merits of a new global agreement.

Throughout 2023, the UNODC collected the views of Member States (a total of 76 States and the European Union responded) and analysed them in report E/CN.15/2024/14, which was released in May 2024. According to the UNODC report, more than half of the responding States "considered that the current international legal framework to prevent and combat illicit trafficking in wildlife contained gaps", with "many respondents" highlighting the lack of harmonisation, as well as the need for more effective law-enforcement and judicial cooperation. Additionally, more than half of the respondents had a positive view of the potential of an additional Protocol to the UNTOC, which could, in their view, "support States in strengthening capacities to prevent and combat illicit trafficking in wildlife, developing the appropriate legislation and improving international and national cooperation and coordination".

Although the next steps in this intergovernmental process remain to be seen, the implementation of CCPCJ resolution 31/1 has shown a promising level of support for an additional instrument addressing wildlife trafficking, which might lead to further developments in the coming years.

Conclusions

Effective prevention and combatting of wildlife trafficking require collaborative international efforts and coordinated action between source and destination countries. A dedicated global agreement would provide the international legal basis for such cooperation, clearly defining the conduct to be criminalised, harmonising national legal regimes, and facilitating mutual legal assistance, at a minimum. As shown with human trafficking, creating a separate Protocol also equals a higher profile for the crime it is dedicated to. While it took time and a significant number of resources to adopt and implement a trafficking in persons Protocol, it triggered "unprecedented levels of action" and firmly placed the issue permanently on the international agenda (Gallagher, 2015). Ultimately, there is no single, all-encompassing solution that will swiftly address all facets of this complex form of criminal activity. Embedding wildlife trafficking in the international legal framework, however, may prove to be a substantial first step towards success. The adoption of a dedicated international instrument, taking the form of an additional Protocol to the UNTOC, would acknowledge the severity of wildlife trafficking and lay the groundwork for a more robust and concerted international response to protect our people and biodiversity.

Note

1 The CCPCJ is the principal policy making body of the United Nations in the field of crime prevention and criminal justice, and it meets annually in Vienna, Austria.

References

American Bar Association. (2023, February 6). *Resolution 508 Adopted by the House of Delegates.* https://www.americanbar.org/content/dam/aba/directories/policy/midyear-2023/508-midyear-2023.pdf

Ayling, J. M. (2015). A Regulatory Approach to Demand Reduction in the Illegal Wildlife Market (2015) *RegNet Research Paper No. 82.* Doi: 10.2139/ssrn.2634303

Coalition to End Wildlife Trafficking Online. (2021). *Coalition to End Wildlife Trafficking Online: 2021 Progress Update.* https://static1.squarespace.com/static/5b53e9789772ae59ffa267ee/t/614c9996d599c17dd162d2b8/1632410006720/Coalition+2021+Progress+Update+2+Pager-English.pdf

Commission on Crime Prevention and Criminal Justice. (2023). Strengthening the international legal framework for international cooperation to prevent and combat illicit trafficking in wildlife, Conference Room Paper 12, November 23. https://www.unodc.org/documents/commissions/CCPCJ/CCPCJ_Sessions/CCPCJ_32Reconvened/ECN152023_CRP12_2323139E.pdf

De Vries, S. (2022, July 19). *Wildlife Trafficking: Is a Fourth Protocol to UNTOC the Answer?*. ICCLR. https://icclr.org/2022/07/19/wildlife-trafficking-is-a-fourth-protocol-to-untoc-the-answer/

Doody, J. S., Reid, J. A., Bilali, K., Diaz, J. and Mattheus, N. (2021). In the post-COVID-19 era, is the illegal wildlife trade the most serious form of trafficking? *Crime Science, 10*(1), 19. Doi: 10.1186/s40163-021-00154-9

Economic and Social Council. (2001). *Illicit trafficking in protected species of wild flora and fauna, Resolution 2001/12, 24 July.* https://www.unodc.org/documents/Wildlife/ECOSOC_resolution_2001-12.pdf

Economic and Social Council. (2003). *Illicit trafficking in protected species of wild flora and fauna, Resolution 2003/27, 22 July.* https://www.un.org/en/ecosoc/docs/2003/resolution%202003-27.pdf

Economic and Social Council. (2008). *International cooperation in preventing and combating illicit international trafficking in forest products, including timber, wildlife and other forest biological resources, Resolution 2008/25, 24 July.* https://www.un.org/en/ecosoc/docs/2008/resolution%202008-25.pdf

Economic and Social Council. (2011). *Crime prevention and criminal justice responses against illicit trafficking in endangered species of wild fauna and flora, Resolution 2011/36, 28 July.* https://www.unodc.org/documents/Wildlife/ECOSOC_resolution_2011-36.pdf

Economic and Social Council. (2013). *Crime prevention and criminal justice responses to illicit trafficking in protected species of wild fauna and flora, Resolution 2013/40, 25 July.* https://www.

unodc.org/documents/commissions/CCPCJ/Crime_Resolutions/2010-2019/2013/ECOSOC/Resolution_2013-40.pdf

Elliott, L. (2017). Cooperation on transnational environmental crime: Institutional complexity matters. *Review of European Comparative International Environmental Law, 26*, 107–117.

Gallagher, A. (2015). Two cheers for the trafficking protocol. *Anti-Trafficking Review*, (4), 14–32. www.antitraffickingreview.org

Global Initiative to End Wildlife Crime. (2020). *Form and content of a possible Protocol on the illicit trafficking of wildlife*. https://endwildlifecrime.org/untoc-wildlife-protocol/

Global Initiative to End Wildlife Crime. (2021). *Joint statement Gabon and Costa Rica call for a new global agreement to prevent and combat wildlife crime*. https://www.endwildlifecrime.org/wp-content/uploads/2021/05/Gabon-and-Costa-Rica-Joint-Statement-on-Wildlife-Crime-Press-Release-and-Statements-May-2021-.pdf

Global Initiative to End Wildlife Crime. (2023). *The Global Initiative to End Wildlife Crime Progress Report 2022*. https://endwildlifecrime.org/#section8

Global Initiative against Transnational Organised Crime. (2018, March 20). *Adopted 18 years ago, why has the UNTOC still not achieved its aim?* https://globalinitiative.net/analysis/adopted-18-years-ago-why-has-the-untoc-still-not-achieved-its-aim/

IPBES. (2019). *Global assessment report on biodiversity and ecosystem services of the Intergovernmental Science-Policy Platform on Biodiversity and Ecosystem Services*. Doi: 10.5281/zenodo.3831673

Legal Atlas. (2023). *Model International Agreement on Repatriation of Trafficked Live Wild Animals*. Doi: 10.13140/RG.2.2.19279.66727

Obokata, T. (2017). The value of international law in combating transnational organised crime in the Asia-Pacific. *Asian Journal of International Law, 7*(1), 39–60. Doi: 10.1017/S2044251315000193

Research Initiative on Organised Crime. (2021, March 14). *Major Challenges Exist as the UN Convention against Transnational Organised Crime Enters its Third Decade*. https://www.rioc.eu/blog/major-challenges-exist-as-the-un-convention-against-transnational-organized-crime-enters-its-third-decade/

Rosen, T. (2020). The Evolving War on Illegal Wildlife Trade. *International Institute for Sustainable Development (IISD)*. https://www.jstor.org/stable/resrep29278

Roux, A. (2015). *Tightening the Net: Toward a Global Legal Framework on Transnational Organised Environmental Crime*. https://www.unodc.org/documents/congress/background-information/NGO/GIATOC-Blackfish/GIATOC_-_Tightening_the_Net.pdf

Schloenhardt, A. and Pitman, M. (2022). A new international instrument against wildlife trafficking? An appraisal of current proposals. *Melbourne Journal of International Law, 23*(2). https://espace.library.uq.edu.au/view/UQ:893e39c

Slobodian, L. (2014). *Addressing transnational wildlife crime through a Protocol to the UN Convention against Transnational Organised Crime: A scoping paper*. IUCN Environmental Law Centre. https://www.informea.org/sites/default/files/imported-literature/MON-092691.pdf

Thomaz, F. (2018). *Illicit Wildlife Markets and the Dark Web: A Scenario of the Changing Dynamics*. Global Initiative against Transnational Organised Crime. https://globalinitiative.net/wp-content/uploads/2018/11/TGIATOC-DarkWebReport-Web.pdf

UNEP. (2019). *Environmental rule of law: First global report*. United Nations Publication. Nairobi, Kenya.

United Nations General Assembly. (2000). *United Nations Convention against Transnational Organised Crime, Resolution 55/25, 15 November*. https://www.un.org/en/development/desa/population/migration/generalassembly/docs/globalcompact/A_RES_55_25.pdf

United Nations General Assembly. (2015). *Tackling illicit trafficking in wildlife, Resolution 69/314, 30 July*. https://cites.org/sites/default/files/eng/news/pr/2015/UNGA_res_illicit_trade_wildlife.pdf

United Nations General Assembly. (2016). *Tackling illicit trafficking in wildlife, Resolution 70/301, 9 September*. https://documents-dds-ny.un.org/doc/UNDOC/GEN/N16/283/36/PDF/N1628336.pdf?OpenElement

United Nations General Assembly. (2017). *Tackling illicit trafficking in wildlife, Resolution 71/326, 11 September*. https://documents-dds-ny.un.org/doc/UNDOC/GEN/N17/282/18/PDF/N1728218.pdf?OpenElement

United Nations General Assembly. (2019). *Tackling illicit trafficking in wildlife, Resolution 73/343, 16 September*. https://documents-dds-ny.un.org/doc/UNDOC/GEN/N19/284/11/PDF/N1928411.pdf?OpenElement

United Nations General Assembly. (2021). *Tackling illicit trafficking in wildlife, Resolution 75/311, 23 July*. https://digitallibrary.un.org/record/3933854
United Nations General Assembly. (2023). *Tackling illicit trafficking in wildlife, Resolution 77/325, 25 August*. https://cites.org/sites/default/files/EST/UNGAN2324389.pdf
UNODC (n.d.). *United Nations Convention against Transnational Organised Crime*. https://www.unodc.org/unodc/en/organised-crime/intro/UNTOC.html
UNODC. (2016). *World wildlife crime report. Trafficking in protected species*. United Nations Publication. Vienna, Austria.
UNODC. (2018). *Guide on drafting legislation to combat wildlife crime*. United Nations Publication. Vienna, Austria.
UNODC. (2020). *World wildlife crime report. Trafficking in protected species*. United Nations Publication. Vienna, Austria.
UNODC. (2021). *Digest of cases: International cooperation in criminal matters involving the United Nations convention against transnational organised crime as a legal basis*. United Nations Publication. Vienna, Austria.
UNODC. (2022). *Illegal wildlife trade and climate change: Joining the dots*. United Nations Publication. Vienna, Austria.
UNODC. (2024). *Global analysis on crimes that affect the environment – Part 1: The landscape of criminalisation*. United Nations Publication. Vienna, Austria.
UNODC. (n.d.). *United Nations Convention against Transnational Organised Crime and the Protocols Thereto: Background Information*. https://www.unodc.org/unodc/en/organized-crime/intro/UNTOC.html
UNODC. (n.d.). *Working groups*. United Nations Office on Drugs and Crime. https://www.unodc.org/unodc/en/organized-crime/intro/COP/working-groups.html
Wyatt, T. (2021). *Is CITES protecting wildlife? Assessing implementation and compliance*. Routledge. London.
Wyatt, T. (2022). *Wildlife trafficking: A deconstruction of the crime, victims and offenders*. Palgrave Macmillan Cham, London. Doi: 10.1007/978-3-030-83753-2
Wyatt, T., Friedman, K. and Hutchinson, A. (2021). Are fish wild? *Liverpool Law Review, 42*(3), 485–492.
Wyatt, T. and Hutchinson, A. (2023). *Plant blindness and the law on international trade in wildlife. Erasmus Law Review, 1*, 14–26. https://assets.budh.nl/tijdschriften/elr/issues/ELR_2210-2671_2023_016_001_totaal.pdf

18

POACHING, ILLEGAL WILDLIFE TRADE, AND SUSTAINABLE DEVELOPMENT

Challenges and opportunity

Rodgers Lubilo, Thelma Nkandu, Bupe Banda, and Cecilia Banda

Introduction

The rapid rise of poaching and illegal wildlife trade (IWT) poses a severe threat to biodiversity, disrupts ecosystems, and undermines sustainable development (UNEP, 2016). Tackling these challenges requires urgent and innovative solutions. Community-Based Natural Resource Management (CBNRM) has emerged as a crucial strategy in this regard. By empowering local communities, promoting conservation incentives, and strengthening governance, CBNRM provides a sustainable approach to mitigating poaching and IWT while fostering socioeconomic development (Roe, 2015).

Building on this, this chapter aims to broaden the understanding of poaching and IWT, highlighting both the challenges they pose and the opportunities for sustainable development. It examines the complexities of these issues for both humans and nature, analyses the impact of IWT on local communities and ecosystems, and discusses potential solutions that balance conservation goals with human needs. Poaching and IWT are escalating at an alarming rate, with iconic species of both fauna and flora being hunted, harvested, and traded illegally (UNEP, 2016). These activities, driven by factors such as poverty, corruption, and the demand for wildlife products in local and international markets (Challender & MacMillan, 2014; Damania & Bulte, 2007; Warchol, 2004), threaten not only biodiversity but also economic stability and human security. Addressing these drivers through CBNRM is essential for promoting conservation and sustainable livelihoods.

Wildlife poaching is directly linked to IWT, which encompasses wildlife-related crimes such as smuggling, poaching, and the illegal collection of endangered species and their parts, including those subject to regulated harvest quotas and permits (Shekhar, 2009). Poaching represents the exploitation of wild fauna and flora in violation of international and national laws, endangering biodiversity, human security, and economic growth (Laurean, 2016). Given these legal and socioeconomic implications, tackling IWT requires a multifaceted strategy that includes stronger enforcement and community-driven conservation efforts. CBNRM is placed to address the root causes of poaching and IWT through

DOI: 10.4324/9781003333906-21

creating employment, stimulating local economies, improving livelihoods, and promoting human–wildlife coexistence.

This chapter places CBNRM at the centre of efforts to combat poaching and IWT, demonstrating how community-driven approaches can enhance conservation and contribute to sustainable development. This chapter is structured as follows: first, we explore the drivers and consequences of poaching and IWT; second, we examine existing challenges in enforcement and policy coordination; and finally, we highlight how CBNRM can serve as a viable solution, aligning conservation efforts with local socioeconomic benefits.

Understanding the broad CBNRM ideology

CBNRM is a people-centred approach to natural resource governance that actively involves indigenous peoples, local communities, and natural resource users in decision-making (Murphee, 2000). It integrates local institutions, customary practices, and traditional knowledge into conservation and natural resource management, regulatory, and enforcement processes (Adams, 2004; Child, 1996). At its core, CBNRM aims to create the right incentives and conditions for local resource users within defined jurisdictions to sustainably profit from the conservation of natural resources. Developed to promote environmental, economic, and social justice, CBNRM combines wildlife conservation with rural development objectives, making it a widely adopted model in wildlife-rich African countries (Getz et al., 2002).

The interplay between social justice, sustainable development, and governance in CBNRM fosters a human rights-based approach to natural resource management, balancing environmental, economic, and social considerations. CBNRM empowers communities by providing rights over land and natural resources, building skills and capacity, establishing community decision-making bodies, promoting community advocacy, and enabling women to take leadership positions (Murphee, 2000). Through community-led development and conservation, local communities are empowered to take ownership of their natural resources while influencing policy and legislative frameworks that promote sustainability and social equity. For CBNRM to be effective, it must be based in a community that can clearly define itself and whose members agree to cooperate in managing resources (Reid, 2016). This management also necessitates that clear rules are understood and respected both within and beyond the community.

Sustainable development in the context of wildlife trade and conservation plays a critical role in protecting endangered species and their habitats while supporting economic growth and sustainable livelihood opportunities. By fostering wildlife-based economies, CBNRM creates jobs, supports ecotourism, and contributes to biodiversity conservation, reducing conflicts over natural resources. CBNRM not only supports conservation but also contributes to rural development by providing income for community projects, diversifying land use, and strengthening local institutions. CBNRM thus complements both national and international regulations to combat illegal wildlife trade and helps to enhance law enforcement institutions to address wildlife crime by addressing many of the key drivers of poaching and wildlife trafficking in Southern Africa, namely poverty and vulnerability (Lunstrum et al., 2023).

Another key component of how CBNRM contributes to addressing poaching and IWT is community-based monitoring and information systems, where indigenous peoples and local community organisations track environmental changes and resource use using a combination of traditional knowledge and modern tools (Berkes, 2008). These monitoring systems enhance CBNRM's role in combating poaching and IWT by strengthening local governance and supporting law enforcement with ground-level information on resource use.

Given its potential to align conservation with community development, CBNRM emerged as a dominant conservation model in the late 1980s. Initially developed to integrate wildlife conservation and rural development, it has since evolved into a leading framework for African conservation, emphasising community organisation for sustainable resource management (DWNP, 1999). Indeed, CBNRM has gained global recognition as an effective strategy for community-led conservation. It is particularly well-established in Southern Africa, including Zambia, where it encourages local ownership and active participation in conservation efforts (Sammy & Opio, 2005; Child & Barnes, 2010).

CBNRM in southern Africa

The history of natural resource management in Southern Africa is a tale of transformation, shaped by indigenous conservation practices, disrupted by colonial policies, and reimagined in the post-colonial era through CBNRM (Child, 1988, 1996; Hulme & Murphree, 2001). This evolution reflects broader socio-political changes and the continuous search for models that balance conservation with human development. The origins of CBNRM in the region are tied to the recognition of the limitations of exclusionary conservation models, which historically restricted local communities from accessing and managing natural resources (Brockington, 2002; Lunstrum et al., 2023; Enns & Bersaglio, 2024; Nabane & Matzke, 1997). As a response, CBNRM became an integral part of the conservation agenda, growing from early successes such as Zimbabwe's Communal Areas Management Programme for Indigenous Resources (CAMPFIRE) and other initiatives. Over time, CBNRM has been institutionalised in policies and legislation, shaping conservation and rural development strategies across the region (Roe et al., 2009; Africa Wildlife Foundation, 2016).

Despite these developments, CBNRM initiatives in Southern Africa have faced varying levels of success and failure. In South Africa, the failure of certain initiatives is often linked to the deep-seated legacy of land expropriation and overcrowding in 'homelands,' which resulted from colonial-era policies that criminalised local wildlife use (Fabricius et al., 2001). Similarly, in Zimbabwe, colonial dispossession disrupted traditional governance systems related to land and wildlife, weakening community engagement in conservation efforts (Murphree, 2009). While CAMPFIRE initially succeeded in restoring some of these traditional governance structures, its impact was later undermined by broader governance challenges, particularly during Zimbabwe's political and economic decline (Balint & Mashinya, 2006).

In contrast, Namibia's CBNRM model stands out as a more resilient approach. Unlike many other regional initiatives, Namibia's programme has been reinforced through strong legal frameworks that grant communities explicit rights over wildlife and tourism revenue (Jones, 2016). These rights, enshrined in national legislation, have enabled communities to directly benefit from conservation efforts, thereby fostering a stronger sense of ownership and long-term sustainability. As studies have highlighted, Namibia's model provides valuable insights into how clearly defined community rights and governance structures can enhance the effectiveness of CBNRM (Roe et al., 2009; Africa Wildlife Foundation, 2016).

Zambia's approach to CBNRM followed a slightly different trajectory, with early efforts centred on the Administrative Management Design for Game Management Areas (ADMADE) programme initiated in 1987 (Lubilo & Child, 2010). The foundations of Zambia's CBNRM can also be traced back to the Joint Lupande Research Project (1979–1984), a collaboration between the New York Zoological Society and Zambia's National Parks and Wildlife Service (NPWS). This project, which focused on elephant management in Lupande Game Management Area (GMA), demonstrated the importance of local community involvement. Findings indicated that local residents possessed significant ecological knowledge, including an understanding of wildlife movements and poaching routes, which was instrumental in shaping future conservation strategies (Mwape, 2003; Foyet, 2024).

Building on these insights, ADMADE expanded to eight additional GMAs and was later implemented in 26 more across Zambia. The programme introduced revenue-sharing mechanisms, where income from wildlife tourism and hunting was distributed to local communities. These funds supported conservation efforts, including hiring community scouts for anti-poaching initiatives, and financed local development projects such as schools and clinics. Crucially, communities were granted decision-making power in allocating revenues for local development priorities (Child, 2009).

The overarching objectives of ADMADE were to create a self-sustaining framework for wildlife conservation, generate long-term community benefits, and foster collaboration between NPWS and local populations. The adoption of Zambia's CBNRM policy in 2007 marked a major milestone, aiming to establish an enabling environment for conservation and sustainable use while reducing human–wildlife conflict (GRZ, 2007). The policy is rooted in the principle that local communities should have equitable access to tourism and conservation benefits, with income from sustainable wildlife utilisation being fairly distributed (DeGeorge & Reilly, 2009).

Beyond Zambia, community-based conservation remains a dominant strategy across Southern Africa, playing a crucial role in reducing illegal wildlife offtake, strengthening governance, and enhancing rural livelihoods (Child & Barnes, 2010; Lubilo, 2018). As CBNRM continues to evolve, its effectiveness depends on how well governance structures, economic incentives, and conservation goals are aligned. The next section explores how these dynamics unfold across Southern Africa, assessing both successes and challenges in implementing CBNRM at different scales.

Complexities, constraints, and challenges in the CBNRM approach

CBNRM's implementation faces multiple complexities, constraints, and challenges spanning technical, economic, cultural, and governance-related aspects. Addressing these issues is critical to ensuring the effectiveness and sustainability of CBNRM initiatives, including their ability to help mitigate poaching and IWT.

Systemic complexities and constraints

One of the fundamental constraints in CBNRM is the absence of an appropriate resource rights framework. Existing legal and institutional structures related to land and resource rights and how communities can engage with them and gain management authority over them are often complex, time-consuming, and costly for rural communities to navigate. These bureaucratic challenges make it difficult for local communities to understand and effectively exercise their rights over natural resources. Moreover, the ability of CBNRM programmes to contribute to local and community economic development and livelihoods largely depends on the value of the resource base, which is not always sufficient to support local communities (Lubilo & Child, 2010).

Another significant, related challenge is the lack of knowledge and capacity in effective resource management techniques. Deficiencies in understanding population dynamics and resource levels make it challenging to determine sustainable yields and implement sound management practices. Furthermore, limited knowledge of techniques to enhance, safeguard, or restore natural productivity further hinders the effectiveness of CBNRM initiatives (Berkes, 2008; Murphree, 2000). This knowledge gap extends to indigenous conservation methods, which are often underappreciated and unsupported.

Governance-related constraints also pose a major challenge. The relationship between local communities and government authorities is often asymmetrical, with communities having minimal influence over decision-making processes (Murphree, 2009). This imbalance can lead to situations where policies, projects, and resource management decisions do not reflect the needs and priorities of the local population, undermining the principles of CBNRM (Fabricius et al., 2001). Additionally, inconsistencies in national policies and laws in ways that favour other stakeholder rights at the expense of community rights further hinder the effectiveness and scalability of CBNRM initiatives (Roe et al., 2009). For example, in Zambia, CBNRM has struggled due to a lack of political will to decentralise decision-making authority and responsibilities over wildlife management to local communities. Weak institutional frameworks and inadequate policies that fail to support local ownership of natural resources have further constrained the potential of CBNRM. Many local communities lack financial resources and face persistent poverty, limiting their ability to engage effectively in conservation activities. As a result, they often resort to illegal resource harvesting from protected areas, which meets immediate household needs but also fuels the IWT and poaching as a means of income generation Roe et al., 2009).

Implementation challenges

Beyond structural and systemic constraints, CBNRM initiatives also face significant implementation challenges that impact their long-term viability. Some of the key elements of these challenges are below.

Financial and capacity constraints

Most communities receive insufficient funds for their programmes, limiting the scalability and sustainability of CBNRM initiatives. Many rural communities rely on external funding and support, which is often inconsistent or inadequate. Additionally, limited technical capacity and expertise among local communities, governments, and NGOs hinder effective implementation. Hiring and retaining conservation experts is often unaffordable for rural communities, making it difficult to develop and implement effective management strategies (Child, 2009; DeGeorge & Reilly, 2009; Hulme & Murphree, 2001; Jones, 2007).

Climate change and environmental pressures

The adverse impacts of climate change exacerbate socio-ecological challenges, disproportionately affecting rural communities that lack the capacity to adapt and mitigate its effects. Deforestation, land degradation, and loss of biodiversity further threaten ecosystem resilience and economic sustainability (UNEP, 2016). The degradation of these natural resources leads to a decline in essential ecosystem services, negatively impacting both conservation efforts and local livelihoods. As valuable natural resources become increasingly scarce, rural communities may struggle to sustain CBNRM initiatives effectively.

Social inequality and community perceptions

Persistent poverty and inequality limit access to resources, markets, and decision-making opportunities. In many cases, local communities feel excluded from conservation benefits, leading to perceptions of injustice. These frustrations can manifest as non-compliance with conservation regulations or even cooperation with poachers. Traditional livelihoods, such as hunting and agriculture, are often perceived as being threatened by conservation initiatives, further reducing local support for conservation and CBNRM (Moreto, 2013). Human–wildlife conflict over crops, livestock, and grazing land exacerbates these tensions, leading to targeted retaliation against wildlife and general resistance to conservation policies (Mugisha & Jacobson, 2004).

Governance and benefit distribution

The distribution of benefits within CBNRM programmes remains a contentious issue. Elite capture—where wealthier or more influential community members disproportionately benefit from CBNRM while poorer members bear the costs—is a common challenge (Shackleton et al., 2002). Even within communities, benefit-sharing is often unequal, reducing incentives for full participation. Additionally, the distribution of benefits between communities and external stakeholders, such as private enterprises or government agencies, is often skewed, limiting the economic impact of CBNRM on local populations (Jones, in press).

Weak monitoring and evaluation (M&E) systems further hinder CBNRM effectiveness. A lack of rigorous data collection and analysis makes it difficult to assess success, adjust strategies, and ensure accountability. Investment in M&E systems is essential for improving resource management, enhancing decision-making, and identifying best practices. With advancements in technology, real-time data collection and knowledge management can significantly improve CBNRM performance, yet investment in these areas remains insufficient.

Opportunities for CBNRM to address poaching and IWT

Despite the above challenges, CBNRM holds significant potential to strengthen conservation, enhance local governance, and promote sustainable development, particularly in combating IWT. Realising this potential, however, requires addressing persistent challenges and strategically leveraging opportunities for improvement. Addressing issues such as weak law enforcement, inadequate community engagement, and ineffective benefit-sharing mechanisms requires a strategic approach that leverages governance reforms (power), economic incentives (wealth), and international cooperation (nature) to achieve a balanced, triple-win outcome. Successful CBNRM thus has simultaneous positive results in three areas—nature (environmental management), wealth (economics and revenue generation), and power (local empowerment and good governance) (Roe et al., 2009). Initiatives that succeed in one or two of these dimensions but fail in another are often unsustainable in the long run. Unlike purely growth-driven or conservation-focused approaches, CBNRM seeks to achieve a balanced, triple-win outcome as a pathway towards realising its full potential.

One of the key opportunities within CBNRM is the mobilisation of stakeholders to drive governance reforms and enhance legal frameworks that support community-led conservation. Strengthening stakeholder networks presents an opportunity to advocate for demand-driven collective management arrangements, ensuring that communities have meaningful decision-making authority over their resources (IIED, 2009). A legislative review that aligns natural resource policies with local realities offers a crucial opportunity to unlock the full potential of CBNRM. In Zambia, for example, advancing policies that formalise benefit-sharing mechanisms and provide legal recognition for community resource rights can empower local communities and create a more sustainable governance framework. One such example is the potential of frameworks and adequate support for communities to have adequate rights over wildlife to participate in the legal game meat economy (Bersaglio et al., 2025). Additionally, capacity-building initiatives provide an opportunity to enhance local institutions, empower civic organisations, and equip communities with skills for effective resource governance. Strengthening M&E indicators further offers a means to measure progress and refine conservation strategies over time (IIED, 2009).

CBNRM offers a significant opportunity to break the poverty-environment degradation trap, and specifically poverty and vulnerability as a driver of poaching and IWT. With appropriate policies and incentives CBNRM has been shown to generate employment in rural areas, improve conditions for marginalised groups such as women and youth, and alleviate pressure on natural resources. By integrating sustainable economic activities, such as ecotourism, small-scale enterprises, and regulated wildlife utilisation, CBNRM creates legitimate income-generating opportunities that reduce reliance on illegal resource extraction, including poaching (Roe et al., 2009). This not only helps curb IWT and poaching but also enhances food security, economic stability, and resilience against climate change impacts (Duffy & Humphreys, 2016). By tapping into these opportunities, communities can transition towards more sustainable economic models that align with conservation priorities.

Furthermore, CBNRM provides an opportunity to strengthen collaboration between rural communities and government institutions, fostering cooperative governance structures that balance conservation with community needs. Developing effective governance frameworks that support local ownership of natural resources ensures that conservation efforts align with socioeconomic development. In Zambia and other regions, the push towards decentralisation offers an opportunity to improve natural resource governance by implementing policies that grant communities greater control over their resources. When fully realised, these institutional arrangements can bridge gaps in resource governance and policy implementation, leading to more effective conservation outcomes.

International cooperation offers additional opportunities to enhance CBNRM initiatives through knowledge exchange, funding, and policy development. Zambia, along with other southern African nations, can leverage international frameworks such as the Convention on Biological Diversity (CBD) and the Convention on International Trade in Endangered Species of Wild Fauna and Flora (CITES) to strengthen its conservation strategies. The CBD presents an opportunity to align national CBNRM policies with global biodiversity targets, ensuring that local conservation efforts contribute to broader international commitments. Similarly, CITES regulations provide an opportunity to improve national enforcement mechanisms to curb IWT and promote legal, sustainable trade in wildlife resources.

Conclusion

CBNRM stands at the intersection of conservation, sustainable development, and social justice in Southern Africa. As the region grapples with the dual challenges of biodiversity loss and rural poverty, CBNRM offers a pathway to harmonise environmental protection, including efforts to address wildlife poaching, with community empowerment. The evolution of CBNRM across Southern Africa, from early initiatives like Zimbabwe's CAMPFIRE to Namibia's robust conservancy model and Zambia's ADMADE programme, demonstrates both the potential and the pitfalls of community-led conservation. Ultimately, the success of CBNRM in Southern Africa demonstrates the transformative potential of community-led conservation efforts. Wildlife-based enterprises, when well-managed, generate significant local revenues that enhance economic and non-economic security in marginalised areas (Child, 2019). Sustainable conservation models that involve local communities offer opportunities not only to protect biodiversity but also to create resilient rural economies. In doing so, CBNRM helps address the root causes of many poaching and IWT challenges by providing rural communities with sustainable livelihoods, stimulating local economic development, and offsetting the potential negative implications of protected areas on rural communities.

As demonstrated, these experiences underscore the critical importance of secure land tenure, clear benefit-sharing mechanisms, and adaptive governance structures in fostering successful CBNRM initiatives. Looking ahead, the future of CBNRM in combating poaching and IWT, and promoting sustainable development, hinges on several key factors including strengthening legal frameworks to enshrine community rights over natural resources, enhancing capacity-building efforts to empower local institutions, fostering innovative partnerships between communities, government agencies, and the private sector, leveraging technology for improved resource monitoring and management, and aligning CBNRM strategies with broader sustainable development goals and climate action plans.

As southern African nations continue to refine their CBNRM approaches, there is a pressing need for cross-border collaboration and knowledge sharing. By learning from both successes and setbacks, countries can develop more resilient and effective CBNRM models that not only curb IWT but also drive rural economic development and biodiversity conservation. Ultimately, the success of CBNRM in addressing poaching and IWT will be measured not just in wildlife population numbers, but in the improved livelihoods, strengthened community governance, and enhanced ecological resilience of Southern Africa's rural landscapes. As we look towards a future where human well-being and environmental health are inextricably linked, CBNRM stands as a beacon of hope for sustainable, community-driven conservation.

Bibliography

African Wildlife Foundation (2016). *African Conservancies: Towards Best Practices*. African Conservancies, African Wildlife Foundation.

Ban, K. M. (2018). *Wildlife Crime is a Threat to Species, Security and Sustainable Development, US*. Wildlife crime initiative.

Bennett, E. et al., (2007). The world can't have wild tigers and eat them. *Conservation Biology, 22*, 222–223.

Berkes, F. (2008). *Sacred Ecology: Traditional Ecological Knowledge and Resource Management* (2nd ed.). Routledge.

Bersaglio, B., Enns, C., Hichoonga, E., Masse, F., Sommerville, M., & van Vliet, N. (2025). *How uneven access shapes the socio-economic and environmental potential of game meat value chains: The case of legal game meat in Zambia. People and Nature, 3*(2), 1–14.

Bond, I. et al., (2006). Community-based natural resources management manual: wildlife management series, s.l: *WWF-World Wild Fund for Nature*.
Brockington, D. (2002). *Fortress Conservation: The Preservation of the Mkomazi Game Reserve*. Tanzania. Indiana University Press.
Challender, D. W. S. & MacMillan, D. C. (2014). 'Poaching is more than an enforcement problem'. *Conservation Letters, 7(5)*, 484–494. https://doi.org/10.1111/conl.12082.
Child, B. (1996). The practice and principles of community based willdife management in Zimbabwe: The Campfire programme. *Biodiversity and Conservation, 5*, 369–398.
Child, B. (2019). *Sustainable Governance of Wildlife and Community-Based Natural Resource Management. From Economic Principles to Practical Governance*. Routledge.
Damania, R. & Bulte, E. H. (2007). *'The economics of wildlife farming and endangered species conservation'. Ecological Economics, 62*(3–4), 461–472. https://doi.org/10.1016/j.ecolecon.2006.07.007.
Danielle, C. (2018). *96 Elephants a Day: A Risk Analysis of Elephant Poaching in Kenya, Kenya*. Cunnings School of Veterinary Medicine.
Duffy, R. & Humphreys, J. (2016*). Poaching, wildlife trafficking and human security. Whitehall Papers, 88*(1), 22–37.
DWNP (1999). *Joint Venture Guideline, Gaborone*. Department of Wildlife and National Parks.
Enns, C. & Bersaglio, B. (2024). *Settler Ecologies: The Enduring Nature of Settler Colonialism in Kenya*. University of Toronto Press.
Foyet, M. (2024). Community-Based Natural Resource Management (CBNRM) in southern Africa: history, principles, evolution and contemporary challenges. *Namibian Journal of Environment, 9*, C: 1–15.
HUlme, D. & Murphree, M. (2001). *Africa Willdife and Livelihoods. The Promises and Performance of Community Conservation*. James Currey. https://www.worldbank.org/en/news/video/2017/11/28/working-together-to-save-wildlife-and-ecosystems#:~:text=The%20World%20Bank-led%20Global%20Wildlife
Jones, B. (2016). *Institutionalised community conservancies in Namibia. In: Conservancies in Africa: towards best practices*. African wildlife foundation., Nairobi
Laurean, M. (2016). *Illegal Dealings in Wildlife and the Law in Tanzania: Examinig Pertinent Issues and Challenges with Respect to Poaching and Illegal Wildlife Trade*. Tanzania wildlife poaching and trafficking workshop.
Lunstrum, E., Givá, N., Massé, F., Mate, F. & Jose, P. L. (2023). The rhino horn trade and radical inequality as environmental conflict. *The Journal of Peasant Studies, 50*(3), 1085–1105.
Matthew, H. et al., (2018). Increase anti-poaching law-enforcement or reduce demand for wildlife products? A framework to guide strategic conservation investments. *A journal of Society for Conservation Biology, 12(4)*, 1.
Meredith, L. et al., (2023). *Advancing interdisciplinary science for disrupting wildlife trafficking network. Proceedings of the National Academy of Sciences, 120(10)*, 1.
Micheal't, S.-R.et al., (2019). Illegal wildlife trade: Scale, processes, and governance. *Annual Review of Environment and Resources, 44*, 201–228.
Moreto, W. (2013). *To Conserve and Protect: Examinig Law Enforcement Ranger Culture and Operations in Queen Elizabeth National Park, Uganda*. Rutgers University-Graduate school-Newark.
Mugisha, A. & Jacobson, S. (2004). *Threat reduction assessment of conventional and community-based conservation approaches to managing protected areas in Uganda. Environmental Conservation, 31*(3), 177–189.
Murphee, M. W. (2000). Community-Based Conservation: Old Ways, New Myths and Enduring Challenges. Key address at the conference on African Wildlife Management in the New Millennium. College of African Wildlife Management, Mweka, Tanzania, 13–15 December 2000.
Murphree, M. W. (2009). The strategic pillars of communal natural resource management: benefit, empowerment and conservation. Biodiversity Conservation, 18, 2551–2562.
Nabane, N. & Matzke, G. (1997). A gender-sensitive analysis *of a community-based wildlife utilization initiative in Zimbabwe's Zambezi Valley. Society and Natural Resources, 10*(6), 519–535.
Neil, H. C. et al., (2016). *A conceptual framework for understanding illegal killing of large carnivores, s.l.* Kungul vetenskaps akademien.
Rabie, M. (2016). *Sustainable Development: A Theory of Sociocultural and Economic Develoment*. London: Palgrave Macmillan.

Reid. (2016). *Ecosystem and community-based adaptation: Learning from community-based natural resource management. Climate and Development, 8*(1), 4–9. https://doi.org/10.1080/17565529.2015.1034233.

Roe, D., Nelson, F. & Sandbrook, C. (2009). *Community Management of Natural Resources in Africa: Impacts, Experiences and Future Directions*. International Institute for Environment and Development (IIED).

Roe, D., Nelson, F. & Sandbrook, C. (eds) (2009). *Community Management of Natural Resources in Africa: Impacts, Experiences and Future Directions, Natural resource issues no. 18*. International Institute for Environment and Development.

Roe, D. (2015). *Conservation, Crime, and Communities: Case Studies of Efforts to Engage Local Communities in Tackling Illegal Wildlife Trade*. International Institute for Environment and Development (IIED). Available at: https://www.iied.org/14640iied

Rolf, D. B. (n.d*). Poaching in Africa: Facts, causes and solutions. African Indaba e-Newsletter, 12*(3), 8

Sammy, J. & Opio, C. (2005). Problems and prospects for conservation and indigenous community development in rural Botswana. *Developmental Southern Africa, 22*(1), 67–85.

Shackleton, S., Campbell, B., Wollenberg, E. & Edmunds, D. (2002). *Devolution and Community Based Natural Resource Management: Creating Space for Local People to Participate and Benefit? ODI Natural Resource Perspectives*. Programme for Land and Agrarian Studies (PLAAS).

Shekhar, K. N. (2009). *Sustainable Development, Poaching and Illegal Wildlife Trade in India, India*. School of Natural Resources.

United Nations Environment Programme (UNEP) and INTERPOL (2016). *The rise of environmental crime: A growing threat to natural resources, peace, development, and security*. Nairobi: UNEP. Available at: https://wedocs.unep.org/20.500.11822/7662 (Accessed: 21 February 2025).

Warchol, G. L. (2004). 'The transnational illegal wildlife trade'. *Criminal Justice Studies, 17*(1), 57–73. https://doi: 10.1080/08884310420001679334.

PART IV

Emerging and ongoing issues in wildlife crime

19
ILLEGAL CAPTIVE BREEDING OF WILDLIFE

José Luis Carpio-Domínguez and Jesús Ignacio Castro-Salazar

Introduction

Captive breeding of wildlife is a human practice of breeding wildlife species in captive conditions under human control. It is characterised by two forms: illegal and legal. Legal breeding of wildlife is a useful practice when a wildlife population is no longer viable in the wild, thus ensuring the preservation of threatened or endangered species (United Nations Office on Drugs and Crime [UNODC], 2020). Many countries promote this practice to protect wild species through breeding centres (farms), research centres, and/or zoos (Castro-Salazar, 2023; Gomez & Nguyen, 2023).

When wildlife breeding is carried out within legal frameworks it can (1) promote the protection and management of wild species as a genetic reserve, (2) provide legal and sustainable economic sources for people, (3) offer scientific data (biological, veterinary and ecological) for the generation of knowledge and environmental education, and (4) promote development in localities through tourism, the pharmaceutical industry, and the sale of derived products, among others (Tensen, 2016; Wang et al., 2019).

However, this practice can be carried out "legally" by exploiting legal loopholes, which often has a detrimental effect on the species (Williams et al., 2015). Legal policies (local, national and international) should be reviewed to amend and clarify regulations. Some studies have reported that market strategies have a strong capacity to identify legal loopholes and gain economic and marketing advantages (see Maher & Sollund, 2016; Wyatt et al., 2018; Musinguzi & Bsheka, 2018; Wyatt, 2020; de Vries, 2022); therefore, it cannot be assumed that everything legal does not cause harm.

Illegal captive breeding of wildlife is a criminal practice of capturing specimens of flora or fauna from the wild or purchasing specimens illegally with the aim of breeding them in captivity to supply a commercial demand for the offspring or products. This practice is closely linked to the illegal wildlife trade in live and dead specimens or products (Tamalene y Kartika, 2019; Sollund & Runhovde, 2020; Wyatt, 2021; Carpio-Domínguez et al., 2021; Gomez & Nguyen, 2023). Although this phenomenon also includes the capture from the wild and commercialisation of plant species (Warchol, 2004; Arroyo-Quiroz & Wyatt, 2019; Wyatt, 2021; Urrego et al., 2022), the illegal breeding and trafficking of fauna has been

DOI: 10.4324/9781003333906-23

mainly made visible due to the high value in the illegal economic sector (Carpio-Domínguez et al., 2018a, 2018b; UNODC, 2020; Gomez & Nguyen, 2023).

Historically, the captive breeding of fauna species has been a practice that has allowed the domestication of species for human activities (Clutton-Brock, 1999; Wang et al., 2019; Widi et al., 2021). For example, for agricultural and livestock production, it was necessary to domesticate and breed cattle, goats, horses, etc. For hunting and security activities, dogs played an important role, and to obtain food sources, techniques such as planting fish species in lakes or streams were implemented, among other examples. Through artificial selection, these practices favoured the development of human civilisation (Wang et al., 2019; Frantz et al., 2020). In addition, captive breeding of wildlife has played an important role as part of human leisure activities (e.g. Roman circuses for fighting, as ornamentation, for domestic entertainment or for commercial purposes such as circuses and zoos) (see Sollund & Runhovde, 2020; Klecel & Martyniuk, 2021; Arroyo-Quiroz et al., 2021; Rozmiarek & Włodarczyk, 2023; Mazorrin & Abatino, 2023).

It has been documented that legal captive breeding in farms has become a common practice to cope with the demand for animal products and offspring in the international and global market (Rizzolo, 2020) and is a focus of attention for environmental regulatory authorities (criminal and administrative) because links have been documented between legal breeding farms and the formation of criminal groups involved in wildlife trafficking through species laundering (Natusch & Lyons, 2014; Auliya et al., 2016; Wyatt, 2022).

Illegal captive breeding of wildlife has been identified as related to four interrelated phenomena analysed in this chapter: (1) wildlife trafficking supplied by illegal captive breeding, (2) species laundering as a criminal practice, (3) inbreeding as a factor not considered in illegal captive breeding, and (4) the pandemic risks of illegal breeding farms.

Wildlife trafficking supplied by illegal captive breeding

Wildlife trafficking involves the capture, confinement and trade of animals and plants in contravention of national and international laws and is considered one of the world's largest black markets (World Bank, 2019). Wildlife traffickers use diverse sources of access to wildlife (e.g. capture from the wild, illegal hunting, illegal captive breeding, laundering of species). The ability of wildlife traffickers to disguise illegal practices as legal practices has made the illegal trade adaptable to government efforts to address it. Among such adaptive practices is illegal captive breeding as a source of wildlife specimens and products.

Several studies have documented that illegal captive breeding of wildlife is carried out in unauthorised species (species in at-risk categories such as endangered and threatened) (Kuhnen et al., 2012; Hogg et al., 2019). This is important because illegal captive breeding of wildlife results in wildlife specimens being captured from the wild to be bred in captivity, often including species in at-risk categories, to supply the illegal market without complying with the legal procedures established by the authority on the management and trade of specimens for captive breeding.

Illegal captive breeding of wildlife has different purposes depending on the place or country. For example, in Mexico, illegal captive breeding of wildlife is largely used to supply the demand for exotic species as pets, which implies that illegal captive breeding supplies the market with specimens that do not have a natural distribution in the country. Other studies in Mexico (e.g. Sosa-Escalante, 2011; Arroyo-Quiroz y Wyatt, 2019; Arroyo et al., 2021) show that fauna is captured from the wild when it is naturally available; however, those species considered "exotic" are often bred illegally in captivity (e.g. *Panthera tigris, Panthera leo, Oryx dammah,* and *Myiopsitta monachus*, among others) (Carpio-Domínguez et al., 2018b, 2021, 2022, 2023). Although the possibility of illegal captive breeding of local or endemic fauna for wildlife trafficking is not ruled out.

In countries such as Laos, Cambodia, and Vietnam, bears (*Ursus thibetanus*) are illegally captured from the wild and bred in captivity to supply the demand for bile and other products (Gomez, Shepherd & Khoo, 2020; Gomez et al., 2021). In other countries such as China, Vietnam, Thailand, and Laos, tiger farms (*Panthera tigris*) have between seven and eight thousand tigers in captivity (Traffic, 2023) with the aim of breeding them and illegally trading live and dead specimens, and products such as skins, skulls, fangs, claws, and bones, among others (Kirkpatric & Emerton, 2010; Rieder, 2023).

In African countries such as South Africa, Uganda, Zambia, and Tanzania, studies have documented that wildlife trafficking involves the trafficking of live and dead specimens, and products of different species such as rhinos, elephants, pangolins, snakes, and lions, among others, originating from different social, legal, economic, cultural and environmental dynamics in these countries (Hübschle, 2016; Hübschle & Gore, 2024). However, (legal) captive breeding of wildlife in Africa is a practice used to supply wildlife parts and products to Asian countries. For example, Williams et al. (2015, 2017) have documented significant markets for African lion bones from breeding farms for transport to Southeast Asian countries, but also from illegal breeding, making it imperative to conduct studies focused on identifying legitimate and illegitimate sources from the wild or from captive breeding.

To summarise the documented cases of illegal captive breeding of wildlife for trafficking purposes would exceed the length of this chapter (e.g. Kirkpatric & Emerton, 2010; Mali et al., 2014; Williams & 't Sas-Rolfes, 2019; Gomez, Shepherd & Khoo, 2020; Gomez et al., 2021; Magliocca et al., 2021; Rieder, 2023; among others), however, illegal captive breeding of specimens captured from the wild allows traffickers to: (1) save economic and time expenses to go into the wild to search for and capture new specimens and (2) reduce the chances of being identified and sanctioned by the authorities for illegal wildlife capture or trafficking activities. Furthermore, illegal captive breeding of wildlife is a source of wildlife supply for wildlife trafficking (see van Uhm, 2016; Carpio-Domínguez et al., 2018a; Arroyo & Wyatt, 2019; Carpio-Domínguez et al., 2023; Sollund & Runhovde, 2020; Sardari et al., 2022; Salas-Picazo et al., 2023; Espejo-Cruz et al., 2023) (Figure 19.1).

Species laundering in the illegal captive breeding of wildlife

With globalisation and the expansion of markets, a number of criminal networks have become involved in the different stages of the illegal wildlife trade: organised criminal groups, corporate criminal groups, and disorganised criminal networks (South & Wyatt, 2011; Wyatt et al., 2020; Carpio-Domínguez et al., 2022; Carpio-Domínguez et al., 2023), as well as collaborations between such groups (in addition to independent individuals who also operate in this market) in different stages of the trafficking process. These stages include cases where illegal wildlife must be processed for sale on the legal market (Wyatt et al., 2020).

Almost any part and product or live and dead specimen of wild fauna can be legally sold, provided it is legally sourced and traded (United Nations, 2016; Tensen, 2016). However, the legal wildlife market may be supplied by illegally traded or bred species (Moyle, 2017). Criminals and criminal organisations regularly use legally registered companies to hide and facilitate their illicit business (van Uhm, 2016).

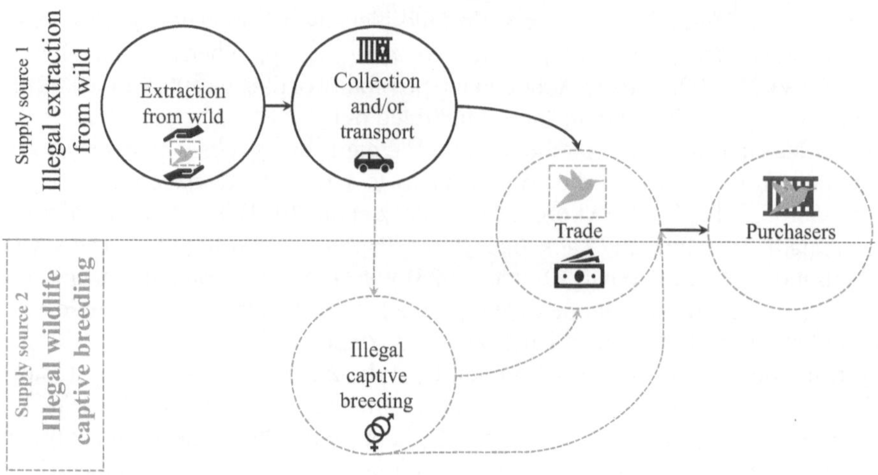

Figure 19.1 Illegal captive breeding in the basic dynamics of illegal wildlife trade.
Source: The authors.

In criminology, the concept of laundering is associated with concealing the proceeds of crime and transferring the (economic) value through legal commercial transactions to legitimise the illicit origin (van Uhm, 2018a; Teichmann, 2020; Castro-Salazar & Carpio-Domínguez, 2023). Although the concept is generally used to explain the illicit origin of money, for green criminology, this phenomenon is of particular interest when wildlife species are laundered. Species laundering conceals the illegal origin of fauna and/or flora specimens, parts or products to facilitate legitimate trade (Wyatt et al., 2020). This means that species laundering is intended to trade illegal wildlife as if it were legal (Wyatt, 2022). Concealing the illegal origin of an animal, plant or its parts and products as legal through documents or marking systems is known as species laundering; it aims to trade illegal wildlife declared as legally obtained in order to obtain documents proving legality (van Uhm, 2018a, 2018b). The laundering of illegal wildlife into the legal supply chain often occurs through intermediaries and distributors that may operate from the international to the local level (Basel Institute on Governance, 2021).

For example, legal wildlife traders may be involved in fraud and the falsification of wildlife marking systems to launder illegal animals (van Uhm, 2018a). Law enforcement and regulatory agency personnel are important participants in the supply chain, albeit in a different way, often through corruption and bribery (Kishor & Lescuyer, 2012). Corruption of government officials contributes to the laundering of illegal species and products (Phelps et al., 2013) when they provide fraudulent permits, of for example illegally bred wildlife (through corrupt processes), to make it seem like an illegal specimen was legally bred in captivity (Wyatt, 2013).

Various stages within the wildlife trade are an opportunity for wildlife laundering (e.g. species registration, issuance of legal documentation, and sale). Illegal captive wildlife breeding centres have the potential to play a key role in supplying illegal species and products to the legal supply chain, to launder illegally sourced wildlife into legal markets. Therefore, identifying and detecting illegal captive breeding of wildlife centres becomes an opportunity to prevent the illegal breeding of specimens that will later be laundered and introduced into the legal market (Castro-Salazar et al., 2022).

Species laundering by breeding centres is often difficult to detect, due to difficulties in differentiating between specimens legally bred in captivity and those from illegal breeding (van Uhm, 2018a). There are few empirical studies documenting the intent and knowledge of legal breeding centres and exotic pet shops, exotic wildlife expos and conventions, etc., where illegal wildlife is known to be mixed and laundered into legal markets and exacerbated by corrupt activities by government officials (see Wyatt & Cao, 2015; van Uhm & Moreto, 2018; Carpio-Domínguez, 2023).

Government attention must focus on traceability in the supply chain of legal wildlife trade to verify that the documentation presented to prove that the specimens come from a legal captive breeding centre and not from illegal captive breeding of wildlife (UNODC, 2020) to ensure that the documentation is correct and corresponds to the specimens declared (Castro-Salazar & Carpio-Domínguez, 2023). Although marking systems are used as a mechanism to individually identify specimens from legal breeding centres (see Knapp & Affre, 2007), it has been identified that the marks can be corrupted and used to launder species (Castro-Salazar & Carpio-Domínguez, 2023). Proposals to ensure the correct declaration of specimens include DNA testing applied during wildlife inspections (see Hogg et al., 2018), but it is also recommended to implement administrative actions such as constant inspections and audits to monitor sources of supply and reduce corruption.

Inbreeding as a consequence of illegal captive breeding

Inbreeding refers to the mating of genetically related individuals, which results in identical inherited genomic regions to their offspring, from both parents to their offspring (Knief et al., 2015). With inbreeding, the harmful recessive mutations of the parents are inherited to the offspring so that the harmful effects of these mutations are fully expressed in the offspring; this is known as inbreeding depression (Keller & Waller, 2002). The consequences of inbreeding include recessive diseases, impoverishment of phenotypic traits and reduced health conditions in offspring (Campbell et al., 2007; Chapman et al., 2009). It also limits the ability to adapt to the environment and environmental changes, produces smaller populations, and reduces the rate of population growth (McPhee, 2004; Willoughby et al., 2017).

It has been documented that invertebrate animals (insects) held in captivity for 50 generations showed inbreeding depression, and when specimens were returned to the wild, 64–86% of the released populations showed a decrease in reproductive fitness (Woodworth et al., 2002). Genetic analysis indicated that inbreeding depression and genetic adaptation were responsible for the genetic deterioration in wild fitness, which demonstrates that long-term captive breeding is a major problem in the reproductive health of populations when they are returned to the wild (Woodworth et al., 2002).

Although inbreeding is unlikely to occur in the wild (Knief et al., 2015), in captivity, the probabilities increase, both "unintentionally" (by mating between inbred males and females within the same cage) and intentionally by humans seeking to breed animals to obtain offspring (often with small breeding populations) for economic gain or specific colouring patterns.

Inbreeding (from anthropogenic origin) has ethically questionable implications when considered as an aesthetic practice for selective hunting (e.g. the preference of animals with variations in colours and physical characteristics that are not normal or natural), the possession of live animals as pets, or animal parts as souvenirs or trophies. It has been documented that lions in South Africa are legally bred in captivity for genetic crossbreeding to produce colour variations that are attractive to hunters and exotic pet owners. For example, the practice called "canned hunting" consists of taking lion cubs from the captive mother to force her to go into heat again in the shortest possible time, and the cubs are used for the "cub-petting" industry for tourists who are unaware of this practice. When the cubs grow up, they are sold to zoos, private collectors or are killed for "special trophies" of lions with "rare" colours (see The Guardian, 2014; Harvey, 2020; Green et al., 2020; Chorney et al., 2022).

In addition, inbreeding can be enhanced by human practices such as disorganised legal hunting, as it can lead to natural inbreeding due to a decrease in the number of individuals in the wild for natural reproduction (see Nilsson, 2004; Hedrick, 2011; Laikre et al., 2013). Among the proposals that have been put forward is to (re)design hunting policies to improve regulatory strategies, considering the minimum viable population (MVP) with the aim of guaranteeing the conservation of the species (Nillson, 2004). However, although regulations for the hunting industry are present (whether or not they are complied with), the truth is that these regulations are not applied to the possession of animals as pets and even less so when breeding is carried out illegally in captivity due to two main (hypothetical) causes: (1) lack of knowledge of the consequences of inbreeding and (2) the prioritisation of economic benefit over the health of the species.

Biological and genetic studies have been exhaustive in analysing the consequences of inbreeding of captive animals. However, from the perspective of the study of wildlife trafficking, and in particular the illegal breeding of captive wildlife, it has been studied to a lesser extent. Despite formal captive breeding programmes such as zoos and wildlife research centres with Zoological Information Management Software (ZIMS) or studbooks as captive breeding programmes (see Farquharson et al., 2021), illegal captive breeding of wildlife is carried out informally and often without keeping records of the breeding process.

The inbreeding within illegal captive breeding in small populations where offspring are illegally traded (e.g. tigers, bears, lions, snakes), should be considered by law enforcement, and in the regulations for the hunting industry and regulatory controls, for the prevention of inbreeding for aesthetic purposes due to the environmental and biological impacts of this practice.

More studies are needed from criminal law and (green) criminology to deepen the study of illegal captive breeding of wildlife, not only to address its direct negative impacts on biodiversity and ecosystems or the ways in which the law is enforced, but also to address the genetic consequences caused by the inbreeding of captive specimens used for illegal breeding that could interfere in natural selective processes and chromosomal translocation and even "dangerous" hybridisations (e.g. deliberate hybridisation of venomous snakes in captivity without any regulation). These are issues that could be considered as an aggravating factor in the crime of illegal wildlife trade.

Pandemic risks of illegal breeding farms

Ecosystems support species diversity and have a disease-regulating effect when the species that inhabit them are linked to the transmission cycle of a disease. In addition, natural food chains favour viruses, bacteria, protozoa and fungi (which can cause zoonotic diseases) to complete their life cycle in a specific ecological niche. Zoonotic diseases are a serious public health problem. The current global focus is on the timely identification of the origin of these diseases, as well as on understanding the human–animal interactions that enhance zoonoses (Centers for Disease Control and Prevention [CDC], 2016). Zoonotic diseases are linked, among other processes like the expansion of urban settlements into wild environments, climate change affecting the behaviour of both hosts and vectors, and even natural animal migrations (Rahman et al., 2020), to the (legal and illegal) inclusion of wild animals in human environments as food, pets, and the use of products derived from wild animals (see Rahman et al., 2020; Mishra et al., 2021; Carpio-Domínguez et al., 2021; Guo et al., 2021; Whitfort, 2021).

Illegal breeding centres, operating clandestinely, in poorly controlled conditions, without authorisation or supervision by the inspection authorities to verify compliance with their legal obligations, are a risk of generating and spreading diseases (epizootic, zoonotic and anthropozoonotic) (see Naguib et al., 2019; Carpio-Domínguez et al., 2021). Diseases can be transmitted between captive specimens of the same species and with other species which, in turn, can spread to and from non-captive animals such as those that may be in the vicinity of illegal breeding sites or to domestic animals (see Naguib et al., 2019; Hilderink and Winter, 2021; Nijman, 2015; Bezerra-Santos et al., 2021; Carpio-Domínguez et al., 2021), with serious and potentially irreversible consequences for biodiversity, animal welfare, the economy, and human health (United Nations, 2020).

Particular concerns have recently been raised about the health and welfare conditions in wildlife breeding farms and the sanitary conditions in captivity in which wildlife is kept (Petry & Liting, 2010; Pickett & Harris, 2015; Environmental Investigation Agency [EIA], 2016; Carpio-Domínguez et al., 2021), due to unhygienic living conditions in captivity which increase the risks of zoonotic diseases that put human health at risk (Diao, 2016; Yu, 2016; Arroyo-Quiroz, 2020). The unsanitary conditions that persist, both in captivity and in the transport of animals during wildlife trafficking, facilitate zoonoses (Morse et al., 2012; CDC, 2016; Welfare & Wright, 2016; Zinsstag et al., 2020; Cardoso et al., 2021). Illegal captive wildlife breeding farms have a zoonotic potential due to the survival and proliferation of disease-transmitting vectors such as ticks, bedbugs, mosquitoes, fleas and mites (Gbogbo & Opoku, 2017; Lorenzo et al., 2018; Sarvi et al., 2018; Gbogbo & Daniels, 2019; Hrnková et al., 2022). This is aggravated when captive wildlife is in constant contact with domestic and other wild animals.

Since the pandemic outbreaks of avian influenza (H1N1) in 2009 and COVID-19 (SARS-CoV-2) in 2019, increased attention has been placed on government agendas regarding the regulation of animal breeding farms (wild and domestic) and markets for the sale of live animals or products (Arroyo-Quiroz, 2020). Green criminology studies report the need for increased regulation of animal breeding farms, exotic animal ownership and wildlife markets for the prevention of zoonotic diseases with epidemic and pandemic potential (Wong, 2019; Beirne, 2020; Arroyo-Quiroz, 2020; Joglekar, 2021; Carpio-Domínguez et al., 2021; Doody et al., 2021). Attention should focus primarily on illegal captive breeding of wildlife because the conditions of captivity (hygienic and safety) are not subject to sanitary regulations by governments or other institutions, and therefore, the pandemic potential is greatest in this criminal practice.

Final reflections and future perspectives

From green criminology, the illegal captive breeding of wildlife is a criminal practice that endangers biodiversity and ecosystems and is a factor that puts human health at risk. Furthermore, the illegal captive breeding of wildlife can be considered a speciesist practice, on the premise that speciesism is the human prejudice towards other species that leads to oppression, discrimination and exploitation for profit (Sollund, 2008). Speciesism is discrimination against those who do not belong to the human species; therefore, it is a systematic negative differentiation based on instrumentalist ideas about other nonhuman species; it is a harmful hegemonic ideological assumption that sustains human violence and the exploitation of wildlife (flora and fauna) (Horta, 2011; Goyes, 2019; Hutchinson, Stephens-Griffin & Wyatt, 2022). From the speciesism paradigm of green criminology, illegal captive breeding of wildlife is a practice in which human needs and interests are prioritised over the physical and reproductive health of species both in captivity and in wildlife populations.

The illegal captive breeding of wildlife is part of the wildlife trafficking chain and is linked to species laundering by acting as a supplier of specimens for legal and illegal wildlife. Furthermore, illegal captive breeding of wildlife promotes biodiversity loss, enhances inbreeding and chromosomal translocation in small captive populations, and is linked to risks of epizootic and zoonotic diseases. This review highlights that illegal wildlife breeding centres are not subject to regulation by the authorities. They thus pose greater biological, environmental, and human health risks because they often fail to meet minimum operational standards and regulatory requirements.

The study of illegal captive breeding of wildlife contributes to learning more about issues related to wildlife trafficking, as reported by other studies (Wyatt et al., 2020; Wyatt, 2022; Carpio-Domínguez et al., 2023). The illegal captive breeding of wildlife not only has a negative impact on the ecosystems due to the loss of (genetic) biodiversity, when specimens are extracted from their natural habitat, but also contributes to the formation and strengthening of criminal groups dedicated to wildlife trafficking that have found in the illegal captive breeding of wildlife a low-risk source of income.

It is important to intensify monitoring by authorities to identify illegal breeding farms. And the study of green crimes and harms should focus on the illegal captive breeding of wildlife to document this practice and better understand the internal processes of illegal breeding farms, the adverse environmental, biological, economic and human impacts, including criminological profiling to understand the motivations of those who illegally breed animals in captivity. This would help to propose strategies for government attention, law enforcement, and prevention of this practice. While illegal captive breeding of wildlife on farms has been documented, more attention should also be paid to illegal captive breeding of wildlife that takes place on a small scale in non-breeding settings (e.g. illegal breeding of exotic animals as pets by private individuals). These are important as they are often the cases that go unnoticed and tend to receive less government attention and consequently fewer safety and care protocols for breeding.

In addition to scientific research and the intensification of government inspections and operatives aimed at locating illegal breeding farms, environmental education efforts must be strengthened to prevent illegal wildlife breeding, prioritise animal welfare, and generate an awareness of prevention in people both to avoid acquiring illegal wildlife specimens and products, and to inform about the risks (criminal, administrative, biological, ecosystem and human–animal health) of the illegal captive breeding of wildlife and the illegal wildlife trade.

References

Arroyo Quiroz, I., Díaz-Rentería, P. y. and Puente-Guerrero, P. (2021). Comercio transatlántico de vida silvestre: un estudio de caso entre México y España. In Zúñiga-Rodríguez, L. y Ballesteros-Sánchez, J. (Eds.) *Nuevos desafíos frente a la criminalidad organizada transnacional y el terrorismo* (pp. 269–299). Madrid: Dykinson. https://digital.casalini.it/4958653

Arroyo-Quiroz, I. (2020). Mercados internacionales de vida silvestre en el contextoeckot del Covid-19. *Notas de Coyuntura del CRIM, 20*, 1–9. https://doi.org/10.22201/crim.001r.2020.20
Arroyo-Quiroz, I., Carpio-Domínguez, J. and Díaz-Rentería, L. (2021). Comercio y posesión de fauna silvestre en la industria de mascotas de México: Una perspectiva criminológica verde. In En Hernández-Jiménez, N. (Ed.) *Criminología y Sistema Penitenciario* (pp. 185–232). Colombia: Tirant Lo Blanch. https://ru.crim.unam.mx/handle/123456789/831
Arroyo-Quiroz, I. and Wyatt, T. (2019). Wildlife trafficking between the European Union and Mexico. *International Journal for Crime, Justice and Social Democracy, 8*(3), 23–37. https://doi.org/10.5204/ijcjsd.v8i3.1243
Auliya, M., Altherr, S., Ariano-Sanchez, D., Baard, E., Brown, C., Brown, R., Cantu, J.C., Gentile, G., Gildenhuys, P., Henningheim, E., Hintz- mann, J., Kanari, K., Krvavac, M., Lettink, M., Lippert, J., Luiselli, L., Nilson, G., Nguyen, T.Q., Nijman, V., Parham, J., Pasachnik, S., Pedrono, M., Rauhaus, A., Cordova, D., Sanchez, M.E., Schepp, U., Schingen, M., Schneeweiss, R., Sy, E., Turkozan, O., Vinke, S., Vinke, T., Vyas, R., Williamson, S. and Ziegler, T. (2016). Trade in live reptiles, its impact on wild populations, and the role of the European market. *Biological Conservation, 204*, 103–119. https://doi.org/10.1016/j.biocon.2016.05.017
Basel Institute on Governance. (2021). *Wildlife Crime-Understanding Risks, Avenues for Action*. Basel: Basel Institute on Governance. https://baselgovernance.org/publications/wildlife-crime-series
Beirne, P. (2020). Wildlife trade and COVID-19: Towards a criminology of anthropogenic pathogen spillover. *The British Journal of Criminology, 61*(3), 607–626. https://doi.org/10.1093/bjc/azaa084
Bezerra-Santos, M., Mendoza-Roldan, J., Thompson, A. and Dantas-Torres, D. (2021). Legal versus illegal wildlife trade: Zoonotic disease risks. *Trends in Parasitology, 37*(1), 360–361. https://doi.org/10.1016/j.pt.2021.02.003
Campbell, H., Carothers, A., Rudan, I., Hayward, C., Biloglav, Z., Barac, L., Pericic, M., Janicijevic, B., Smolej-Narancic, N., Polasek, O., Kolcic, I., Weber, J., Hastie, N., Rudan, P. and Wright, A. (2007). Effects of genome-wide heterozygosity on a range of biomedically relevant human quantitative traits. *Human Molecular Genetics, 16*(2), 233–241. https://doi.org/10.1093/hmg/ddl473
Cardoso, P., Amponsah-Mensah, K., Barreiros, J., Bouhuys, J., Cheung, H., Davies, A., Kumschick, S., Longhorn, S., Martínez-Muñoz, C., Morcatty, T., Peters, G., Ripple, W., Rivera-Téllez, E., Stringham, O., Toomes, A., Tricorache, P. and Fukushima, C. (2021). Scientists' warning to humanity on illegal or unsustainable wildlife trade. *Biological Conservation, 263*(2021), 109341. https://doi.org/10.1016/j.biocon.2021.109341
Carpio-Domínguez, J. (2023). Posesión de animales exóticos como mascotas en México. Una aproximación criminológica verde desde los reporrtes hemerográficos durante 2008–2018. *Constructos Criminológicos, 3*(5), 127–154. https://doi.org/10.29105/cc3.5-56
Carpio-Domínguez, J., Arroyo-Quiroz, I., Villarreal-Martínez, M. and Castro-Salazar, I. (2022). Tigers of Mexican drug traffickers, national statistical review and a case study through network analysis of the possession of a tiger (Panthera tigris) in northeastern Mexico. *Forensic Science International: Animals and Environments, 2*(2022), 100039. https://doi.org/10.1016/j.fsiae.2021.100039
Carpio-Domínguez, J., Vargas-Orozco, C., Meraz-Esquivel, M. and Villarreal-Sotelo, K. (2018a). Las redes sociales como factor criminógeno de la venta ilegaleckot de especies en Tamaulipas (México): el caso de Facebook. *CienciaUAT, 13*(1), 9–34. https://www.scielo.org.mx/scielo.php?script=sci_arttext&pid=S2007-78582018000200019
Carpio-Domínguez, J., Vargas-Orozco, C. and Villarreal-Sotelo, K. (2023). Redes de tráfico ilegal de vida silvestre en el noreste mexicano, un análisis desde la criminología verde. *Revista Española De Investigación Criminológica, 21*(2), e789. https://doi.org/10.46381/reic.v21i2.789
Carpio-Domínguez, J., Vargas-Orozco, C., Villarreal-Sotelo, K., Santillana-Cantú, R. and Hernández-Rodríguez, I. (2018b). Percepción Criminológica de la Posesión de Animales por el Narcotráfico en Tamaulipas, el Zoológico del Crimen Organizado. *Letras Jurídicas, 26*, 1–18. https://dialnet.unirioja.es/servlet/articulo?codigo=6954735
Carpio-Domínguez, J., Villarreal-Martínez, M. and Hernández-Jiménez, M. (2021). Posesión de animaleseckot exóticos y enfermedades zoonóticas: una aproximación social desde el contextoeckot mexicano. *Sociedad y Ambiente, 24*, e29. https://doi.org/10.31840/sya.vi24.2414
Castro-Salazar, I. and Carpio-Domínguez, J. (2023). Lavado de vida silvestre y forestaleckotal desde el discurso legal y la autoridad en México. *DIKE, 32*(2022), 1–32. https://www.apps.buap.mx/ojs3/index.php/dike/article/view/2474

Castro-Salazar, J. (2023). Estrategias legales e institucionales para asegurar y trasladar fauna silvestre a centros de rehabilitación. *Regiones y Desarrollo Sustentable, 23*(44), e2023. http://www.elcoltlax.edu.mx/openj/index.php/ReyDS/article/view/291

Castro-Salazar, J., Carpio-Domínguez, J. and Arroyo-Quiroz, I. (2022). Criminología verde, esfuerzo de aplicación de las regulaciones sobre vida silvestre en México entre el 2006 y 2020. *Constructos Criminológicos, 2*(2), 109–134. https://doi.org/10.29105/cc2.2-21

Centers for Disease Control and Prevention (CDC). (2016). *(zoonotic) Animals*. Global Health, Division of Parasitic Diseases and Malaria. https://www.cdc.gov/parasites/es/animals.html

Chapman, J., Nakagawa, S., Coltman, D., Slate, J. and Sheldon, B. (2009). A quantitative review of heterozygosity-fitness correlations in animal populations. *Molecular Ecology, 18*(13), 2746–2765. https://doi.org/10.1111/j.1365-294X.2009.04247.x

Chorney, S., DeFalco, A., Jacquet, J., LaFrance, C., Lary, M., Pirker, H. and Franks, B. (2022). Poor welfare indicators and husbandry practices at lion (Panthera Leo) "cub-petting" facilities: Evidence from public youtube videos. *Animals, 12*, 2767. https://doi.org/10.3390/ani12202767

Clutton-Brock, J. (1999). *A Natural History of Domesticated Mammals*. United Kingdom: Cambridge University Press. https://www.degruyter.com/document/doi/10.1515/pz-1990-0115/html

de Vries, S. (2022). Wildlife law and policy loopholes and canadian species of concern: Practitioners' perspectives. *Journal of International Wildlife Law and Policy, 25*(1), 62–83. https://doi.org/10.1080/13880292.2022.2077388

Diao, F. (2016). With four mentions of "utilization" at the start, the revised draft of the Wildlife Protection Law has attracted controversy. The Paper. https://eia-international.org/wp-content/uploads/Diao-Fanchao-With-four-mentions-of-utilisation-at-the-start1.pdf.

Doody, J., Reid, J., Bilali, K., Diaz, J. and Mattheus, N. (2021). In the post-COVID-19 era, is the illegal wildlife trade the most serious form of trafficking? *Crime Science, 10*, 19. https://doi.org/10.1186/s40163-021-00154-9

Environmental Investigation Agency (EIA). (2016). *Comments on draft implementation regulations for the revised law of the People"s Republic of China on the protection of wildlife*. EIA. https://eia-international.org/wp-content/uploads/EIA-comments-on-WPL-draft-implementation-regulations-bilingual.pdf.

Espejo-Cruz, A., Espejo-Martínez, A., Chávez-Ángeles, M., Lagunez-Rivera, L. and Solano, R. (2023). Deficiencies in compliance with environmental regulation for orchid trade via social networks in Mexico. *Botanical Sciences, 101*(2), 400–416. https://doi.org/10.17129/botsci.3159

Farquharson, K., Hogg, C. and Grueber, C. (2021). Offspring survival changes over generations of captive breeding. *Nature Communications, 12*, 3045. https://doi.org/10.1038/s41467-021-22631-0

Frantz, L., Bradley, D., Larson, G. and Orlando, L. (2020). Animal domestication in the era of ancient genomics. *Nature Review Genetics, 21*, 449–460. https://doi.org/10.1038/s41576-020-0225-0

Gbogbo, F. and Daniels, J. (2019). Trade in wildlife for traditional medicine in ghana: Therapeutic values, zoonoses considerations, and implications for biodiversity Conservation. *Human Dimensions of Wildlife, 24*(3), 296–300. https://doi.org/10.1080/10871209.2019.1605637

Gbogbo, F. and Opoku, K. (2017). Knowledge, perceptions and attitude of a community living around a colony of straw Coloured fruitBats (*Eidolon helvum*) in ghana after ebola virus disease outbreak in west Africa. *Zoonoses and Public Health, 64*, 628–635. https://pubmed.ncbi.nlm.nih.gov/28371424/

Gomez, L. and Nguyen, M. (2023). A rapid assessment of the illegal otter trade in Vietnam. *European Journal of Wildlife Research, 69*, 77, 69–77. https://doi.org/10.1007/s10344-023-01707-w

Gomez, L., Shepherd, C. and Khoo, M. (2020). Illegal trade of sun bear parts in the Malaysian states of Sabah and Sarawak. *Endangered Species Research, 41*, 279–287. https://doi.org/10.3354/esr01028

Gomez, L., Wright, B., Shepherd, C. and Joseph, T. (2021). An analysis of the illegal bear trade in India. *Global Ecology and Conservation, 27*, e01552. https://doi.org/10.1016/j.gecco.2021.e01552

Goyes, D. (2019). Ecological discrimination: A science to end discrimination and the science of the discriminated against. In Goyes, D. (Ed) *Southern Green Criminology: A Science to End Ecological Discrimination* (Perspectives on Crime, Law and Justice in the Global South) (pp. 13–20). Leeds: Emerald Publishing Limited. https://doi.org/10.1108/978-1-78769-229-920191002

Green, J., Jakins, C., Asfaw, E., Bruschi, N., Parker, A., de Waal, L. and D'Cruze, N. (2020). African iions and Zoonotic diseases: Implications for commercial lion farms in South Africa. *Animals, 10*(9), 1692. https://doi.org/10.3390/ani10091692

Guo, Y., Li, N., Feng, Y. and Xiao, L. (2021). Zoonotic parasites in farmed exotic animals in China: Implications to public health. *International Journal for Parasitology: Parasites and Wildlife*, *14*(2021), 241–247. https://doi.org/10.1016/j.ijppaw.2021.02.016

Harvey, R. (2020). Towards a cost-benefit analysis of South Africa's captive predator breeding industry. *Global Ecology and Conservation*, *23*, e01157. https://doi.org/10.1016/j.gecco.2020.e01157

Hedrick, P. (2011). Rapid decrease in horn size of bighorn sheep: Environmental decline, inbreeding depression, or evolutionary response to trophy hunting? *Journal of Heredity*, *102*(2), 770–781. https://doi.org/10.1093/jhered/esr082

Hilderink, M. and Winter, I. (2021). No need to beat around the bushmeat–The role of wildlife trade and conservation initiatives in the emergence of zoonotic diseases. *Heliyon*, *7*(29021), e07692. https://www.cell.com/heliyon/pdf/S2405-8440(21)01795-3.pdf

Hogg, C., Dennison, S., Frankham, G., Hinds, M. and Johnson, R. (2018). Stopping the spin cycle: genetics and bio-banking as a tool for addressing the laundering of illegally caught wildlife as 'captive-bred'. *Conservation Genetics Resources*, *10*, 237–246. https://doi.org/10.1007/s12686-017-0784-3

Horta, O. (2011). La cuestión de la personalidad legal más allá de la especie humana. *Isonomía. Revista de Teoría y Filosofía del Derecho*, *34*, 55–83. https://www.redalyc.org/pdf/3636/363635640004.pdf

Hrnková, J., Golovchenko, M., Sadiq, A., Needham, T., Italiya, J., Caecero, F., Kotrba, R., Grubhoffer, L., Rudenko, N. and Cerny, J. (2022). *Borrelia spirochetes* in European exotic farm animals. *Frontiers in Veterinary Science*, *9*(2022), 1–10. https://doi.org/10.3389/fvets.2022.996015

Hübschle, A. (2017). The social economy of rhino poaching: Of economic freedom fighters, professional hunters and marginalized local people. *Current Sociology*, *65*(3), 427–447. https://doi.org/10.1177/0011392116673210

Hübschle, A. and Gore, M. (2024). Lessons in resilience from the illegal wildlife trade during COVID-19 lockdowns. *Science of the Total Environment*, *916*, 170365. https://doi.org/10.1016/j.scitotenv.2024.170365

Hutchinson, A., Stephens-Griffin, N. and Wyatt, T. (2022). Speciesism and the Wildlife Trade: Who gets listed, downlisted and uplisted in CITES? *International Journal for Crime, Justice and Social Democracy*, *11*(2), 191–209. https://doi.org/10.5204/ijcjsd.1945

Joglekar, Y. (2021). Environmental crimes: Effect of COVID-19 on non-human victims. *Journal of Victimology and Victim Justice*, *4*(1), 33–40. https://doi.org/10.1177/25166069211031140

Keller, L. and Waller, D. (2002). Inbreeding effects in wild populations. *Trends in Ecology and Evolution*, *17*(5), 230–241. https://doi.org/10.1016/S0169-5347(02)02489-8

Kirkpatric, R. and Emerton, L. (2010). Killing tigers to save them: Fallacies of the farming argument. *Conservation Biology*, *24*(3), 655–659. https://doi.org/10.1111/j.1523-1739.2010.01468.x

Kishor, N. and Lescuyer, G. (2012). Controlling illegal logging in domestic and international markets by harnessing multi-level governance opportunities. *International Journal of the Commons*, *6*(2), 255–270. https://doi.org/10.18352/ijc.327

Klecel, W. and Martyniuk, E. (2021). From the Eurasian Steppes to the Roman circuses: A review of early development of Horse breeding and management. *Animals*, *11*(7), 1859. https://doi.org/10.3390/ani11071859

Knapp, A. and Affre, A. (2007). *A briefing paper on marking techniques used in the control of wildlife in the European Union*. A TRAFFIC Europe Report for the European Commission, Brussels, Belgium. https://www.traffic.org/site/assets/files/10076/marking-techniques-used-in-the-eu.pdf

Knief, U., Hemmrich-Stanisak, G., Wittig, M., Franke, A., Griffith, S., Kempanaers, B. and Forstmeier, W. (2015). Quantifying realized inbreeding in wild and captive animal populations. *Heredity*, *114*, 397–403. https://doi.org/10.1038/hdy.2014.116

Kuhnen, V., Remor, J. and Lima, R. (2012). Breeding and trade of wildlife in Santa Catarina state, Brazil. *Bazilian Journal of Biology*, *72*(1), 59–64. https://www.scielo.br/j/bjb/a/65ZktHj586Gz4xRn73jvJ3g/?format=pdf&lang=en

Laikre, L., Jansson, M., Allendorf, F., Jakobsson, S. and Ryman, N. (2013). Hunting effects on favourable conservation status of highly inbred Swedish wolves. *Conservation Biology*, *27*(2), 248–253. https://doi.org/10.1111/j.1523-1739.2012.01965.x

Lorenzo, C., Rioja, T., Carrillo, A., Bolaños, J., Sántiz, E. y. and Navarrete, D. (2018). Enfermedades zoonóticas virales emergentes. Importancia ecológica y su evaluación en el sureste de México. *Sociedad y Medio Ambiente*, *5*(15), 131–146. https://revistas.ecosur.mx/sociedadyambiente/index.php/sya/article/view/1791

Magliocca, N., Torres, A., Margulies, J., McSweeney, K., Arroyo-Quiroz, I., Carter, N., Curtin, K., Easter, T., Gore, M., Hübschle, A., Masse, F., Rege, A. and Tellman, E. (2021). Comparative analysis of illicit supply network structure and operations: Cocaine, wildlife, and sand. *Journal of Illicit Economies and Development, 3*(1), 50–73. https://nrl.northumbria.ac.uk/id/eprint/49236/1/76_1139_3_PB.pdf
Maher, J. and Sollund, R. (2016). Law enforcement of the illegal wildlife trafficking: a comparative strengths, weaknesses, opportunities and threats analysis of the UK and Norway. *Journal of Trafficking, Organized Crime and Security, 2*(1), 82–99. https://pure-test.southwales.ac.uk/ws/files/269627/55_193_1_PB.pdf
Mali, I., Vandewege, M., Davis, S. and Forstner, M. (2014). Magnitude of the freshwater turtle exports from the US: Long term trends and early effects of newly implemented harvest management regimes. *PloS ONE, 9*(1), e86478. https://doi.org/10.1371/journal.pone.0086478
Mazorrin, J. and Abatino, C. (2023). The exotic animal trade from Roman provinces. The results of animal remains study from the Roman town of Thamusida (Kénitra, Morocco). *Journal of Archaeological Science: Reports, 51*, 104116. https://doi.org/10.1016/j.jasrep.2023.104116
McPhee, M. (2004). Generations in captivity increase behavioral variance: considerations for captive breeding and reintroduction programs. *Biological Conservation, 115*(1), 71–77. https://doi.org/10.1016/S0006-3207(03)00095-8
Mishra, J., Mishra, P. and Arora, N. (2021). Linkages between environmental issues and zoonotic diseases: with reference to COVID-19 pandemic. *Environmental Sustainability, 4*, 455–467. https://doi.org/10.1007/s42398-021-00165-x
Morse, S., Mazet, J., Woolhouse, M., Parrish, C., Carroll, D., Karesh, W., Zambrana-Torrelio, C., Lipkin, W. and Daszak, P. (2012). Prediction and prevention of the next pandemic zoonosis. *Lancet, 380*(9857), 1956–1965. https://doi.org/10.1016/S0140-6736(12)61684-5
Moyle, B. (2017). Wildlife markets in the presence of laundering: A comment. *Biodiversity and Conservation, 26*, 2979–2985. https://doi.org/10.1007/s10531-017-1396-7
Musinguzi, J. and Bsheka, B. (2018). The efficacy of the legal framework in the management of wildlife trade in uganda. *Sabinet African Journals, 26*(4), 220–238. https://hdl.handle.net/10520/ejc-adminpub-v26-n4-a13
Naguib, M., Verhagen, J., Samy, A., Eriksson, P., Fife, M., Lundkvist, Å., Ellström, P. and Järhult, J. (2019). Avian influenza viruses at the wild–domestic bird interface in Egypt. *Infection Ecology & Epidemiology, 9*(1), 1–9. https://doi.org/10.1080/20008686.2019.1575687
Natusch, D. and Lyons, J. (2014). *Assessment of python breeding farms supplying the international high-end leather industry. A report under the 'Python Conservation Partnership' programme of research. Occasional Paper of the IUCN Species Survival Commission No. 50*. Gland, Switzerland: IUCN.
Nijman, V. and Shepherd, C. (2015). *Adding Up the Numbers. An Investigation into Commercial Breeding of Tokay Gekos in Indonesia*. Malaysia: Traffic Report. https://www.traffic.org/site/assets/files/6060/adding-up-the-numbers.pdf
Nilsson, T. (2004). Integrating effects of hunting policy, catastrophic events, and inbreeding depression, in PVA simulation: the Scandinavian wolf population as an example. *Biological Conservation, 115*(2), 227–239. https://doi.org/10.1016/S0006-3207(03)00120-4
Petry, M. and Liting, B. (2010). *China- Peoples Republic of Fur animals and products*. Global Agricultural Information Network Report CH10031. U.S. Department of Agriculture Foreign Agricultural Service. https://gain.fas.usda.gov/Recent%20GAIN%20Publications/Fur%20Animals%20and%20Products_Beijing_China%20-%20Peoples%20Republic%20of_5-25-2010.pdf.
Phelps, J., Carrasco, L. and Webb, E. (2013). A framework for assessing supply-side wildlife conservation. *Conservation Biology, 28*(1), 244–257. https://doi.org/10.1111/cobi.12160
Pickett, H. and Harris, S. (2015). *The case against fur factory farming: a scientific review of animal welfare standards and "WelFur"*. Respect for Animals, Fighting the International Fur Trade. https://www.furfreealliance.com/wp-content/uploads/2015/11/Case-against-fur-farming.pdf.
Rahman, M., Sobur, M., Islam, M., Ievy, S.; Hossain, M., El Zowalaty, M., Rahman, A. and Ashour, H. (2020). Zoonotic diseases: Etiology, impact, and control. *Microorganisms*, 8, 1405. https://doi.org/10.3390/microorganisms8091405
Rieder, E. (2023). *Using Participatory System Dynamics Modeling to Address Complex Conservation Problems: The Case of Tiger Farms*. [Thesis Degree of Doctor of Philosophy, Faculty of North Carolina State University]. Raleigh: North Carolina State University Repository. https://repository.lib.ncsu.edu/bitstream/handle/1840.20/41240/etd.pdf?sequence=1

Rizzolo, J. (2020). Wildlife Farms, Stigma and Harm. *Animals, 10*(10), 1783. https://doi.org/10.3390/ani10101783

Rozmiarek, M. and Włodarczyk, A. (2023). From british Menageries and hippodromes to the olympic circus in the grand duchy of posen: The origins of the use of ostriches in european sport and entertainment. *European Review, 31*(3), 291–307. https://doi.org/10.1017/S1062798723000054

Salas-Picazo, R., Ramírez-Bravo, E., Meza-Padilla, I. and Camargo-Rivera, E. (2023). The role of social media groups on illegal wildlife trade in four Mexican states: A year-long assessment. *Global Ecology and Conservation, 45*, e02593. https://doi.org/10.1016/j.gecco.2023.e02539

Sardari, P., Felfelian, F., Mohammadi, A., Nayeri, D. and Davis, E. (2022). Evidence on the role of social media in the illegal trade of Iranian wildlife. *Conservation Science and Practice, 4*(7), e12725. https://doi.org/10.1111/csp2.12725

Sarvi, S., Daryani, A., Sharif, M., Rahimi, M., Kohansal, M., Mirshafiee, S., Siyadatpanah, A., Hosseini, S. and Gholami, S. (2018). Zoonotic intestinal Parasites of Carnivores: A systematic review in Iran. *Veterinary World, 11*(1), 58–65. https://www.ncbi.nlm.nih.gov/pmc/articles/PMC5813513/

Sollund, R. (2008). Causes for speciesism: Difference, distance and denial. In R. Sollund (Ed.) *Global arms: Ecological crime and speciesism* (pp. 109–131). Nova Science. London: Routledge.

Sollund, R. and Runhovde, S. (2020). Reponses to wildlife crime in post-colonial times. Who fares best? *British Journal of Criminology, 60*, 1014–1033. https://doi.org/10.1093/bjc/azaa005

Sosa-Escalante, J. (2011). Aplicación de la ley para el combate del tráfico eckot de vida silvestre en México: El caso de Charco Cercado. *Therya*,(3), 245–262. https://www.scielo.org.mx/pdf/therya/v2n3/v2n3a5.pdf

South, N. and Wyatt, T. (2011). Comparing illicit trades in wildlife and drugs: An exploratory study. *Deviant Behavior, 32*(6), 538–561. https://doi.org/10.1080/01639625.2010.483162

Tamalene, N. and Kartika, S. (2019). Local knowledge and community behavior in the exploitation of parrot in surrounding area of Aketajawe Lolobata National Park. *Biosfer: Jurnal Pendidikan Biologi, 12*(1), 24–33. https://journal.unj.ac.id/unj/index.php/biosfer/article/view/10435/7097

Teichmann, F. (2020). Recent trends in money laundering. *Crime Law and Social Change, 73*, 237–247. https://doi.org/10.1007/s10611-019-09859-0

Tensen, L. (2016). Under what circumstances can wildlife farming benefit species conservation? *Global Ecology and Conservation, 6*, 286–298. https://doi.org/10.1016/j.gecco.2016.03.007

The Guardian. (March 11, 2014). *Canned hunting' of white lions is despicable – and it must stop*. The Guardian. Retrieved from https://www.theguardian.com/environment/blog/2014/mar/11/canned-hunting-white-lions-despicable

United Nations. (2016). *Combating illegal wildlife trafficking. Report of the Secretary General*. https://documents-dds-ny.un.org/doc/UNDOC/GEN/N16/283/40/PDF/N1628340.pdf?OpenElement

United Nations. (2020). [Spanish] *UNODC Global Report on Wildlife and Forest Crime 2020: The COVID-19 pandemic has shown that these crimes threaten not only the environment and biodiversity, but also human health*. United Nations. https://www.unodc.org/lpomex/es/noticias/julio-2020/07_10_WLFC_UNODC.html

United Nations Office on Drugs and Crime (UNODC). (2020). *World Wildlife Crime Roport. Trafficking in Protected Species*. Vienna: United Nations. https://www.unodc.org/documents/data-and-analysis/wildlife/2020/World_Wildlife_Report_2020_9July.pdf

Urrego, D., Pérez-Lara, M., Polanco, C. and Pérez-Poveda, M. (2022). Control del aprovechamiento ilegaleckot de flora a partir de la anatomía de siete especies maderables en estado de amenaza. Aportes a la política nacional ambientaleckotal. *Revista Criminalidad, 54*(1), 259–281. https://biblat.unam.mx/hevila/Revistacriminalidad/2012/vol54/no1/5.pdf

van Uhm, D. (2016). *The Illegal Wildlife Trade. Inside the World of Poachers, Smugglers and Traders*. Switzerland: Springer International Publishing. https://doi.org/10.1007/978-3-319-42129-2

van Uhm, D. (2018a). Wildlife and laundering: Interaction between the under and upper world. In Spapens, T., White, R., van Uhm, D. and Huisman, W. (Eds.) *Green Crimes and Dirty Money* (pp. 197–214). London: Routledge. https://doi.org/10.4324/9781351245746

van Uhm, D. (2018b). Wildlife Crime and Security. In Reichel, P. and Randa, R. (Eds) *Transnational Crime and Global Security* (pp. 73–96). United States: Praeger. https://api.pageplace.de/preview/DT0400.9781440843181_A31981018/preview-9781440843181_A31981018.pdf

van Uhm, D. and Moreto, W. (2018). Corruption within the illegal wildlife trade: A symbiotic and antithetical enterprise. *The British Journal of Criminology, 58*(4), 864–885. https://doi.org/10.1093/bjc/azx032

Wang, W., Yang, L., Wronski, T., Chen, S., Hu, Y. and Huang, S. (2019). Captive breeding of wildlife resources-China"s revised supply-side approach to conservation. *Wildlife Society Bulletin, 43,* 425–435. https://doi.org/10.1002/wsb.988

Warchol, G. (2004). The Transnational illegal wildlife trade. *Criminal Justice Studies, 17*(1), 57–73. https://doi.org/10.1080/0888431042000l679334

Welfare, W. and Wright, E. (2016). Planning for the unexpected: Ebola virus, Zika virus, what"s next? *British Journal of Hospital Medicine, 77*(12), 704–707. https://doi.org/10.12968/hmed.2016.77.12.704

Whitfort, A. (2021). COVID-19 and wildlife farming in china: Legislating to protect wild animal health and welfare in the wake of a global pandemic. *Journal of Environmental Law, 33*(1), 57–84. https://doi.org/10.1093/jel/eqaa030

Widi, T., Udo, H., Oldenbroek, K., Budisatria, I., Baliarti, E. and van der Zijpp, A. (2021). Designing genetic impact assessment for crossbreeding with exotic beef breeds in mixed farming systems. *Outlook on Agriculture, 50*(1), 34–45. https://doi.org/10.1177/0030727020915206

Williams, V., Loveridge, A., Newton, D. and Macdonald, D. (2017). A roaring trade? The legal trade in Panthera leo bones from Africa to East-Southeast Asia. *PLoS ONE, 12*(10), e0185996. https://journals.plos.org/plosone/article?id=10.1371/journal.pone.0185996

Williams, V., Newton, D., Loveridge, A. and Macdonald, D. (2015). Bones of Contention: An Assessment of the South African trade in African Lion Panthera Leo Bones and Other Body Parts. United Kingdom: TRAFFIC. https://policycommons.net/artifacts/1935592/bones-of-contention/2687362/

Williams, V. and 't Sas-Rolfes, M. (2019). Born captive: A survey of the lion breeding, keeping and hunting industries in South Africa. *PLoS ONE, 14*(5): e0217409. https://doi.org/10.1371/journal.pone.0217409

Willoughby, J., Ivy, J., Lacy, R., Doyle, J. and De Woody, J. (2017). Inbreeding and selection shape genomic diversity in captive populations: Implications for the conservation of endangered species. *PloS ONE, 12*(4), e0175996. https://doi.org/10.1371/journal.pone.0175996

Wong, R. (2019). *The Illegal Wildlife Trade in China. Understanding The Distribution Networks*. London: Palgrave Macmillan. https://doi.org/10.1007/978-3-030-13666-6

Woodworth, L., Montgomery, M., Briscoe, D. and Frankham, R. (2002). Rapid genetic deterioration in captive populations: Causes and conservation implications. *Conservation Genetics, 3,* 277–288. https://doi.org/10.1023/A:1019954801089

World Bank. (2019). *Illegal logging, fishing and wildlife trade: The costs and how to combat it*. World Bank. https://thedocs.worldbank.org/en/doc/482771571323560234-0120022019/original/WBGReport1017Digital.pdf

Wyatt, T. (2021). Canada and the convention on international trade in endangered species of wild fauna and flora (CITES): Lessons learned on implementation and compliance. *Liverpool Law Review, 42,* 143–159. https://doi.org/10.1007/s10991-020-09267-8

Wyatt, T. (2022). *Wildlife Trafficking. A Deconstruction of the Crime, Victims and Offenders*. London: Palgrave Macmillan. https://doi.org/10.1007/978-3-030-83753-2

Wyatt, T. and Cao, A. (2015). *Corruption and wildlife trafficking*. U4 Anti-Corruption Resource Centre. https://open.cmi.no/cmi-xmlui/handle/11250/2475047

Wyatt, T., Johnson, K., Hunter, L., George, R. and Gunter, R. (2018). Corruption and wildlife trafficking: Three case studies involving asia. *Asian Criminology, 13,* 35–55. https://doi.org/10.1007/s11417-017-9255-8

Wyatt, T., van Uhm, D. and Nurse, A. (2020). Differentiating criminal networks in the illegal wildlife trade: organized, corporate and disorganized crime. *Trends in Organized Crime, 23,* 350–366. https://doi.org/10.1007/s12117-020-09385-9

Yu, F. (2016). *What conservationists expect from a Wildlife Protection Law*. https://eia-international.org/wp-content/uploads/Yu-Fengqin-What-conservationists-expect-from-a-Wildlife-Protection-Law2.pdf.

Zinsstag, J., Utzinger, J., Probst-Hensch, N., Shan, L. and Zhou, X. (2020). Towards integrated surveillance-response systems for the prevention of future pandemics. *Infectious Diseases of Poverty, 9*(1), 140. https://doi.org/10.1186/s40249-020-00757-5

20

SHARK FIN TRADE IN HONG KONG

Forensic and investigative solutions

Hannah Bethany Tilley
Astrid A. Andersson

Introduction

Hong Kong SAR ('Hong Kong') is a major hub for wildlife trafficking (Lau, 2014; Andersson et al., 2021). Wildlife is traded here for local consumption, as well as re-export, as various products including ornaments, exotic pets, traditional medicine, fashion, and food (Mozer & Prost, 2023). Hong Kong is one of the world's largest consumers and trade ports of seafood (Clark, 2004a; Thomas, 2011). The high export value and sizable local demand mean wild stocks for species traded as seafood are decreasing worldwide, including live reef fishes, abalone, bêche-de-mer, fish maws, and shark fins (Clarke, 2004b; Lau, 2018; Purcell et al., 2018; Sadovy de Mitcheson et al., 2019). Due to the volume of luxury seafood trade in the city, illegally harvested species are often traded illegally or laundered in legal markets (Sadovy, 2010; Wu & Sadovy, 2016; Shea & To, 2017; Sadovy de Mitcheson et al., 2018; Lau, 2018). Hong Kong is one of the largest trade territories for shark fin, a luxury seafood commodity, as well as being a transit hub for one of the largest consumers, mainland China (Vannuccini, 1999). It is unsurprising, therefore, that Hong Kong has been previously recorded to trade the most shark fin globally (Clark et al., 2006b; Shea & To, 2017). The volume of shark fin imported, exported, and re-exported from Hong Kong means that it is challenging to monitor and regulate this trade, and as a result, much of the trade includes endangered or illegal shark species (Clarke et al., 2006a; Cardeñosa et al., 2018a, 2019). This illegal and unsustainable trade in shark fin is threatening sharks worldwide due to the difficulties associated with enforcing and regulating this trade (Worm et al., 2024).

In this chapter, we provide an overview of the trade of shark fin-related products in Hong Kong and explore how crimes committed against sharks are being addressed. We explore the environmental importance of sharks and the cultural significance of fin consumption in Hong Kong. We then discuss legal issues with shark trade, and forensic and investigative scientific methods to aid enforcement in regulating the shark fin trade.

DOI: 10.4324/9781003333906-24

Sharks: their importance and threats

Elasmobranchii is a large subclass of cartilaginous predatory fish which encompasses sharks, skates, rays, chimaeras, and sawfish (Nair & Zacharia, 2015). There are >950 species of elasmobranchs with representatives in every type of environment, from freshwater estuaries to marine deep-sea trenches and in polar and tropical ecosystems (Camhi et al., 1998).

Large-bodied predators, such as sharks, are vital in ecosystems for food web management. They regulate prey dynamics through direct predation, consuming smaller or weaker animals, and indirectly through modifying prey behaviour (Roff et al., 2016). For example, tiger sharks in Western Australia have positive impacts on seagrass beds by modifying turtle movement and grazing (Heithaus et al., 2007). Large, un-predated turtle populations have the potential to overgraze seagrass leading to habitat destruction and elimination of other species (Heithaus et al., 2014; Gangal et al., 2021). The removal of sharks by overfishing, by-catch, fin trade, and destruction of marine ecosystems can result in community-wide trophic cascades and may have irreversible effects on oceanic systems and the human communities that rely on these systems (Myers et al., 2007; Heithaus et al., 2008).

Widespread exploitation of sharks has occurred over the past few decades (Rose, 1996; Lack, 2006), resulting in the dramatic global decline of shark species (Baum & Myers, 2004; Ferretti et al., 2008). Rising demand for shark products has led to increased fishing of adults, resulting in reduced numbers of offspring (Momigliano & Harcourt, 2014). Sharks are particularly susceptible to exploitation as they have long life histories and produce few offspring (Last & Stevens, 2009). It is estimated that only one-third of species are currently considered safe from extinction (Dulvy et al., 2014), with Chondrichthyes (sharks, rays, and chimaeras) being one of the most threatened vertebrate classes assessed on the International Union for Conservation of Nature (IUCN) Red List (Dulvy et al., 2021).

The rate of shark catch is often unsustainable, with many populations now threatened through overfishing (Clarke, 2006b; Worm, 2013, 2024). In 2010, the global catch of elasmobranchs was ~1.8 million tonnes (Sadovy et al., 2018), significantly contrasting with the sustainable catch estimates of 200,000 tonnes annually (Simpfendorfer & Dulvy, 2017). As a result, more shark species are now being listed as species of conservation concern on the Convention on International Trade in Endangered Species of Wild Fauna and Flora (CITES) appendices[1] (Figure 20.1, Sadovy et al., 2018).

Elasmobranchs are commonly grouped together under the general term 'shark' or 'shark fin' in the trade; therefore, we will discuss shark fins from this perspective (unless specified). Shark meat, fins, oil, cartilage, and jaws are diverse in their usage and value (Rose, 1996; Hareide et al., 2007). The largest growing threat to sharks is the increasing demand for shark fins, with an estimated 26–73 million sharks traded for their fins annually before 2006 (Clarke et al., 2006b). Shark fins are mainly harvested for use in shark fin soup, a delicacy historically popular among the Chinese elite class (Clarke et al., 2007; Fabinyi, 2012). This dish is still regularly consumed on significant occasions, e.g. at business functions, Lunar New Year, and wedding banquets (Cheung & Chang, 2011). The materials consumed in this dish are anatomically known as 'ceratotrichia' (shark fin needles) and are made of collagen (Kemp, 1977). Traditional medicine suggests that consuming these needles provides health benefits, such as extended lifespan and increased virility (Vannuccini, 1999). However, recent studies suggest that shark products may be unhealthy due to the accumulation of heavy metals such as mercury and neurotoxins within shark tissues (Choy & Wainwright, 2022; García Barcia et al., 2023), substances which are known to pose health risks to humans (World Health Organization, 2018; 2024).

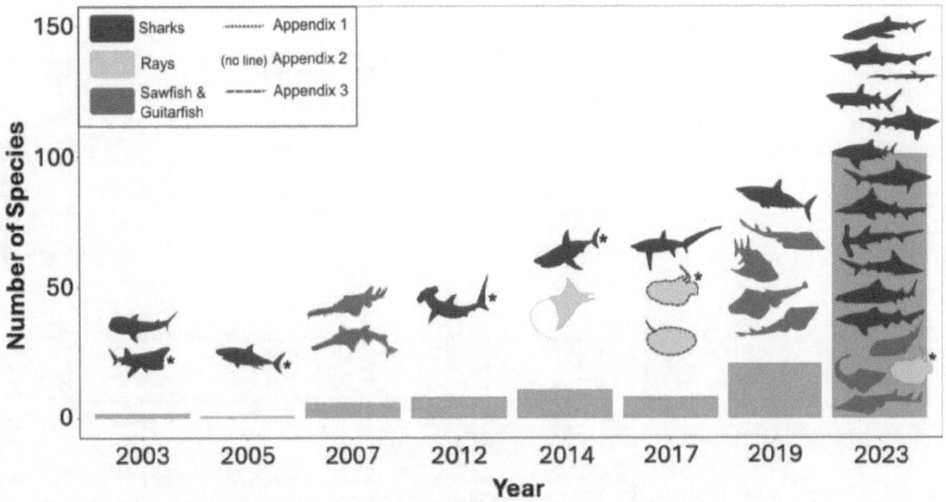

Figure 20.1 Elasmobranchs genera listed on CITES by year.

Silhouettes depict one representative species from each of the different genera listed in CITES that year. Black: sharks (superorder Selachimorpha), dark grey: sawfish & guitarfish (order Rhinopristiformes) and light grey: rays (order Myliobatiformes). Dotted outline: Appendix I species, no outline: Appendix II species, dashed outline: Appendix III species. Asterisks (*) indicates where CITES-listing changed or was listed at different times for different countries. For further details, please see Notes: Appendix Table 1.

Despite potential health risks, total fishing mortality for sharks increased from 76 to 80 million from 2012 to 2019 (Worm et al., 2024). As the market demand increases, the volume of this trade is becoming increasingly unsustainable (Camhi et al., 1998; Clarke et al., 2007). Globally, the largest markets and consumption centres are in Asia, predominantly in mainland China and Hong Kong (Cardeñosa et al., 2020b). However, Hong Kong lies at the centre of the trade, accounting for 50–85% of the global exports (Clarke, 2004a, 2006b).

Shark fin trade in Hong Kong

Hong Kong has historically contributed ~50% of dried shark fin trade volumes globally (Fong & Anderson, 2000; Shea & To, 2017). The industry boomed between 1960 and 1990s, during which time volumes increased six-fold (Sadovy et al., 2019). In the early 2000s, the industry started to decline, attributable to an overall reduction in shark populations caused by unsustainable harvests, transport bans by multiple private companies, public awareness campaigns, CITES listings for sharks, and governmental removal of shark fin from official functions (Vallianos et al, 2018; Eriksson & Clarke, 2015; Fabinyi & Liu, 2014). Imports then recovered slightly, increasing to 5,700 and 6,000 tonnes in 2015 and 2016, respectively – over 90% coming in by sea and the rest by air (CSD, 2022). These volumes alone are staggeringly large, but still only constitute 50% of what was annually imported during Hong Kong's shark fin industry's heyday in the late 1990s (Sadovy et al., 2018). According to the Hong Kong Census & Statistics Department and CITES trade data combined, total legal imports reached 189,823 tonnes between 1998 and 2023 (Figure 20.2). After commercially important elasmobranch species began to be listed on CITES in 2014, Hong Kong saw 11,242 kg of illegal trade in shark fin coming through its borders, with the highest amount included in 2019, and the most commonly identified sharks being hammerhead species,

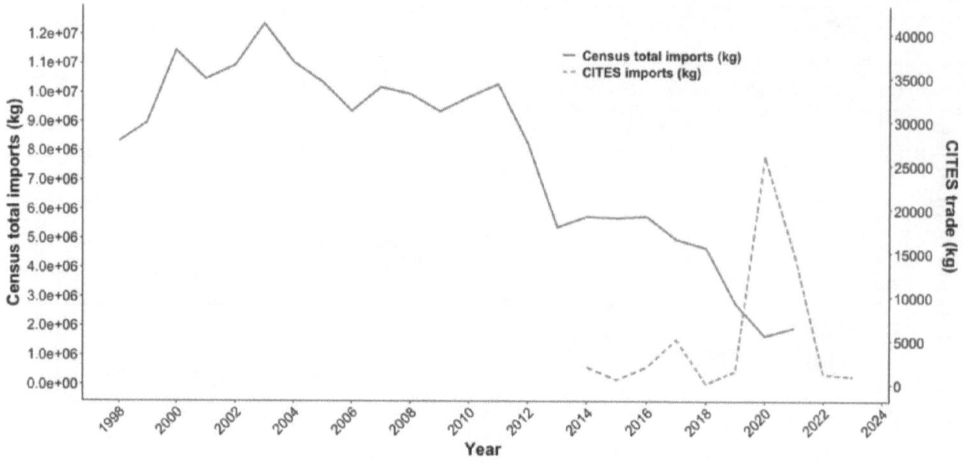

Figure 20.2 Legal trade in shark fins in Hong Kong.

As reported by the Hong Kong Census & Statistics Department (Census total imports (kg); CSD, 2022) and listed under the CITES trade database (CITES trade (kg), CITES Trade Database, 2022) by year.

which are restricted from trade (HKSAR press release, 2020). However, true trade volumes are likely much more than this, considering different data sources.

Hong Kong retains its position as one of the top global importers of shark fins, and ~80% of celebratory banquet menus still include shark fin (Shea & Ho, 2014; Ho & Shea, 2021). However, studies have shown consumption may be shifting (ADMCF et al., 2018) with other regional hubs emerging such as Macau and Vietnam (Vallianos et al., 2018; Sadovy et al., 2019). A huge amount of shark fins consumed in Hong Kong are legally imported, with Hong Kong recorded as the world's top legal importer of fins from CITES-listed sharks (Cardeñosa et al., 2018a). However, black market trade operates in parallel, with customs regularly seizing large shipments of undeclared and illegally harvested/traded shark fins, with most of this representing unsustainable harvest (Vannuccini, 1999). For example, in 2017, Hong Kong Customs seized a total of 2,646 kg of oceanic white-tip and hammerhead dried shark fin from over nine countries (Sadovy et al., 2018). Much of this harvesting also goes largely unmonitored, with >50% of shark fins in Hong Kong sourced from oceanic species found in the high seas, in vast areas of international jurisdiction (Clarke et al., 2006b; Fields et al., 2018).

After the most recent 19th meeting of the Conference of the Parties (CoP19) in Panama (2022), many new species were proposed for listing, and as of 2023, CITES now regulates 171 species of sharks and rays (Figure 20.1). Under CITES, Hong Kong is a dependent territory to China, which is a party to CITES. Within this party status, Hong Kong law Cap. 586 regulates shark fin trade under CITES. As such, only those species listed on CITES that are traded with permits are allowed entry into the local market. However, in practice, the number of species listed on CITES, as well as the sheer volume of fins traded into and out of Hong Kong, makes this incredibly challenging to monitor. Increasing the number of species regulated under CITES means that trading them will become more difficult due to the additional administration processes involved, including permits required by law for import, export, and re-export, as well as adhering to new quotas and regulations that traders are required to follow. However, it also means that enforcement has more species to regulate, and capacity needs to keep up to date, in terms of knowledge and forensic traceability methods. Put simply, the resources and capacity of enforcers will need to increase alongside the number of additional species they are required to monitor and regulate trade of, otherwise the additional regulations are futile.

Tackling crimes against sharks in Hong Kong

Wildlife forensic and investigative science

Scientific techniques can be leveraged to support law enforcement through various methods of analysis of biological life. Wildlife forensic science is an applied discipline of forensic science that concerns nonhuman biological species and provides analysis to support law enforcement. This typically differs from pure research or even investigative/traceability tools by means of validating and standardising the methods of analysis, providing quality management systems, having a stringent chain of custody, and being driven by investigative questions (SWFS Technical Working Group, 2018). Traceability methods can aid enforcement by gathering information on a case or about a trade, e.g. to assess trade patterns, but without the above measures, these methods may not be suitable for supporting evidence in court (Ogden et al., 2009).

The most common questions asked in wildlife forensic and investigative applications include: What is the specimen? Where did it come from? What individual did it come from? (Ogden et al., 2009). By far the most common question is that of species identification. This is particularly important when dealing with crimes against laws that enact CITES. However, a species that has been processed for the consumer market can be very difficult to identify, i.e. if it is not in its raw, natural state, but has been modified into pieces, powder, or medicine, depending on the final consumer product. See the book by Huffman and Wallace (2012) for more information.

Forensic and investigative analysis of shark fins

In the most recent CITES CoP19, proposals were submitted for all species within two shark families to be listed on CITES. Partially because of DNA research conducted in Hong Kong (Cardeñosa et al., 2017, 2018b, 2019, 2020b), it was decided that 54 species all from the same family of requiem sharks (Carcharhinidae) would be listed on the CITES App. II in CoP19, due to their 'look-alike' nature with other protected requiem sharks (CoP19 Prop. 37, 2022). These look-alikes closely resemble the most traded critically endangered and endangered requiem species, and to minimise the risk of misidentification and therefore mis-regulation, all were listed to ensure effective trade control. This listing came into force in November 2023, so a total of 56 requiem sharks are now listed on CITES App. II. Furthermore, 37 species of guitarfishes were listed under CITES App. II, along with the remaining species within the hammerhead shark family (Sphyrnidae, to now include all ten species). With many species in these families being traded into Hong Kong (Cardeñosa et al., 2017, 2018b, 2019, 2020b), this means that scientific tests will need to be updated alongside the new CITES listings. Technically, for these new listings, only identification of sharks to family level is required to prove that a crime has been committed, with the process of elimination likely being the easiest method (Shea, *pers. comms.*).

Identifying regulated species in the trade is important for Hong Kong law enforcement to ensure that a crime has not been committed. Most CITES listings are species, genus, or family-specific, apart from the rare population-specific listings (e.g. some Potamotrygon ray species in South America). Shark fins are typically traded detached from the shark's body; they are often imported dried and unprocessed, and sold without their skins, and bleached, in retail outlets. Although products from the same species or origin may be traded together, these changes to shape, size, colour, tissue, and DNA of the fin create difficulties when identifying the species of origin. To assess whether products are regulated species or not, various forensic and investigative scientific tests can be implemented. The most commonly used methods for investigative and forensic applications are DNA and morphology-based, but others such as the use of stable isotopes, x-ray fluorescence (XRF), and elemental analysis are also used, with the method often changing depending on the investigative question, suspected taxa involved, and other data available for the analysis to be successful.

Morphology taxonomic identification

First responders, customs officers, police, prosecutors, and law enforcement officials are often non-biological specialists. Therefore, the easiest method (cost-efficient, technical, and capacity-wise) to identify traded wildlife products is through visual, morphological guides. Identifying species using morphological methods depends on the state of that specimen, e.g. specimens with key-taxonomic features on show are easier to identify than dried shark fins. Even then, many fins are morphologically similar between species, making taxonomic identification difficult and only reliable for a few species.

Morphological shark identification guides have been developed for use by government and customs officials globally. In Hong Kong, NGOs conduct capacity building workshops to help local and international enforcement identify species of shark fins, focusing on fin shape, body ratio, colouration, distinctive traits, and size (BLOOM Association Hong Kong, 2016). Similarly, morphological guides recommended by CITES use the external skin markings and fin shape to identify species, and focus on easily identifiable skin markings (e.g. using the white tip of a dorsal fin to identify an oceanic white-tip shark) (Seret, 2006; Abercrombie & Chapman, 2012; Ahmad & Lim, 2012, https://citessharks.org/resources). Following the need for rapid identification and to minimise human error in the identification process, advanced computational techniques, such as artificial intelligence (AI), have been used to aid enforcement in shark fin identification. These include using taxonomic differences between the shape of shark denticles (small 'teeth' covering shark skin) to distinguish between certain shark species (Sheng, 2020), and the use of smartphone applications such as iSharkFin (FAO, 2020) and Fin Finder (CITES News, 2022). These tools and guides are useful for on-site identification at the point of landing if data is available for a particular species, and if the identifiable part of the fin is visually intact. The problem arises when the fin is processed, in pieces, or devoid of skin (typical in the trade). Although recent genetic tools can provide rapid diagnosis of fins from small cuttings (processed fins) (Cardeñosa et al., 2017), further research is needed to morphologically identify processed shark fins to enhance traceability. Many of the AI-morphology methods are still under development and not yet validated for forensic casework in court. Other limitations include that, currently, only morphological differences between primary fins (first dorsal fin, pectoral fins, and caudal fin) of sharks and guitarfish can be ascertained through guides (Jabado & Morata, 2022). Furthermore, fins may change morphologically when processed, making it nearly impossible to identify to a taxonomically informative level. However, traditional visual identification has been key in aiding on-site identification at the point of landing.

DNA taxonomic identification

In instances where fins cannot be reliably identified visually, genetic analysis can help identify shark species. The use of genetic analysis in taxonomic identification is a commonly used method utilising segments of DNA to compare organisms. Scientists can analyse sequences of DNA to distinguish individuals, populations, species, or even higher taxonomic levels. DNA analysis can also be used to test for mislabelling (Hobbs et al., 2019), providing a rapid assessment of seizures (Cardeñosa et al., 2017), screening markets for illegal species (Cardeñosa et al., 2019), or to identify a product suspected of being CITES-listed as evidence in a case.

DNA barcoding methods – using short, standardised DNA segments to identify specimens – are particularly useful, rapid, and cost-efficient, and can identify shark products along the supply chain (Shivji et al., 2002; Chapman et al., 2003; Abercrombie et al., 2005; Caballero et al., 2012; Fields et al., 2015). Shark fin products that are processed or modified often have degraded DNA that can be difficult to analyse. Recent work has looked to overcome this, using shorter fragments of the genome via creating mini-DNA barcoding assays (Cardeñosa et al., 2017). This work has since been utilised to assess the species diversity between markets in Hong Kong and Guangzhou using shark fin trimmings, finding many CITES-listed species being sold openly across both locations (Cardeñosa et al., 2020b).

When confiscations are made by enforcement, rapid turnarounds for evidence analysis are often required (e.g. <24 hours). Recent work in Hong Kong and Indonesia has led to the development and implementation of real-time PCR (rtPCR) methods to identify shark species by increasing the speed of detection and flexibility of location, as analyses can be conducted outside of the laboratory (Cardeñosa et al., 2018b; Prasetyo et al., 2023). In Hong Kong, this technique has been used to enhance the capacity for enforcement to detect illegally traded shark fin products, detect CITES-listed species outside of the government's morphological identification guides, and provide a near real-time genetic analysis detection of species' fins after initial inspection. With new species added to CITES, future genetic studies are underway to update DNA tests accordingly (Shea, *pers. comms.*). Many of these studies provide good examples of international collaboration between scientists, NGOs, and the government in aiding enforcement efforts (Figure 20.3).

Recommendations to advance wildlife forensic capacity in HK

Hong Kong is a pioneer in the research and development of tools and methods to help tackle the illegal shark fin trade. In recent years, there have been considerable investments into method development and training to improve the quality of existing inspection and enforcement capacity, including workshops to train enforcement in visual shark fin identification by local NGOs (Shea, *pers. comm.*) and rapid species identification methods developed by scientists and implemented by government (Cardeñosa et al., 2018b).

Such methods for identifying shark fins help enforcement to assess species at the point of landing, borders, or in markets. However, to be used as evidence in court, these methods need to meet certain standards. This includes using appropriate methods, e.g. in genetic analyses: a suitable DNA 'marker' (segment of genome), or in morphological analyses: a

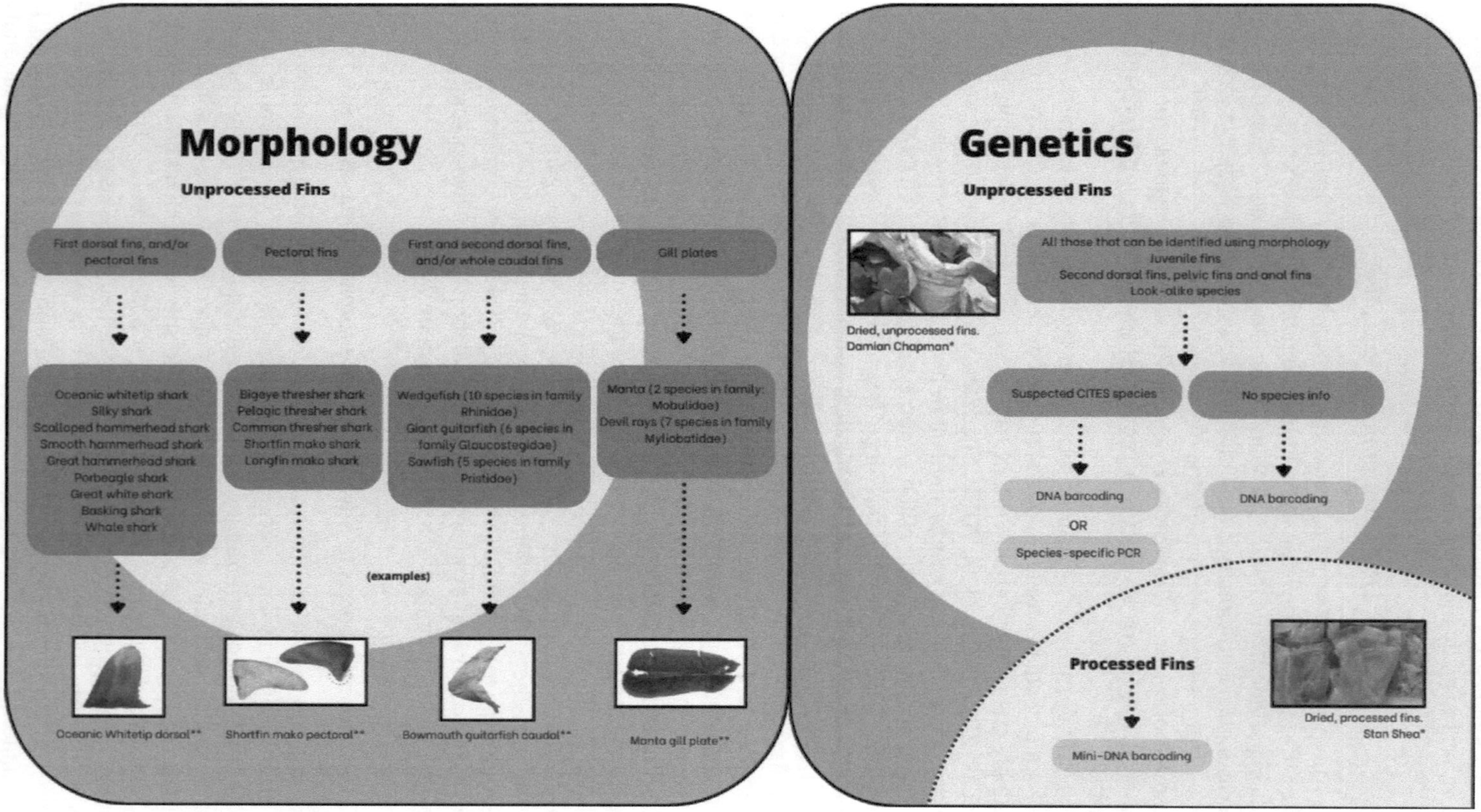

Figure 20.3 Current wildlife forensic and investigative solutions for shark fin products. Adapted from: (Abercrombie et al., 2018; Abercrombie & Jabado, 2022).

suitable morphological identifier (Trail, 2021); ensuring sufficient capacity for conducting analyses; and following comprehensive reference databases (Ogden et al., 2009; SWFS Technical Working Group, 2018). The marker/identifier of choice will depend on the question (e.g. is this shark fin from a CITES-regulated species or family?) and information/research available (e.g. reference databases might be scant for less commonly studied species). For methods to be admissible in court, they should be validated to meet certain scientific standards, ensuring the test's specificity, sensitivity, reproducibility, and robustness (e.g. Hatten et al., 2023; Webster et al., 2024).

It is important that tests are standardised and applicable for a range of species or taxa and can be implemented worldwide. For example, an average of 16,177 Mt of shark fin was imported globally per year from 2000 to 2016 (Okes & Sant, 2019), with Hong Kong importing from an average of 83 nations (Shea & To, 2017). We recommend that, as far as possible, forensic/investigative tests and databases be standardised globally, and that in-country wildlife forensic capacity be built. This is important to ensure that tests are applicable across jurisdictions, laboratories, and personnel. Though scientific facilities exist globally, it is important to ensure special measures are exercised for forensic tests to be admissible in court, e.g. maintaining the chain of custody of evidence, and having appropriate quality management systems in place (SWFS Technical Working Group, 2018). Many different types of facilities are able to conduct forensic or investigative analyses if they keep the appropriate requirements (Ogden & Mailley, 2016).

Unlike many other countries, Hong Kong does not have a dedicated wildlife forensic laboratory, and forensic case work is tendered out by the government to competing institutions (Hatten et al., 2020). Instead, NGOs and academics work with the Hong Kong government to develop and implement scientific research tools to identify shark fins coming through Hong Kong's borders. Therefore, further work is essential to support new CITES listings of elasmobranchs and to ensure the methods for shark fin identification meet the standards for forensic use in court, to support sentencing.

Conclusion – going further to protect sharks

The past decade has seen a concerted effort to regulate the shark fin trade and mitigate its impact on dwindling shark populations. The number of shark species receiving protection and trade quotas under CITES has increased substantially, from just 5 shark and 22 ray species in 2014 (the first year a commercially important shark species was listed), to 171 species of sharks and rays in 2023. However, sharks will not reap the benefits of this increased protection unless we also provide tools to confidently identify species, increasing support and capacity for law enforcement. The recent advancement of shark fin identification techniques and regulations is demonstrative of the impact that can be achieved when scientific communities, NGOs, and government collaborate to protect species' subject to unsustainable and illegal trade. However, it is crucial that data used in forensic and investigative identification are available as evidence in a court of law to prosecute illegal traders, to support sentencing and to ensure there are notable consequences associated with illegal trade of shark-related products. For this to occur, the facilities, procedures and equipment used in the forensic identification of shark fins in Hong Kong need to be improved to comply with the standards of forensic science internationally. In addition, in order to ensure effective dismantling of criminal networks involved in illegal, unregulated and unsustainable trade of shark-related products, police need to leverage the investigative powers that are available to them under the Hong Kong SAR Organised and Serious Crime Ordinance (OSCO), to collect cross-boundary evidence(s) that implicates the high-level 'kingpins' of these networks. With confidently produced forensic data and evidence of organised criminal activity, punishments associated with illegally trading shark fins are likely to be handled more appropriately, reflecting the seriousness of the crime. Finally, reducing demand and changing our collective consumptive behaviour for these products will ultimately reduce the trade. Only with these measures will unsustainably depleting populations of this threatened group be reduced and a more sustainable trade management system be achieved.

Acknowledgements

We thank Stan Shea and Kathleen Ho at BLOOM Hong Kong for allowing us to use their collated data on Hong Kong Census imports for shark products (Figure 20.2) and for taking the time to review our manuscript. Thank you also to Sam Inglis and Sophie le Clue at ADM Capital Foundation for sharing their data on illegal shark fin trade (Figure 20.2). Figure 20.1 Elasmobranchs genera listed on CITES by year. Figure 20.2 Legal trade in shark fins in Hong Kong. Figure 20.3 Current wildlife forensic and investigative solutions for shark fin products. Adapted from: (Abercrombie et al., 2018; Abercrombie & Jabado, 2022).

Note

1 The Convention on International Trade in Endangered Species of Wild Fauna and Flora (CITES) exists to help regulate international trade and is split into three appendices (see https://cites.org/ for more details). All need permits to trade. Appendix I trade is permitted only under exceptional circumstances, and not for commercial purposes (6 elasmobranch species in 2 genera). Appendix II species are those whose trade needs regulating, including 'look-alike' species (147 elasmobranch species in 28 genera). Appendix III species are protected in at least one country, which has asked other CITES Parties for help to regulate their trade (18 elasmobranch species in 2 genera).

References

Abercrombie, D. L., Cardeñosa, D., & Chapman, D. D. (2018) Genetic approaches for identifying Shark Fins and Other Products: A Tool for International Trade monitoring and enforcement. Abercrombie & Fish, Marine Biological Consulting, Suffolk County, NY, 13. URL: https://shorturl.at/DHHc8

Abercrombie, D. L., & Chapman, D. D. (2012) Identifying Shark Fins: Oceanic Whitetip, Porbeagle, and Hammerheads. *Pew Environment Group and the Roe Foundation*. URL: https://cites.org/sites/default/files/sharks_id_material/identifyingsharkfins_guide.pdf

Abercrombie, D. L., Clarke, S. C., & Shivji, M. S. (2005) Global-scale genetic identification of hammerhead sharks: Application to assessment of the international fin trade and law enforcement. *Conservation Genetics, 6*, 775–788. https://doi.org/10.1007/s10592-005-9036-2

Abercrombie, D. L., & Hernandez, S. (2017) Identifying shark fins: Implementing and enforcing CITES. Abercrombie & Fish, Marine Biological Consulting, Suffolk County, NY. 21. URL: https://shorturl.at/Rdnyr

Abercrombie, D. L., & Jabado, R. W. (2022) CITES Sharks and Rays - Implementing and Enforcing Listings: Volume III - Dried Product ID. URL: https://shorturl.at/ZntJ4

ADMCF, Globe Scan, WWF. (2018) Shark Fin Soup Consumption Reduction Survey 2018 Hong Kong. In-Market Report Drivers of demand and behavior change in Hong Kong report. URL: https://www.admcf.org/wp-content/uploads/2021/09/Shark-Fin-Soup-Consumption-Reduction_Hong-Kong-Report_31072018DraftFinal_compressed.pdf

Ahmad, A., & Lim, A. P. K. (2012) Field guide to sharks of the Southeast Asian region. *Kuala Terengganu*: SEAFDEC/MFRDMD/SP/18:210. URL: https://cites.org/sites/default/files/sharks_id_material/FieldGuideToSharks.pdf

Andersson, A. A., Tilley, H. B., Lau, W., Dudgeon, D., Bonebrake, T. C., & Dingle, C. (2021) CITES and beyond: Illuminating 20 years of global, legal wildlife trade. *Global Ecology and Conservation, 26*, e01455. https://doi.org/10.1016/j.gecco.2021.e01455

Baum, J. K., Myers, R. A., Kehler, D. G., Worm, B., Harley, S. J., & Doherty, P. A. (2003) Collapse and conservation of shark populations in the Northwest Atlantic. *Science, 299*(5605), 389–392. https://doi.org/10.1126/science.1079777

Baum, J. K., & Myers, R. A. (2004) Shifting baselines and the decline of pelagic sharks in the Gulf of Mexico. *Ecology Letters, 7*(2), 135–145. https://doi.org/10.1111/j.1461-0248.2003.00564.x

BLOOM Association Hong Kong (2016) *Annual Report 2015–2016*. URL: https://www.bloomassociation.org/en/wp-content/uploads/2016/04/Annual-report-2015_16_spread.pdf.

Caballero, S., Cardenosa, D., Soler, G., & Hyde, J. (2012) Application of multiplex PCR approaches for shark molecular identification: Feasibility and applications for fisheries management and conservation in the Eastern Tropical Pacific. *Molecular Ecology Resources, 12*(2), 233–237. https://doi.org/10.1111/j.1755-0998.2011.03089.x

Camhi, M., Fowler, S. L., Musick, J. A., Bräutigam, A., & Fordham, S. V. (1998) *Sharks and their Relatives–Ecology and Conservation. IUCN/SSC Shark Specialist Group*. IUCN, Gland, Switzerland and Cambridge. iv + 39 pp.

Cardeñosa, D., Fields, A., Abercrombie, D., Feldheim, K., Shea, S. K. H., & Chapman, D. D. (2017) A multiplex PCR mini-barcode assay to identify processed shark products in the global trade. *PLoS One, 12*, e0185368–9. https://doi.org/10.1371/journal.pone.0185368.

Cardeñosa, D., Fields, A. T., Babcock, E. A., Zhang, H., Feldheim, K., Shea, S. K., Fischer, G. A., & Chapman, D. D. (2018a) CITES-listed sharks remain among the top species in the contemporary fin trade. *Conservation Letters, 11*(4), e12457. https://doi.org/10.1111/conl.12457.

Cardeñosa, D., Fields, A. T., Babcock, E., Shea, S. K. H., Feldheim, K. A., Kraft, D. W., Hutchinson, M., Herrera, M. A., Caballero, S., & Chapman, D. D. (2020a) Indo-Pacific origins of silky shark fins in major shark fin markets highlights supply chains and management bodies key for conservation. *Conservation Letters, 14*, 1–11. https://doi.org/10.1111/conl.12780

Cardeñosa, D., Fields, A. T., Babcock, E. A., Shea, S. K., Feldheim, K. A., & Chapman, D. D. (2020b) Species composition of the largest shark fin retail-market in mainland China. *Scientific Reports, 10*(1), 12914. https://doi.org/10.1038/s41598-020-69555-1

Cardeñosa, D., Shea, K. H., Zhang, H., Feldheim, K., Fischer, G. A., & Chapman, D. D. (2019) Small fins, large trade: A snapshot of the species composition of low-value shark fins in the Hong Kong markets. *Animal Conservation, 23*, 203–211. https://doi.org/10.1111/acv.12529.

Cardeñosa, D., Quinlan, J., Shea, K. H., & Chapman, D. D. (2018b) Multiplex real-time PCR assay to detect illegal trade of CITES-listed shark species. *Scientific Reports, 8*(1), 16313. https://doi.org/10.1038/s41598-018-34663-6

Census & Statistics Department (CSD). (2022) *Hong Kong Imports and Exports Classification List (Harmonized System) 2022 Edition Volume One: Commodity Section I – X*. Census and Statistics Department, Hong Kong SAR.

Chapman, D. D., Abercrombie, D. L., Douady, C. J., Pikitch, E. K., Stanhopen, M. J., & Shivji, M. S. (2003) A streamlined, bi-organelle, multiplex PCR approach to species identification: Application to global conservation and trade monitoring of the great white shark, *Carcharodon carcharias*. *Conservation Genetics, 4*, 415–425. https://doi.org/10.1023/A:1024771215616

Cheung, G. C. K., & Chang, C. Y. (2011) Cultural identities of Chinese business: Networks of the shark-fin business in Hong Kong. *Asia Pacific Business Review, 17*(3), 343–359. https://doi.org/10.1080/13602380903461623

Choy, C. P. P., & Wainwright, B. J. (2022) What is in your shark fin soup? Probably an endangered shark species and a bit of mercury. *Animals, 12*(7), 802. https://doi.org/10.3390/ani12070802.

CITES News (2022) *AI helps to fight the illegal trade in shark and ray fins. Convention on International Trade in Endangered Species of Wild Fauna and Flora*. URL: https://cites.org/eng/news/ai-helps-fight-illegal-trade-shark-ray-fins

CITES Trade Database (2022) *Compiled by UNEP-WCMC for the CITES Secretariat*. URL: trade.cites.org

Clarke, S. (2004a) *Shark Product Trade in Hong Kong and Mainland China, and Implementation of the Shark CITES Listings*. TRAFFIC East Asia, Hong Kong. URL: https://www.traffic.org/site/assets/files/9513/shark-product-trade-in-hong-kong-and-mainland-china.pdf)

Clarke, S. (2004b) Understanding pressures on fishery resources through trade statistics: A pilot study of four products in the Chinese dried seafood market. *Fish and Fisheries, 5*(1), 53–74. https://doi.org/10.1111/j.1467-2960.2004.00137.x.

Clarke, S. C., Magnussen, J. E., Abercrombie, D. L., McAllister, M. K., & Shivji, M. S. (2006a) Identification of shark species composition and proportion in the Hong Kong shark fin market based on molecular genetics and trade records. *Conservation Biology, 20*(1), 201–211. https://doi.org/10.1111/j.1523-1739.2005.00247.x

Clarke, S. C., McAllister, M. K., Milner-Gulland, E. J., Kirkwood, G. P., Michielsens, C. G., Agnew, D. J., Pikitch, E. K., Nakano, H., & Shivji, M. S. (2006b) Global estimates of shark catches using trade records from commercial markets. *Ecology Letters, 9*(10), 1115–1126. https://doi.org/10.1111/j.1461-0248.2006.00968.x

Clarke, S., Milner-Gulland, E. J., & Bjørndal, T. (2007) Social, economic, and regulatory drivers of the shark fin trade. *Marine Resource Economics, 22*(3), 305–327. https://doi.org/10.1086/mre.22.3.42629561

CoP19 Prop. 37. (2022) Consideration of Proposals for Amendment of Appendices I and II. Nineteenth meeting of the Conference of the Parties Panama City (Panama), 14–25 November 2022. Convention on International Trade in Endangered Species of Wild Fauna and Flora (CITES). URL: https://cites.org/sites/default/files/documents/E-CoP19-Prop-37.pdf [1]

Dulvy, N. K., Fowler, S. L., Musick, J. A., Cavanagh, R. D., Kyne, P. M., Harrison, L. R., … & White, W. T. (2014) Extinction risk and conservation of the world's sharks and rays. *eLife, 3*, e00590. https://doi.org/10.7554/eLife.00590.

Dulvy, N. K., Pacoureau, N., Rigby, C. L., Pollom, R. A., Jabado, R. W., Ebert, D. A., … & Simpfendorfer, C. A. (2021) Overfishing drives over one-third of all sharks and rays toward a global extinction crisis. *Current Biology, 31*, 4773–4787. https://doi.org/10.1016/j.cub.2021.08.062.

Eriksson, H., & Clarke, S. (2015) Chinese market responses to overexploitation of sharks and sea cucumbers. *Biological Conservation, 184*, 163–173. https://doi.org/10.1016/j.biocon.2015.01.018

Fabinyi, M. (2012) Historical, cultural and social perspectives on luxury seafood consumption in China. *Environmental Conservation, 39*(1), 83–92. https://doi.org/10.1017/S0376892911000609

Fabinyi, M., & Liu, N. (2014) Seafood banquets in Beijing: Consumer perspectives and implications for environmental sustainability. *Conservation and Society, 12*(2), 218–228. https://doi.org/10.4103/0972-4923.138423

FAO. (2020) Food and Agriculture Organization of the United Nations - iSharkFin. In: Food and Agriculture Organization of the United Nations [online]. Rome. URL: https://www.fao.org/ipoa-sharks/tools/software/isharkfin

Ferretti, F., Myers, R. A., Serena, F., & Lotze, H. K. (2008) Loss of large predatory sharks from the mediterranean sea. *Conservation Biology, 22*(4), 952–964. https://doi.org/10.1111/j.1523-1739.2008.00938.x

Fields, A. T., Abercrombie, D. L., Eng, R., Feldheim, K., & Chapman, D. D. (2015) A novel mini-DNA barcoding assay to identify processed fins from internationally protected shark species. *PloS one, 10*(2), e0114844. https://doi.org/10.1371/journal.pone.0114844

Fields, A. T., Fischer, G. A., Shea, S. K., Zhang, H., Abercrombie, D. L., Feldheim, K. A., … & Chapman, D. D. (2018) Species composition of the international shark fin trade assessed through a retail-market survey in Hong Kong. *Conservation Biology, 32*(2), 376–389. https://doi.org/10.1111/cobi.13043

Fong, Q. S. W., & Anderson, J. L. (2000) *Assessment of the Hong Kong Shark Fin Trade*. INFOFISH International1/2000: 28–32.

Gangal, M., Gafoor, A. B., D'Souza, E., Kelkar, N., Karkarey, R., Marbà, N., Arthur, R., & Alcoverro, T. (2021) Sequential overgrazing by green turtles causes archipelago-wide functional extinctions of seagrass meadows. *Biological Conservation, 260*, 109195 https://doi.org/10.1016/j.biocon.2021.109195

Garcia Barcia, L., Valdes, A. E., Wothke, A., Fanovich, L., Mohammed, R. S., Shea, S., … & Chapman, D. (2023) Health risk assessment of globally consumed shark-derived products. *Exposure and Health, 15*(2), 409–423. https://doi.org/10.1007/s12403-022-00500-5

Hareide, N. R., Carlson, J., Clarke, M., Clarke, S., Ellis, J., Fordham, S., … & Polti, S. (2007) European shark fisheries: A preliminary investigation into fisheries, conversion factors, trade products, markets and management measures. *European Elasmobranch Association, 27*.

Hatten, C. E. R, Fitriana, Y. S., Prigge, T. L., Irham, M., Sutrisno, H., & Dingle, C. (2023) DNA analysis and validation for species identification of seized helmeted hornbill (*Rhinoplax vigil*) casques.

Forensic Science International: Animals and Environments, 3, 100058. https://doi.org/10.1016/j.fsiae.2022.100058.
Hatten, C. E. R., Whitfort, A., Baker, D. M., & Dingle, C. (2020) Wildlife forensic science in Hong Kong. *Wiley Interdisciplinary Reviews: Forensic Science, 2*(6), e1376. https://doi.org/10.1002/wfs2.1376.
Heithaus, M. R., Alcoverro, T., Arthur, R., Burkholder, D. A., Coates, K. A., Christianen, M. J. A., Kelkar, N., Manuel, S. A., Wirsing, A. J., Kenworthy, W. J., & Fourqurean, J. W. (2014) Seagrasses in the age of sea turtle conservation and shark overfishing. *Frontiers in Marine Science, 1*, 1–6. https://doi.org/10.3389/fmars.2014.00028
Heithaus, M. R., Frid, A., Wirsing, A. J., & Worm, B. (2008) Predicting ecological consequences of marine top predator declines. *Trends in Ecology & Evolution*, *23*(4), 202–210. https://doi.org/10.1016/j.tree.2008.01.003
Heithaus, M. R., Wirsing, A. J., Dill, L. M., & Heithaus, L. I. (2007) Long-term movements of tiger sharks satellite-tagged in Shark Bay, Western Australia. *Marine Biology, 151*, 1455–1461. https://doi.org/10.1007/s00227-006-0583-y
HKSAR press release. (2020) Smuggling of shark fins. The Government of Hong Kong Special Administrative Region Press Releases. URL: https://www.info.gov.hk/gia/general/202005/27/P2020052700507.htm
Ho, K. Y., & Shea, K. H. (2021) Comparison Report: The 2009/10, 2014/15 and 2019/20 Surveys on the shark consumption habits and attitudes in Hong Kong. Bloom Association Hong Kong. URL: https://www.bloomassociation.org/en/wp-content/uploads/2021/06/2019-20-Shark-fin-sociological-survey-comparison-report.pdf
Hobbs, C. A., Potts, R. W., Bjerregaard Walsh, M., Usher, J., & Griffiths, A. M. (2019) Using DNA barcoding to investigate patterns of species utilisation in UK shark products reveals threatened species on sale. *Scientific Reports, 9*(1), 1028. https://doi.org/10.1038/s41598-018-38270-3.
Huffman, J. E., & Wallace, J. R. (Eds.) (2012) *Wildlife Forensics: Methods and Applications*, John Wiley & Sons, Ltd., West Sussex, 201–236. https://doi.org/10.1002/9781119953142.
Jabado, R. W., & Morata, A. Z. A. (2022) Identifying sharks from their fins. Elasmo Project. Dubai, United Arab Emirates, 21. URL: https://static1.squarespace.com/static/5be1cec125bf028361db95dc/t/63701e55a05c9d1cc183fc5b/1668292220371/Web_Identifying+sharks+from+their+fins_EN.pdf
Kemp, N. E. (1977) Banding pattern and fibrillogenesis of ceratotrichia in shark fins. *Journal of Morphology, 154*(2), 187–203. https://doi.org/10.1002/jmor.1051540202
Lack, M. (2006) Conservation of Spiny Dogfish Squalus Acanthias: A Role for CITES? TRAFFIC International. URL: https://www.traffic.org/species-reports/traffic_species_fish5.pdf
Last, P. R., & Stevens, J. D. (2009) Sharks and Rays of Australia. *Copeia*, 1994(4), 1055. https://doi.org/10.2307/1446735
Lau, W. (2014) Taking from the wild: Moving towards sustainability in Hong Kong's wildlife trade. pp 1–88. URL: https://www.civic-exchange.org
Lau, W. (2018) An assessment of South African dried abalone *Haliotis midae* consumption and trade in Hong Kong. TRAFFIC International, Cambridge. URL: https://www.traffic.org/site/assets/files/2235/traffic-hk-abalone-report.pdf.
Momigliano, P., & Harcourt, R. (2014) Shark conservation, governance and management: The science–law disconnect. In *Sharks: Conservation, Governance and Management*, Techera, E. J., & Klein, N. (Eds). Routledge. London. https://doi.org/10.4324/9780203750292
Mozer, A., & Prost, S. (2023) An Introduction to illegal wildlife trade and its effects on biodiversity and society. *Forensic Science International: Animals and Environments, 3*, 100064. https://doi.org/10.1016/j.fsiae.2023.100064.
Myers, R. A., Baum, J. K., Shepherd, T. D., Powers, S. P., & Peterson, C. H. (2007) Cascading effects of the loss of apex predatory sharks from a coastal ocean. *Science, 315*(5820), 1846–1850. https://doi.org/10.1126/science.1138657
Nair, R. J., & Zacharia, P. U. (2015) An introduction to the classification of elasmobranchs, 118–133. URL: https://core.ac.uk/download/pdf/33020498.pdf
Okes, N., & Sant, G. (2019) An overview of major shark traders, catchers and species. TRAFFIC, Cambridge. URL: https://www.traffic.org/site/assets/files/12427/top-20-sharks-web-1.pdf.
Ogden, R., Dawnay, N., & McEwing, R. (2009) Wildlife DNA forensics—bridging the gap between conservation genetics and law enforcement. *Endangered Species Research, 9*(3), 179–195. https://doi.org/10.3354/esr00144.

Ogden, R., & Mailley, J. (2016) A review of wildlife forensic science and laboratory capacity to support the implementation and enforcement of CITES. CITES CoP17 Doc. 25 Annex 4. A review commissioned by the Secretariat of the Convention on International Trade in Endangered Species of Wild Fauna and Flora and undertaken by the United Nations Office on Drugs and Crime. URL: https://cites.org/sites/default/files/eng/cop/17/WorkingDocs/E-CoP17-25-A4.pdf.

Prasetyo, A. P., Cusa, M., Murray, J. M., Agung, F., Muttaqin, E., Mariani, S., & McDevitt, A. D. (2023) Universal closed-tube barcoding for monitoring the shark and ray trade in megadiverse conservation hotspots. *iScience, 26*(7), 107065. https://doi.org/10.1016/j.isci.2023.107065.

Purcell, S. W., Williamson, D. H., & Ngaluafe, P. (2018) Chinese market prices of beche-de-mer: Implications for fisheries and aquaculture. *Marine Policy, 91*, 58–65. https://doi.org/10.1016/j.marpol.2018.02.005.

Roff, G., Doropoulos, C., Rogers, A., Bozec, Y.-M., Krueck, N. C., Aurellado, E., Priest, M., Birrell, C., & Mumby, P. J. (2016) The ecological role of sharks on coral reefs. *Trends in Ecology & Evolution, 31*(5), 395–407. https://doi.org/10.1016/j.tree.2016.02.014.

Rose, D. (1996) An overview of world trade in sharks and other cartilaginous fishes, IUCN: International Union for Conservation of Nature. TRAFFIC International, Center for Marine Conservation, US, IUCN Species Survival Commission (SSC), Shark Specialist Group, WWF, IUCN. URL: https://policycommons.net/artifacts/1373118/an-overview-of-world-trade-in-sharks-and-other-cartilaginous-fishes/1987339/

Sadovy, Y. (2010) Humphead wrasse and illegal, unreported and unregulated fishing. *SPC Live Reef Fish Information Bulletin, 19*, 19–21. URL: https://www.spc.int/DigitalLibrary/Doc/FAME/InfoBull/LRF/19/LRF19_19_Sadovy.pdf

Sadovy de Mitcheson, Y., Andersson, A. A., Hofford, A., Law, C. S., Hau, L. C., & Pauly, D. (2018) Out of control means off the menu: The case for ceasing consumption of luxury products from highly vulnerable species when international trade cannot be adequately controlled; shark fin as a case study. *Marine Policy, 98*, 115–120. https://doi.org/10.1016/j.marpol.2018.08.012

Sadovy de Mitcheson, Y., To, A. W. L., Wong, N. W., Kwan, H. Y., & Bud, W. S. (2019) Emerging from the murk: Threats, challenges and opportunities for the global swim bladder trade. *Reviews in Fish Biology and Fisheries, 29*, 809–835. https://doi.org/10.1007/s11160-019-09585-9

Seret, B. (2006) Identification Guide of the Main Shark and Ray Species of the Eastern Tropical Atlantic: For the Purpose of the Fishery Observers and Biologists. FIBA. URL: https://horizon.documentation.ird.fr/exl-doc/pleins_textes/divers18-05/010072286.pdf

Shea, K. H., & To, A. W. L. (2017) From boat to bowl: Patterns and dynamics of shark fin trade in Hong Kong—implications for monitoring and management. *Marine Policy, 81*, 330–339. https://doi.org/10.1016/j.marpol.2017.04.016.

Sheng, V. (2020) A World Without Sharks: Forensic methods, trade networks, and impacts of predator removal. PhD thesis. The University of Hong Kong. URL: https://hub.hku.hk/handle/10722/311101

Shivji, M., Clarke, S., Pank, M., Natanson, L., Kohler, N., & Stanhope, M. (2002) Genetic identification of pelagic shark body parts for conservation and trade monitoring. *Conservation Biology, 16*(4), 1036–1047. https://doi.org/10.1046/j.1523-1739.2002.01188.x

Simpfendorfer, C. A., & Dulvy, N. K. (2017) Bright spots of sustainable shark fishing. *Current Biology, 27*(3), R97–R98. https://doi.org/10.1016/j.cub.2016.12.017

SWFS Technical Working Group. (2018) Standards and guidelines for wildlife forensic analysis, version 3. Ed. Lucy M. I. Webster. Published by the Society for Wildlife Forensic Science, 19th November 2018, pp.21. URL: https://www.wildlifeforensicscience.org/wp-content/uploads/2018/11/SWFS-Standards-and-Guidelines_Version-3_19-11-18.pdf

Thomas, N. (2011) The impact of seafood consumption on endangered marine species on Hong Kong. *International Journal of Environmental Sciences, 1*(7), 2048–2085. https://www.indianjournals.com/ijor.aspx?target=ijor:ijes&volume=1&issue=7&article=058

Trail, P. W. (2021) Morphological analysis: A powerful tool in wildlife forensic biology. *Forensic Science International: Animals and Environments, 1*, 100025. https://doi.org/10.1016/j.fsiae.2021.100025.

Vannuccini, S. (1999) Shark utilization, marketing and trade. FAO Fisheries Technical Paper 389. Rome. URL: https://www.fao.org/documents/card/fr/c/1448b1bf-3354-540f-926b-/

Vallianos, C., Sherry, J., Hoford, A., & Baker, J. (2018) *Sharks in Crisis: Evidence of Positive Behavioral Change in China as New Threats Emerge*. WildAid Report.

Webster, L. M., Prigge, T. L., & Frankham, G. J. (2024) A guide for the validation of DNA based species identification in forensic casework. *Forensic Science International: Animals and Environments, 5*, 100080. https://doi.org/10.1016/j.fsiae.2023.100080

World Health Organization. (2018, February 15) Arsenic. URL: https://www.who.int/news-room/fact-sheets/detail/arsenic

World Health Organization. (2024, Oct 24) Mercury. URL: https://www.who.int/news-room/fact-sheets/detail/mercury-and-health

Worm, B., Davis, B., Kettemer, L., Ward-Paige, C. A., Chapman, D., Heithaus, M. R.,... & Gruber, S. H. (2013) Global catches, exploitation rates, and rebuilding options for sharks. *Marine Policy, 40*, 194–204. https://doi.org/10.1016/j.marpol.2012.12.034

Worm, B., Orofino, S., Burns, E. S., D'Costa, N. G., Manir Feitosa, L., Palomares, M. L. D., Schiller, L., & Bradley, D. (2024) Global shark fishing mortality is still rising despite widespread regulatory change. *Science, 383*(6679), 225–230. https://doi.org/10.1126/science.adf8984

Wu, J., & Sadovy, Y. (2016) Humphead (Napoleon) Wrasse *Cheilinus undulatus* trade into and through Hong Kong. *Hong Kong: TRAFFIC*. URL: https://www.traffic.org/site/assets/files/2426/humphead-wrasse-hong-kong-trade.pdf

21
LEGALISATION AND SUSTAINABLE USE AS AN ALTERNATIVE APPROACH TO TACKLING WILDLIFE CRIME

Michael 't Sas-Rolfes and Shylock Muyengwa

Introduction

Amidst growing academic and practical interest in the topic of wildlife crime, value judgements over human actions that impact the environment remain salient; views differ on the extent to which such actions should be regarded as 'crime' and the extent to which legal and enforcement measures should be employed to deter them (Wellsmith, 2011; White, 2003). Growing international interest in wildlife crime during the early 21st century has included a geopolitical dimension: consumer demand for wildlife products has surged in rapidly growing East Asian countries, while many affected species occur in developing African, Asian, and Latin American countries, prompting international political interventions from Western countries such as the US, UK, and EU member states. Prominent interventions include security-related foreign aid packages from the United States (Massé & Margulies, 2020) and a series of international meetings on illegal wildlife trade convened by the UK government and royal family (Massé et al., 2020). However, these initiatives have drawn criticism for promoting a violent approach to wildlife conservation, harming local people and potentially losing their critical longer-term support (Duffy, 2022; Duffy et al., 2019), prompting some scholars to propose an alternative 'socioecological harm reduction approach' (Hübschle & Margulies, 2024).

Contemporary global policy towards wildlife crime is historically rooted in the experience of the USA and colonial European countries during the 19th and 20th centuries and reflects their persistent hegemonic influence. Concepts of wildlife preservation and conservation emerged from North American philosophy and practice, shaped by influential individuals such as John Muir, Gifford Pinchot, and Aldo Leopold (Callicot, 1990). Muir and Pinchot differed over whether nature should be protected from humans or sustainably managed for human use, with Leopold adopting a more nuanced view. The consequent compromise 'North American model' of wildlife management relies heavily on a legalistic regulatory approach in which the state plays a dominant role and market activity is suppressed, but citizen hunting and fishing of non-threatened species for personal use is permitted (see Geist, 1994). This 'command and control' regulatory logic is evident in two significant (and somewhat related) legal documents shaped in the early 1970s in Washington, DC:

DOI: 10.4324/9781003333906-25

the US Endangered Species Act (ESA) and CITES, the Convention on International Trade in Endangered Species of Wild Fauna and Flora (Epstein, 2006). The logic is also of a time that coincided with US President Richard Nixon's launch of the similarly framed 'war on drugs' aimed at suppressing illicit drug use through enforced regulation.

CITES' emergence is also linked to the 1972 United Nations (UN) Conference on the Human Environment (held in Stockholm), which was followed in 1992 by the UN Conference on Environment and Development (held in Rio de Janeiro). The latter meeting spawned two further significant treaties, the UN Framework Convention on Climate Change and the Convention on Biological Diversity (CBD). Both treaties significantly broaden the remit of conservation from protecting wild species to more inclusive global management of multiple interacting environmental concerns. The CBD adopted a different approach from CITES in several respects, being less legalistic, more consensus-driven, and more oriented towards the concerns of the Global South. Consequently, by 1994, it had already achieved more support through ratifications than CITES (Escobar-Pemberthy & Ivanova, 2020).

Subsequently, the UN has also developed the widely supported Sustainable Development Goals (SDGs), which include addressing environmental concerns while ending poverty, reducing inequality, and building more peaceful, prosperous societies by 2030. In 2022, the Parties to the CBD approved the Kunming-Montreal Global Biodiversity Framework, which lists 23 global targets to be implemented by 2030 in alignment with the SDGs. The tone of these targets, in which the rights of indigenous peoples and local communities feature strongly, contrasts sharply with the current approach to large terrestrial mammals and many other species for which CITES is increasingly establishing a global 'prohibition regime' (Nadelmann, 1990) following repeated calls from conservation scientists and others to intensify trade restrictions (e.g. Hughes, 2021; Macdonald et al., 2021). Given the (necessarily) strong people-centred orientation of inclusive global governance policies and following the realisation that the heavy-handed prohibitionist approach to drug regulation has failed and is steadily giving way to more nuanced approaches (Babor et al., 2018; Stevens et al., 2022), prohibitionist approaches to wildlife protection warrant re-evaluation.

A recent systematic review of scientific evidence on the effectiveness of trade restrictions to protect elephants, rhinos, lions, and pangolins in southern Africa reveals ambiguous results at best (Hiller & 't Sas-Rolfes, 2024). Other recent research on ideas influencing global wildlife trade policy towards charismatic African megafauna identifies three broad and somewhat conflicting policy narratives ('t Sas-Rolfes & Gooden, 2024). CITES remains primarily rooted in a conventional legalistic approach that assumes policy decisions informed by science will be adequately implemented by its members through enforcement, which continues to constitute the dominant narrative. However, the design of CITES leaves it vulnerable to the influence of special interest groups pursuing a prohibitionist agenda that draws from ideology more than science, especially in relation to the more charismatic species ('t Sas-Rolfes et al., 2024). An assessment of varying institutional approaches to African rhinoceros (hereafter rhino) conservation suggests that countries adopting more nuanced decentralised approaches have significantly outperformed those depending more exclusively on state control ('t Sas-Rolfes & Emslie, 2024). Intensifying the prohibitionist approach may eventually fail to achieve species conservation goals in many countries ('t Sas-Rolfes, 2024).

In view of the thrust of contemporary international sustainable development policy and the evidence that prohibitionist approaches towards wildlife use and trade may deliver deficient social and conservation outcomes, we explore the potential of legalisation as an alternative approach. Our chapter proceeds as follows: After discussing the historical development of wildlife crime, we draw on institutional economic literature to provide an analytic frame to relate its relevance to conservation. This leads us to discuss the rationale for legalisation and the success of certain past initiatives. Next, we discuss concerns raised by those opposed to or sceptical of legalisation. We then suggest ways to address those concerns and present two illustrative African case studies. We conclude by briefly contemplating the prospects for successfully implementing this approach and some broader implications.

Origins of wildlife crime

Humanity has enabled a scaled-up social order and galvanised collective progress through 'institutions.' Following Hodgson (2006, p.18), we define institutions as 'systems of established and embedded social rules that structure social interactions.' Building off language as a base institution (Searle, 2005), early humans developed informal institutions such as customs, taboos, and traditions to modulate their behaviour, including their interactions with nature. Institutions governing wildlife developed within varying contexts, with correspondingly varying degrees of formality, from traditional localised forms of hunting regulation to the development of statutory conservation law.

The conceptual emergence of wildlife crime as a more serious issue can be linked to historical developments around the time of global colonisation by Europeans and developments in firearm technology, which greatly facilitated the efficient elimination of wild animals through hunting ('t Sas-Rolfes et al., 2025). By the early 19th century, European colonists in the Cape region of South Africa had hunted an endemic antelope species, the bluebuck, to extinction. In the 1870s, American bison were hunted to near extinction by European settlers who had dismantled traditional Native American institutions that had previously ensured their conservation (Lueck, 2002). The near loss of the bison species and the subsequent hunting to extinction of the once-abundant passenger pigeon added to a growing 19th-century environmental consciousness, leading to the creation of state-protected areas and the 1900 Lacey Act in the USA, which outlawed cross-border wildlife trade. In the American mind, 'market hunting' was to blame for wildlife demise rather than institutional failure (Tober, 1981).

In African and Asian colonies, European settlers similarly ignored local institutions and effectively appropriated legal hunting rights for themselves, albeit also distributing more sophisticated weaponry among indigenous people (MacKenzie, 1997; Beinart & Hughes, 2009). 'Sport hunting' by white settlers was elevated as a noble pursuit, whereas traditional hunting by indigenous people became progressively outlawed and consequently labelled as 'poaching' (Steinhart, 2005). During the 20th century, as populations of key species started to decline and countries gained independence, some outlawed hunting entirely (e.g. India in 1972 and Kenya in 1977). However, many others, notably in Africa, continued to allow financially lucrative foreign 'safari' hunting while continuing to prohibit locals from hunting and harvesting wildlife, having also frequently evicted rural communities from their traditional lands to create state-protected areas. Consequently, and partly in further rebellion against imposed post-colonial regimes, pockets of well-armed Africans throughout the continent started poaching high-value species such as rhinos and elephants to benefit from lucrative international markets for rhino horn and ivory. This poaching escalated and spread significantly in the 1970s and 1980s, bolstering the arguments for CITES interventions.

In its founding documents, CITES concerns itself with the welfare of animals in transport but otherwise focuses on species conservation as its primary objective. However, with rapidly growing concern for animal welfare since the 1970s, animal protection activists have increasingly influenced CITES policy and practice, frequently conflating welfare and conservation concerns ('t Sas-Rolfes & Gooden, 2024) and seeking to frame wild animal welfare violations as wildlife crime (van Uhm, 2015; Wyatt, 2021). CITES decisions thus increasingly reflect the preferences of affluent Western society over the incentive-driven approaches better suited to countries in the Global South (Abensperg-Traun, 2009).

Viewing the emergence of wildlife crime through the lenses of history, geography, and political ecology, we can observe that, as with other forms of crime, wildlife 'crime' is a social construct (van Uhm, 2016), as is the term 'poaching' ('t Sas-Rolfes et al., 2025), with both frequently contested by a range of actors in society (Hübschle, 2016; Lavadinović et al., 2021). The framing of crime is somewhat contextual, political, and dynamic in nature, as elucidated by recent literature on illicit geographies and contested environments (Margulies et al., 2023). Approaches to tackling crime should consider the role that crime classification itself plays in bolstering power through exclusionary environmental protection and securitisation.

Institutional economic analysis

We briefly review literature that explores the relationship between institutional arrangements, especially property rights (including legal ownership of wild animals), and conservation outcomes. Whereas the 'Tragedy of the Commons' framing by Hardin (1968) raised the profile of this issue among environmental scientists, its basic premise is somewhat simplistic (Ostrom, 1990) and obscures a wealth of insights obtained from prior and subsequent theoretical developments and empirical research in the broader social science literature. The discipline of resource economics examines optimal harvesting rates and affirms the significance of property rights regimes in influencing these and impacting resource stocks ('t Sas-Rolfes et al., 2019). Initially focused on forestry and fisheries, extending this approach to terrestrial wild animals highlights the need to address habitat conversion pressures and the opportunity costs of active wildlife management in addition to harvesting incentives when crafting wildlife trade policy (Swanson, 1994; Barbier & Shulz, 1997). These analyses provide a rationale for sustainable wildlife use as a conservation policy towards many terrestrial species.

Parallel literature affirms property as a meaningful and universal human custom (Wilson, 2020), examines the economics of property rights evolution (e.g. Demsetz, 1967; Anderson & Hill, 1975; Lueck & Torrens, 2020), and demonstrates powerful motivations for people to create, strengthen, and obtain property rights. Property rights are strengthened by clearer definition, stronger government support through enforcement, easier transferability, and greater durability (Alchian & Demsetz, 1973). Empirical analysis of institutional changes in southern Africa demonstrates how these brought about demonstrable changes in conservation outcomes, with strengthened property rights generating both economic and conservation benefits and weakened rights subsequently reversing some of the gains (e.g. Child & Child, 2015; Taylor et al., 2020; 't Sas-Rolfes & Emslie, 2024). The American bison recovery is similarly linked to strengthened property rights (Lueck, 2002).

Other parallel literature examines motivations for crime from perspectives of neoclassical economics (e.g. Becker, 1968), economic sociology (e.g. Beckert & Dewey, 2017), and criminology (e.g. Nagin, 2013) and applies the insights to the analysis of wildlife crime (e.g. Milner-Gulland & Leader-Williams, 1992; Hübschle, 2017; Kurland et al., 2017).

A common theme that emerges from this literature is that wildlife crime is contextual and socioeconomically motivated. It frequently occurs within a context of social resistance to externally imposed ethical norms, and decisions to conduct specific criminal activities can be explained as circumstantially economically rational after assessing the risks and expected cost of apprehension and punishment relative to the expected gain.

Finally, applying broader principles of institutional analysis, the work of Berkes (1989), Ostrom (1990, 2005, 2009), and others (e.g. Cooney et al., 2017; Gavin et al., 2018) points to the development of polycentric adaptive governance systems for effective conservation and natural resource management. This literature strongly emphasises engaging relevant local communities, integrating multiple stakeholder objectives and pluralistic worldviews to form partnerships, and tailoring interventions to specific socioecological contexts. In this context, definitions of crime may be reassessed, and legalisation considered as an alternative.

Rationale for legalisation

The rationale for legalisation is threefold. The first reason is practical and relates to the limitations of the regulatory approach. Supporters of the legalistic CITES approach to wildlife conservation argue that it has been inadequately implemented by governments, and this is due to a 'lack of political will' (e.g. Reeve, 2002; Wyatt, 2021). However, governments of many developing countries likely do not and cannot prioritise sufficient resource allocation towards effective implementation and enforcement because they lack support from their constituent citizens, who have other more pressing socioeconomic needs. Wildlife conservation is perpetually underfunded relative to conservationists' prescriptions (James et al., 2001; Waldron et al., 2013; Lindsey et al., 2018). Although governments often accept financial and other assistance from foreign governments and non-governmental organisations, many are also justifiably wary of the implications (Duffy, 2022; Massé & Margulies, 2020).

Second, if appropriately implemented, legalisation can unlock significant long-term benefits for people who live with wildlife and currently mostly incur socioeconomic costs, both direct (through human-wildlife and crime-related conflict) and indirect (the opportunity costs of maintaining wildlife habitat, as per Barbier & Shulz [1997]). It can furthermore benefit governments by reducing the costs of management and enforcement and providing new sources of income from direct sales of wildlife products and taxation of private wildlife trading activity.

Third, appropriately implemented legalisation can also benefit conservation, possibly very significantly, by directing meaningful new benefit streams to relevant local people, thereby incentivising them to maintain wildlife populations and habitats. Relevant past examples of conservation success stories include the legalisation of white and black rhino safari hunting in South Africa and Namibia ('t Sas-Rolfes et al., 2022), the legalisation of vicuña wool shearing and trade in South America (Kasterine & Lichtenstein, 2018), and the worldwide legalisation of trade in crocodilian skins (Hutton & Webb, 2002). In all these cases, the previously overexploited species initially benefitted from strict protection but then thrived after legalisation.

Legalisation concerns

Objections towards legalisation vary from outright ideological opposition to concerns around effective implementation ('t Sas-Rolfes & Gooden, 2024) and somewhat align with concerns about the roles of non-state actors in land conservation (Gooden & 't Sas-Rolfes, 2020).

Ideological opposition is grounded in normative ideas that view the commodification of nature and wild species, especially animals, as morally reprehensible. Such anti-commodification sentiments are also frequently linked to animal welfare concerns and deontological notions of animal rights, which reject the notion of non-human animals as human property (Francione, 1995). Whereas extreme expressions of such sentiments tend to be confined to pockets of (predominantly urbanised) Western affluence (Epstein, 2004), a somewhat more lenient 'compassionate conservation' movement still largely opposes the notion of sustainable extractive use of wild animals, thereby confronting the original intent of the 'conservation' approach towards nature (see Bekoff, 2013).

In addition to normative arguments against legal wildlife use and trade, sceptics raise other concerns about the potential destigmatisation of products and practices deemed risky or culturally unacceptable. Conservationists fear that reopening legal product markets may unleash undesirable and uncontrollable market forces that ultimately stimulate further illegal or ecologically unsustainable activity ('t Sas-Rolfes & Gooden, 2024; Hughes et al., 2023). These concerns include the potential role of increased wildlife husbandry practices (i.e. farming) in adversely affecting conservation and animal welfare and, in turn, increasing the risk of zoonotic disease proliferation (Roe et al., 2020). Pathways to these adverse effects include the role of wildlife farms in exploiting wild sources to replenish stocks, laundering illegally sourced stocks into legal supply channels, and causing undesirable genetic spillover effects from domesticated or otherwise ecologically modified species into wild populations (Meeks et al., 2024).

Solutions

Fundamental and uncompromising ideological objections to legalisation leave no room for solutions beyond political confrontation. However, by carefully considering the specific circumstances under which legalising previously banned wildlife products may produce undesirable effects (Fischer, 2004), valid practical concerns can be mitigated through thoughtful institutional design and the implementation of appropriate governance and management mechanisms. A critical aspect of any legalisation attempt would be to ensure that reformed institutions channel benefit flows to the local custodians of wildlife, thereby providing them with the added resources and incentives to conserve more effectively. Therefore, stronger property rights over in situ (in relation to captive) wildlife sources should constitute a cornerstone of any legalisation strategy.

More specific measures can include using existing technology to establish traceability systems that already effectively mitigate product laundering (Williams et al., 2021) and the setting of wildlife product industry standards (for both goods and services) with respect to ecological and animal welfare criteria. Such measures can be supported through certification systems in consultation with targeted consumer groups, following appropriate engagement and messaging, to ensure effective competition with illegal supply channels. If deemed necessary, certification supported by blockchain technology could be used to further direct and track the reinvestment of financial proceeds into effective in situ conservation programmes (that include direct mitigation of wildlife crime, such as anti-poaching measures).

Case study 1: African citizen hunting

Hunting has deep traditional roots in African society and was regulated through informal institutions throughout precolonial history (Muposhi et al., 2016). Whereas African society is highly culturally diverse in response to varied local circumstances, there are many localities in which wildlife hunting has been—and remains—an integral part of culture and a critical food source, with many indigenous languages using the same word for 'wildlife' and 'meat.' Consequently, many local African hunters reject being labelled as 'poachers' (Lubilo & Hebinck, 2019). Recognising those traditions and the legacy of colonial confiscation of rights, the thoughtful restoration of legal citizen hunting could significantly benefit local people and conservation.

Existing citizen hunting in Africa takes various forms, from subsistence to cultural hunting. While some forms, like institutionalised community-based hunting, can contribute positively to conservation efforts and local livelihoods, others, like uncontrolled poaching, pose significant threats to wildlife. Managing hunting through appropriate regulations and community involvement is key to ensuring the sustainable use of Africa's wildlife resources. Most African countries continue to deem most forms of citizen hunting illegal, resulting in frequent arrests and social disruption to communities. African governments can and should explore restoring traditional rights to relevant local communities, providing mechanisms for subsistence hunting (for food security purposes) and cultural hunting (for cultural preservation).

Hunting small game is a critical source of protein for many rural communities in sub-Saharan Africa, particularly in Tanzania and Zimbabwe. However, the legal and financial barriers imposed on this activity can significantly burden households that rely on bushmeat to meet their nutritional needs. While such regulations aim to protect wildlife populations and ecosystems, their unintended consequences disproportionately affect impoverished families, making the case for carefully calibrated reforms to ensure food security while preserving biodiversity.

The costs imposed for hunting small game vary widely across the region, often depending on the species involved and the country's legal framework. In Tanzania, for example, fines for hunting species such as the Lesser Kudu range between 500,000 Tanzanian Shillings (approximately USD 315) to 2 million TZS (about USD 1,260), with potential imprisonment of five to ten years. For Greater Kudu, penalties are somewhat lower but still substantial, ranging from USD 13 to USD 31.[1] Similarly, in Zimbabwe, fines for hunting small game are structured by gender, with penalties set at USD 2,000 for males and USD 2,500 for females.[2] These costs are significant in countries where rural families often subsist on minimal incomes.

The enforcement of such fines creates a multifaceted burden on households, especially those in rural areas where hunting small game is a traditional and essential practice for meeting protein needs. High fines or imprisonment deprives families of essential financial resources, further exacerbating poverty. Restrictions on small game hunting reduce access to affordable protein sources, increasing malnutrition risks. Additionally, families face criminal charges for engaging in survival-driven hunting, which disrupts community structures and livelihoods. In Tanzania and Zimbabwe, the inability to afford alternative protein sources, such as meat from livestock or fish, drives communities to risk fines or imprisonment by hunting small game illegally. This highlights the tension between wildlife conservation objectives and human survival needs.

To balance conservation goals with community well-being, introducing legally sanctioned quotas for small game hunting in regions where bushmeat is a vital dietary component is essential. Establishing community-managed hunting zones that allow sustainable use of small game within regulated quotas could mitigate the financial and legal pressures on rural families. Implementing low-cost hunting permits and wildlife monitoring systems to adapt quotas based on local wildlife populations would help ensure conservation efforts do not undermine the food security and livelihoods of the most vulnerable populations.

Case study 2: Rhino horn and elephant ivory trade

International trade in rhino horn has been banned since 1977, and in elephant ivory since 1989 (except for two subsequent controversial once-off sale events). For many reasons, neither of these bans can be considered an unqualified success; at best, they have delivered mixed results ('t Sas-Rolfes, 1997, 2000; Di Minin et al., 2022). They are only very reluctantly supported—and in some cases opposed—by several African countries, especially those in the SADC region, which have been most successful at conserving rhinos and elephants. Stockpiles of rhino horn and ivory, worth hundreds of millions of dollars, continue to accumulate in multiple African countries, collected from rhino dehorning (for security reasons) and natural mortalities of both species. These stockpiles amount to stranded assets, which would otherwise potentially inject substantial funding into cash-strapped African conservation agencies.

Trade legalisation for African rhino horn and elephant ivory has been frequently debated but remains contentious. Since 2007, all attempts by range states to lift CITES trade restrictions on these products have failed. This failure reflects the prohibitionist turn in wildlife trade policy of the recent era, spurred by the surge in Asian consumer demand for endangered wildlife products in the early 21st century. Demand for both products no longer appears to be growing rapidly but certainly persists as a driver of continued poaching pressure in Africa (Balfour et al., 2024; Vigne et al., 2025; Kuiper et al., 2023). A substantial body of research has also emerged during the last decade, covering a range of trade and conservation aspects concerning both species. These factors enable a more measured and nuanced approach to tackling the question of trade policy.

The circumstances and challenges facing African rhinos and elephants vary with species and geography. Viewed at the continental scale, the species are disproportionately distributed, with relatively few countries accounting for the largest proportions of wild rhino (especially) and elephant populations. In parts of Africa with high human population density, these species will tend to be confined to smaller parks that require a different management approach than the larger free-ranging populations that occur in more sparsely settled areas (such as the KAZA region in southern Africa). Rhinos furthermore currently require significantly more intensive security measures to protect them from poachers than elephants. Considering these factors, geopolitical considerations, and further relevant differences in the commodity characteristics of rhino horn and elephant ivory, the legalisation question for the two should be approached quite differently.

The species conservation case for legalising rhino horn trade appears more compelling than that for elephant ivory. Most rhino custodians in South Africa and Namibia (which jointly account for more than 80% of Africa's wild rhinos) support this approach (see Rubino & Pienaar, 2020), with most also routinely dehorning their animals for added security. Stockpiles of horn continue to accumulate and are known to be substantial, amounting to multiple times the current annual illegal supply from poaching. In both countries, existing institutional arrangements already ensure that the proceeds from sustainable use are likely to be reinvested in conservation ('t Sas-Rolfes et al., 2022), and these could be further strengthened within existing constitutional frameworks. There is significant potential to realign the rhino horn economy in favour of legitimate rhino custodians rather than criminals and thereby realise substantial benefits for both conservation and local people.

Regarding the ivory trade, a few key range state governments could gain significant economic benefits from an appropriately designed legal trading system. However, given existing institutional arrangements, it is less clear how these would be redirected to reduce crime and support conservation more broadly—more significant institutional reform may be needed. Two previously permitted 'one-off sales' were poorly conceived to achieve meaningful conservation benefits and were sufficiently controversial to ensure that all subsequent attempts to trade would be blocked ('t Sas-Rolfes et al., 2014; 2024).

In assessing the potential benefits of legalisation of a species product, we can refer to insights from economics regarding supply and demand elasticities, which refer to the sensitivity of supply and demand to changes in prices (Marshall, 2009). For products with relatively inelastic supply and demand, trade bans typically underperform, and legalisation accompanied by taxation is a more advisable approach (Becker et al., 2006). For a product with relatively inelastic consumer demand, as appears to be the case for rhino horn, introducing a competing substitute (e.g. legal horn) is likely to reduce the level of poaching significantly (Chen & 't Sas-Rolfes, 2021). However, for elephant conservation, an empirical analysis of historical ivory prices suggests that poaching supply is relatively inelastic, and that legalising trade would have less 'crowding out' potential against illegal trade (Do et al., 2020).

Attempts to legalise either the international rhino horn or ivory trade through CITES will continue to face certain challenges. The first is convincing the governments of key Asian consumer countries to re-establish legal domestic markets after having been politically pressured to outlaw them. They are unlikely to do this without overwhelming support from the majority of rhino and elephant range states, as well as influential Parties to CITES such as the USA and EU, which are strongly influenced by non-governmental organisations. Importantly, some range states currently oppose trade legalisation. In the case of rhinos, the number of opposing range states is small but includes Asian countries with critically threatened species (Indonesia) or ideological opposition to sustainable use (India). For elephants, the number of opposing range states appears much larger, and the current geopolitical environment suggests that legalising the international ivory trade is not an imminent possibility. The geopolitical aspects of such trade policy decisions remain salient.

The future?

Following others before us, we have argued that wildlife' crime' is socially constructed and subject to global hegemonic influences. The contemporary narrative around wildlife crime remains under the strong influence of predominantly North American philosophical views on wildlife conservation and a legalistic regulatory approach to governance. Decisions reached and implemented through the global regulatory regime established by CITES also remain strongly influenced by Western values as expressed through the influential views of the United States, European Union, and other developed countries, which increasingly also reflect an animal welfare orientation in conservation decision-making. This, in turn, is contributing towards a drive against wildlife commodification and consequent support of prohibition. However, this is negatively impacting people-centred and incentive-driven approaches to conservation in developing countries in the Global South.

We argue that this prohibitionist thrust can and must evolve into a more nuanced approach—one that involves carefully considered and conceived legalisation through institutional reform. Such reform would align with the prevailing more inclusive thrust of global sustainable development and biodiversity policy, which highlights the critical role of indigenous people and local communities. It is also supported by the thrust of much of the evolving literature at the intersection of social-ecological systems thinking and institutional design. What are the prospects for such reform? We believe that such reform must eventually take place as its unassailable long-term socioecological logic takes hold. However, the pace of reform will be determined by geopolitical shifts that are difficult to predict. We remain hopeful that, sooner rather than later, society will develop and move closer towards more inclusive and universal value systems and institutional arrangements that produce better outcomes for both people and wildlife.

Notes

1 "Trafficking and Poaching: Tanzania." Library of Congress. Retrieved from https://tile.loc.gov/storage-services/service/ll/llglrd/2013417642/2013417642.pdf

2 Zimbabwe Parks and Wildlife Authority, "Statutory Instrument 56 of 2012." Retrieved from https://www.zimparks.org.zw/wp-content/uploads/2024/04/Statutory-Instrument-56-of-2012-Trapping-of-Animals.pdf.

References

Abensperg-Traun, M. (2009). CITES, Sustainable Use of Wild Species and Incentive-Driven Conservation in Developing Countries, with an Emphasis on Southern Africa. *Biological Conservation*, 142(5), 948–963. https://doi.org/10.1016/j.biocon.2008.12.034

Alchian, A. A., & Demsetz, H. (1973). The Property Right Paradigm. *The Journal of Economic History*, 33(1), 16–27.

Anderson, T. L., & Hill, P. J. (1975). The Evolution of Property Rights: A Study of the American West. *The Journal of Law & Economics*, *18*(1), 163–179.

Babor, T., Caulkins, J., Fischer, B., Foxcroft, D., Humphreys, K., Medina-Mora, M. E., Obot, I., Rehm, J., Reuter, P., & Room, R. (2018). *Drug Policy and the Public Good* (2nd edition). Oxford University Press.

Balfour, D., Makoma, K., & Ferreira, S. M. (2024). African Rhino Specialist Group Chair report/Rapport du Groupe de Spécialistes du Rhinocéros d'Afrique. *Pachyderm*, *65*, 20–34. https://doi.org/10.69649/pachyderm.v65i.1311

Barbier, E. B., & Schulz, C.-E. (1997). Wildlife, Biodiversity and Trade. *Environment and Development Economics*, *2*(02), 145–172. https://doi.org/10.1017/S1355770X97000144

Becker, G. S. (1968). Crime and Punishment: An Economic Approach. *Journal of Political Economy*, *76*(2), 176–177. https://doi.org/10.1086/259394

Becker, G. S., Murphy, K. M., & Grossman, M. (2006). The Market for Illegal Goods: The Case of Drugs. *Journal of Political Economy*, *114*(1), 38–60. https://doi.org/10.1086/498918

Beckert, J., & Dewey, M. (2017). *The Architecture of Illegal Markets: Towards an Economic Sociology of Illegality in the Economy*. Oxford University Press.

Beinart, W., & Hughes, L. (2009). *Environment and Empire* (1st edition). Oxford University Press.

Bekoff, M. (Ed.). (2013). *Ignoring Nature No More: The Case for Compassionate Conservation*. University Of Chicago Press.

Berkes, F. (Ed.). (1989). *Common Property Resources: Ecology and Community-Based Sustainable Development*. Belhaven Press.

Callicott, J. B. (1990). Whither Conservation Ethics? *Conservation Biology*, *4*(1), 15–20. https://doi.org/10.1111/j.1523-1739.1990.tb00261.x

Chen, F., & 't Sas-Rolfes, M. (2021). Theoretical Analysis of a Simple Permit System for Selling Synthetic Wildlife Goods. *Ecological Economics*, *180*, 106873. https://doi.org/10.1016/j.ecolecon.2020.106873

Child, G., & Child, B. (2015). The conservation movement in Zimbabwe: An early experiment in devolved community based regulation. *African Journal of Wildlife Research*, *45*(1), 1–16.

Cooney, R., Roe, D., Dublin, H., Phelps, J., Wilkie, D., Keane, A., Travers, H., Skinner, D., Challender, D. W. S., Allan, J. R., & Biggs, D. (2017). From Poachers to Protectors: Engaging Local Communities in Solutions to Illegal Wildlife Trade. *Conservation Letters, 10*(3), 367–374. https://doi.org/10.1111/conl.12294

Demsetz, H. (1967). Toward a Theory of Property Rights. *The American Economic Review, 57*(2), 347–359. https://www.jstor.org/stable/1821637

Di Minin, E., 't Sas-Rolfes, M., Selier, J., Louis, M., & Bradshaw, C. J. A. (2022). Dismantling the Poachernomics of the Illegal Wildlife Trade. *Biological Conservation, 265*, 109418. https://doi.org/10.1016/j.biocon.2021.109418

Do, Q.-T., Levchenko, A. A., Ma, L., Blanc, J., Dublin, H., & Milliken, T. (2020). The Price Elasticity of African Elephant Poaching. *The World Bank Economic Review, 35*(3), 545–562. https://doi.org/10.1093/wber/lhaa008

Duffy, R. (2022). *Security and Conservation: The Politics of the Illegal Wildlife Trade*. Yale University Press.

Duffy, R., Massé, F., Smidt, E., Marijnen, E., Büscher, B., Verweijen, J., Ramutsindela, M., Simlai, T., Joanny, L., & Lunstrum, E. (2019). Why We Must Question the Militarisation of Conservation. *Biological Conservation, 232*, 66–73. https://doi.org/10.1016/j.biocon.2019.01.013

Epstein, C. (2006). The Making of Global Environmental Norms: Endangered Species Protection. *Global Environmental Politics, 6*(2), 32–54. https://doi.org/10.1162/glep.2006.6.2.32

Epstein, R. A. (2004). Animals as Objects, or Subjects, of Rights. In C. R. Sunstein & M. C. Nussbaum (Eds.), *Animal Rights: Current Debates and New Directions* (pp. 143–161). Oxford University Press. https://www.oxfordscholarship.com/view/10.1093/acprof:oso/9780195305104.001.0001/acprof-9780195305104

Escobar-Pemberthy, N., & Ivanova, M. (2020). Implementation of Multilateral Environmental Agreements: Rationale and Design of the Environmental Conventions Index. *Sustainability, 12*(17), Article 17. https://doi.org/10.3390/su12177098

Fischer, C. (2004). The complex interactions of markets for endangered species products. *Journal of Environmental Economics and Management, 48*(2), 926–953. https://doi.org/10.1016/j.jeem.2003.12.003

Francione, G. L. (1995). *Animals, Property, and the Law*. Temple University Press.

Gavin, M. C., McCarter, J., Berkes, F., Mead, A. T. P., Sterling, E. J., Tang, R., & Turner, N. J. (2018). Effective Biodiversity Conservation Requires Dynamic, Pluralistic, Partnership-Based Approaches. *Sustainability, 10*(6), 1846. https://doi.org/10.3390/su10061846

Geist, V. (1994). Wildlife Conservation as Wealth. *Nature, 368*(6471), 491–492. https://doi.org/10.1038/368491a0

Gooden, J., & 't Sas-Rolfes, M. (2020). A Review of Critical Perspectives on Private Land Conservation in Academic Literature. *Ambio, 49*(5), 1019–1034. https://doi.org/10.1007/s13280-019-01258-y

Hardin, G. (1968). The Tragedy of the Commons. *Science, 162*(3859), 1243–1248.

Hiller, C., & 't Sas-Rolfes, M. (2024). Systematic review of the impact of restrictive wildlife trade measures on conservation of iconic species in southern Africa. *Conservation Biology, 39*(1), e14262. https://doi.org/10.1111/cobi.14262

Hübschle, A. (2016). *A Game of Horns: Transnational Flows of Rhino Horn*. IMPRS-SPCE. https://globalinitiative.net/wp-content/uploads/2016/05/Huebschle-A-Game-of-Horns-2016.pdf

Hübschle, A., & Margulies, J. (2024). The Need for a Socioecological Harm Reduction Approach to Reduce Illegal Wildlife Trade. *Conservation Biology, 38*(5), e14335. https://doi.org/10.1111/cobi.14335

Hübschle, A. M. (2017). The Social Economy of Rhino Poaching: Of Economic Freedom Fighters, Professional Hunters and Marginalised Local People. *Current Sociology, 65*(3), 427–447. https://doi.org/10.1177/0011392116673210

Hughes, A., Auliya, M., Altherr, S., Scheffers, B., Janssen, J., Nijman, V., Shepherd, C. R., D'Cruze, N., Sy, E., & Edwards, D. P. (2023). Determining the sustainability of legal wildlife trade. *Journal of Environmental Management, 341*, 117987. https://doi.org/10.1016/j.jenvman.2023.117987

Hughes, A. C. (2021). Wildlife Trade. *Current Biology, 31*(19), R1218–R1224. https://doi.org/10.1016/j.cub.2021.08.056

Hutton, J., & Webb, G. (2002). Crocodiles: Legal Trade Snaps Back. In S. Oldfield (Ed.), *The Trade in Wildlife: Regulation for Conservation* (illustrated edition, pp. 108–120). Earthscan.

James, A., Gaston, K. J., & Balmford, A. (2001). Can We Afford to Conserve Biodiversity? *BioScience, 51*(1), 43–52. https://doi.org/10.1641/0006-3568(2001)051[0043:CWATCB]2.0.CO;2

Kasterine, A., & Lichtenstein, G. (2018). *Trade in Vicuña: Implications for Conservation and Rural Livelihoods*. International Trade Centre. https://www.intracen.org/resources/publications/trade-in-vicuna-implications-for-conservation-and-rural-livelihoods

Kuiper, T., Altwegg, R., Beale, C., Carroll, T., Dublin, H. T., Hauenstein, S., Kshatriya, M., Schwarz, C., Thouless, C. R., Royle, A., & Milner-Gulland, E. J. (2023). Drivers and Facilitators of the Illegal Killing of Elephants across 64 African Sites. *Proceedings of the Royal Society B: Biological Sciences, 290*(1990), 20222270. https://doi.org/10.1098/rspb.2022.2270

Kurland, J., Pires, S. F., McFann, S. C., & Moreto, W. D. (2017). Wildlife Crime: A Conceptual Integration, Literature Review, and Methodological Critique. *Crime Science*, 6(1), 4. https://doi.org/10.1186/s40163-017-0066-0

Lavadinović, V. M., Islas, C. A., Chatakonda, M. K., Marković, N., & Mbiba, M. (2021). Mapping the Research Landscape on Poaching: A Decadal Systematic Review. *Frontiers in Ecology and Evolution, 9*. https://doi.org/10.3389/fevo.2021.630990

Lindsey, P. A., Miller, J. R. B., Petracca, L. S., Coad, L., Dickman, A. J., Fitzgerald, K. H., Flyman, M. V., Funston, P. J., Henschel, P., Kasiki, S., Knights, K., Loveridge, A. J., Macdonald, D. W., Mandisodza-Chikerema, R. L., Nazerali, S., Plumptre, A. J., Stevens, R., Zyl, H. W. V., & Hunter, L. T. B. (2018). More Than $1 Billion Needed Annually to Secure Africa's Protected Areas with Lions. *Proceedings of the National Academy of Sciences*, 201805048. https://doi.org/10.1073/pnas.1805048115

Lubilo, R., & Hebinck, P. (2019). 'Local hunting' and community-based natural resource management in Namibia: Contestations and livelihoods. *Geoforum, 101*, 62–75. https://doi.org/10.1016/j.geoforum.2019.02.020

Lueck, D. (2002). The Extermination and Conservation of the American Bison. *The Journal of Legal Studies, 31*(S2), S609–S652. https://doi.org/10.1086/340410

Lueck, D., & Torrens, G. (2020). Property Rights and Domestication. *Journal of Institutional Economics, 16*(2), 199–215. Cambridge Core. https://doi.org/10.1017/S1744137419000390

Macdonald, D. W., Harrington, L. A., Moorhouse, T. P., & D'Cruze, N. (2021). Trading Animal Lives: Ten Tricky Issues on the Road to Protecting Commodified Wild Animals. *BioScience, biab035, 71*(8), 846–860. https://doi.org/10.1093/biosci/biab035

MacKenzie, J. M. (1997). *The Empire of Nature: Hunting, Conservation and British Imperialism (Studies in Imperialism): 7*. Manchester University Press.

Margulies, J. D., Massé, F., & VandeBerg, B. (2023). Illicit Geographies and Contested Environments: An Introduction. *Political Geography, 101*, 102818. https://doi.org/10.1016/j.polgeo.2022.102818

Marshall, A. (2009). *Principles of Economics: Unabridged Eighth Edition* (8th edition). Cosimo Classics.

Massé, F., Dickinson, H., Margulies, J., Joanny, L., Lappe-Osthege, T., & Duffy, R. (2020). Conservation and Crime Convergence? Situating the 2018 London Illegal Wildlife Trade Conference. *Journal of Political Ecology, 27*(1), 23–42. https://doi.org/10.2458/v27i1.23543

Massé, F., & Margulies, J. D. (2020). The Geopolitical Ecology of Conservation: The Emergence of Illegal Wildlife Trade as National Security Interest and the Re-Shaping of US Foreign Conservation Assistance. *World Development, 132*, 104958. https://doi.org/10.1016/j.worlddev.2020.104958

Meeks, D., Morton, O., & Edwards, D. P. (2024). Wildlife farming: Balancing economic and conservation interests in the face of illegal wildlife trade. *People and Nature*, 6(2), 446–457. https://doi.org/10.1002/pan3.10588

Milner-Gulland, E. J., & Leader-Williams, N. (1992). A Model of Incentives for the Illegal Exploitation of Black Rhinos and Elephants: Poaching Pays in Luangwa Valley, Zambia. *Journal of Applied Ecology, 29*, 388–401.

Muposhi, V. K., Gandiwa, E., Bartels, P., & Makuza, S. M. (2016). Trophy Hunting, Conservation, and Rural Development in Zimbabwe: Issues, Options, and Implications. *International Journal of Biodiversity, 2016*, e8763980. https://doi.org/10.1155/2016/8763980

Nadelmann, E. A. (1990). Global Prohibition Regimes: The Evolution of Norms in International Society. *International Organisation, 44*(4), 479–526.

Nagin, D. S. (2013). Deterrence in the Twenty-First Century. *Crime and Justice: A Review of Research, 42*, 199.

Ostrom, E. (1990). *Governing the Commons: The Evolution of Institutions for Collective Action*. Cambridge University Press.
Ostrom, E. (2005). *Understanding Institutional Diversity*. Princeton University Press.
Ostrom, E. (2009). A General Framework for Analysing Sustainability of Social-Ecological Systems. *Science, 325*(5939), 419–422. https://doi.org/10.1126/science.1172133
Reeve, R. (2002). *Policing International Trade in Endangered Species: The CITES Treaty and Compliance*. Earthscan Ltd.
Roe, D., Dickman, A., Kock, R., Milner-Gulland, E. J., Rihoy, E., & 't Sas-Rolfes, M. (2020). Beyond Banning Wildlife Trade: COVID-19, Conservation and Development. *World Development, 136*, 105121. https://doi.org/10.1016/j.worlddev.2020.105121
Rubino, E. C., & Pienaar, E. F. (2020). Rhinoceros Ownership and Attitudes Towards Legalisation of Global Horn Trade within South Africa's Private Wildlife Sector. *Oryx, 54*(2), 244–251. https://doi.org/10.1017/S0030605318000030
Searle, J. R. (2005). What is an institution? *Journal of Institutional Economics, 1*(1), 1–22. Cambridge Core. https://doi.org/10.1017/S1744137405000020
Steinhart, E. I. (2005). *Black Poachers, White Hunters: A Social History of Hunting in Colonial Kenya* (1st edition). James Currey.
Stevens, A., Hughes, C. E., Hulme, S., & Cassidy, R. (2022). Depenalisation, Diversion and Decriminalisation: A Realist Review and Programme Theory of Alternatives to Criminalisation for Simple Drug Possession. *European Journal of Criminology, 19*(1), 29–54. https://doi.org/10.1177/1477370819887514
Swanson, T. M. (1994). The Economics of Extinction Revisited and Revised: A Generalised Framework for the Analysis of the Problems of Endangered Species and Biodiversity Losses. *Oxford Economic Papers, 46*, 800–821.
't Sas-Rolfes, M. (1997). Elephants, Rhinos and the Economics of Illegal Trade. *Pachyderm, 24*, 23–29.
't Sas-Rolfes, M. (2000). Assessing CITES: Four Case Studies. In J. Hutton & B. Dickson (Eds.), *Endangered Species, Threatened Convention* (pp. 69–87). Earthscan Publications Ltd.
't Sas-Rolfes, M. (2024). *Conflicting Framings in Global Conservation Governance: Consequences for African megafauna* [Https://purl.org/dc/dcmitype/Text, University of Oxford]. https://ora.ox.ac.uk/objects/uuid:dac4c238-2114-4750-aa1b-aa1ce76e0c6d
't Sas-Rolfes, M., Challender, D. W. S., Hinsley, A., Veríssimo, D., & Milner-Gulland, E. J. (2019). Illegal Wildlife Trade: Scale, Processes, and Governance. *Annual Review of Environment and Resources, 44*(1), 201–228. https://doi.org/10.1146/annurev-environ-101718-033253
't Sas-Rolfes, M., Challender, D. W. S., & Wainwright, L. (2024). Playing the CITES Game: Lessons on Global Conservation Governance from African Megafauna. *Environmental Policy and Governance, n/a*(n/a). https://doi.org/10.1002/eet.2123
't Sas-Rolfes, M., & Emslie, R. (2024). African Rhino Conservation and the Interacting Influences of Property, Prices, and Policy. *Ecological Economics, 220*, 108123. https://doi.org/10.1016/j.ecolecon.2024.108123
't Sas-Rolfes, M., Emslie, R., Adcock, K., & Knight, M. (2022). Legal Hunting for Conservation of Highly Threatened Species: The Case of African Rhinos. *Conservation Letters, 15*(3), e12877. https://doi.org/10.1111/conl.12877
't Sas-Rolfes, M., & Gooden, J. (2024). A Conflict of Visions: Ideas Shaping Wildlife Trade Policy toward African Megafauna. *People and Nature*, 6(5), 2029–2045. https://doi.org/10.1002/pan3.10705
't Sas-Rolfes, M., Moyle, B., & Stiles, D. (2014). The Complex Policy Issue of Elephant Ivory Stockpile Management. *Pachyderm, 55*, 62–77.
't Sas-Rolfes, M., Rademeyer, J., Vigne, L., Emslie, R., Knight, M., Gaymer, J., & Talukdar, B. (2025). The Impact of Poaching on Rhino Conservation. In M. Melletti, B. Talukdar, & D. Balfour (Eds.), *Rhinos of the World: Ecology, Conservation and Management*. Springer. https://link.springer.com/book/9783031671685
Taylor, W. A., Lindsey, P. A., Nicholson, S. K., Relton, C., & Davies-Mostert, H. T. (2020). Jobs, Game Meat and Profits: The Benefits of Wildlife Ranching on Marginal Lands in South Africa. *Biological Conservation, 245*, 108561. https://doi.org/10.1016/j.biocon.2020.108561
Tober, J. (1981). *Who Owns the Wildlife? Political Economy of Conservation in Nineteenth-century America: 37 (Contributions in Economics & Economic History): The... Of Conservation in Nineteenth-Century America*. Praeger.

van Uhm, D. P. (2015). Towards Moral Principles Regarding Non-Human Animals: A Green Criminological Perspective. In F. De Jong (Ed.), *Overarching Views of Crime and Deviancy: Rethinking the Legacy of the Utrecht School*. (pp. 565–587). Eleven International Publishing.
van Uhm, D. P. (2016). *The Illegal Wildlife Trade: Inside the World of Poachers, Smugglers and Traders*. Springer International Publishing. https://www.springer.com/gb/book/9783319421285
Vigne, L., 't Sas-Rolfes, M., Milliken, T., Rademeyer, J., & Talukdar, B. (2025). The Impact of Rhino Horn Trafficking on Conservation. In M. Melletti, B. Talukdar, & D. Balfour (Eds.), *Rhinos of the World: Ecology, Conservation and Management*. Springer. https://link.springer.com/book/9783031671685
Waldron, A., Mooers, A. O., Miller, D. C., Nibbelink, N., Redding, D., Kuhn, T. S., Roberts, J. T., & Gittleman, J. L. (2013). Targeting Global Conservation Funding to Limit Immediate Biodiversity Declines. *Proceedings of the National Academy of Sciences*, *110*(29), 12144–12148. https://doi.org/10.1073/pnas.1221370110
Wellsmith, M. (2011). Wildlife Crime: The Problems of Enforcement. *European Journal on Criminal Policy and Research*, *17*(2), 125–148. https://doi.org/10.1007/s10610-011-9140-4
White, R. (2003). Environmental Issues and the Criminological Imagination. *Theoretical Criminology*, *7*(4), 483–506. https://doi.org/10.1177/13624806030074005
Williams, V. L., Coals, P. G., Bruyn, M. de, Naude, V. N., Dalton, D. L., & Kotzé, A. (2021). Monitoring Compliance of CITES Lion Bone Exports from South Africa. *PLOS ONE*, *16*(4), e0249306. https://doi.org/10.1371/journal.pone.0249306
Wilson, B. J. (2020). *The Property Species: Mine, Yours, and the Human Mind*. Oxford University Press.
Wyatt, T. (2021). *Is CITES Protecting Wildlife? Assessing Implementation and Compliance* (1st edition). Routledge.

22

"DENIAL OF INJURY" AND "DENIAL OF THE VICTIM" IN THE (ILLEGAL) WILDLIFE TRADE

The case of the European eel

Mònica Pons Hernández

Introduction

Skyes and Matza's (1957) neutralisation theory was developed as a promising line of research to further understand juvenile delinquency. In accordance with the theory, individuals who consider engaging in harmful or criminal actions are frequently hindered by the possibility of feeling guilty or forming an unfavourable self-perception. To carry out criminal acts successfully, it becomes essential to surpass these internal constraints (Copes & Maruna, 2017). One of the theories developed to explain how people surpass moral constraints is the neutralisation theory (Skyes & Matza, 1957). Over the years, many scholars have drawn on it to gain a deeper understanding of how these neutralisation techniques enable all kinds of deviant behaviour. It is also true for scholars researching environmental crimes.

Among others, the neutralisation theory has been used for studying animal abuse (Sollund, 2013), climate change (Wyatt & Brisman, 2016), or polluting behaviours (Mir Mohamad Tabar, Rizzolo et al., 2022; Mir Mohamad Tabar, South et al., 2022). It has also been employed to analyse how Animal Rights activists rationalise their actions (Liddick, 2013) or to examine the messages of the Climate Change Counter Movement (McKie, 2019). And, most importantly, within the frame of this research, the neutralisation theory has also been used to study wildlife crimes. For example, the use of neutralisation techniques among precious orchids collectors (Mackenzie & Yates, 2016), deer poachers (Eliason & Dodder, 1999), illegal hunters (Rytterstedt, 2016), farmers engaging in badger culling (Enticott, 2011), those engaging in the illegal killing of large carnivores (Pohja-Mykrä, 2017), and on other wildlife law violations (Eliason, 2003).

In this chapter, I examine the use of the neutralisation techniques eel fishers and traders use to overcome their guilt and engage in illegal, unreported, and unregulated (IUU) eel fishing and their trafficking, as well as in persisting in fishing and trading a critically endangered species. The European eel is one of the many species facing extinction. Among other threats, European eel numbers are in decline due to their overexploitation (Drounieau et al., 2018; Gutierrez & Duffy, 2023). Accordingly, in 2007, the species was listed in Appendix II of the Convention on International Trade in Endangered Species of Wild Fauna and Flora (CITES), and, in 2010, the European Union (EU) set an export and import ban. Despite the

 DOI: 10.4324/9781003333906-26

fishing and trading restrictions imposed, the criminalisation under certain circumstances, and the precarious situation of the European eel, fishers and traders across the eel distribution area persist in engaging in this activity (see Europol, 2022; Pons-Hernandez, 2023; UNODC, 2020).

Using the philosophical underpinning of green criminology (White & Heckenberg, 2014) and employing the neutralisation theory (Skyes & Matza, 1957) applied to the case study of the European eel, this piece aims to examine the role of neutralisation techniques in enabling wildlife trafficking. To begin, this chapter presents an overview of green criminology and the neutralisation theory. Next, the methods used for this research are presented, followed by a presentation of the case study. This chapter concludes by highlighting the *denial of injury* and the *denial of the victim* as the key techniques of neutralisation eel fishers and traders employ and the need for further research applied to different case studies to better grasp the use of neutralisation techniques in enabling wildlife trafficking and wildlife crimes more generally.

Philosophical and theoretical underpinning

Green criminology: an overview

The scholarship in the field of green criminology concerning wildlife crimes has increased over the years (Sollund, 2020). Wildlife crimes—which encompass the illegal wildlife trade—are criminal acts against nonhuman animals (UNODC, 2016). However, because each country protects wildlife inhabiting its territory differently, there is no universal definition of wildlife crime, which means that some acts are deemed criminal in some countries but, despite being equally harmful, are allowed in others (UNODC, 2016). Accordingly, to overcome this conceptual and legal barrier, this research is framed within green criminology.

Green criminology not only broadens the scope of what is considered criminal to include those actions that are harmful, yet overlooked by traditional criminal law and the criminal justice system (Ruggiero & South, 2010; White, 2011; White & Heckenberg, 2014), but it also expands the understanding of who can be considered a victim, including human and nonhuman animals, ecosystems, and the planet itself (Beirne, 2018; Brisman, 2014; Nurse, 2017; Nurse & Wyatt, 2020; Wyatt, 2022). This more comprehensive recognition of actions causing harm to nonhumans and their consideration as victims is what makes green criminology particularly relevant in studying wildlife trafficking and other crimes against wildlife within the field of criminology.

Neutralisation theory: an overview

The neutralisation theory asserts that those who commit harmful or criminal acts employ neutralisation techniques to avoid culpability. As opposed to subcultural theorists, who claim that delinquency is the product of the rejection of the dominant social order (Copes & Maruna, 2017), Skyes and Matza's (1957) neutralisation theory argues that delinquents are committed to the dominant social order and, for this reason, they need to neutralise the guilt associated with violating the social norms. Initially, Skyes and Matza (1957) differentiated among five techniques of neutralisation: denial of responsibility, denial of injury, denial of the victim, condemnation of the condemner, and appeal to higher loyalties. Nowadays, as other types of offending have been defined, other techniques have been described (Copes &

Maruna, 2017). For instance, Liddick (2013, p. 618) refers to the "appeal to a higher moral principle" used by Animal Rights Activists.

Nevertheless, this chapter is built around the use of two techniques, the denial of injury and the denial of the victim, and how both techniques intertwine to enable illegal trade in wildlife. On the one hand, the denial of injury centres on the harm involved in the delinquent act (Skyes & Matza, 1957). On the other, the denial of the victim refers to the movement of the delinquent in an avenger position and the victim in the position of a wrongdoer (Skyes & Matza, 1957). The existence of the victim can be denied in a different sense, however. As Skyes and Matza (1957, p. 668) explain, "when the victim is physically absent, unknown, or a vague abstraction, the awareness of the victim's existence is weakened". As explored using the case of the European eel, the denial of injury and the denial of the victim are not only of key importance but are also intertwined when analysing wildlife trade and traffic.

Methods

Like most research employing this criminological theory, this research uses qualitative methods to examine how deviant individuals utilise neutralisation tactics to deactivate the guilt associated with committing wildlife crimes (Copes & Maruna, 2017). This research draws on 14 open-ended semi-structured interviews with key stakeholders, including eel fishers, eel traders, environmental law enforcement agents, and members of the scientific community. I utilised purposive sampling to choose participants based on their potential contributions in an iterative manner. Subsequently, I employed snowball sampling, following the guidelines outlined in Ritchie et al. (2014). All interviews underwent recording, transcription, anonymisation, and analysis using NVivo 12 software. Ethical approval for the research was granted by the Universitat Rovira i Virgili Ethics Committee concerning Research into People, Society and the Environment on 29/10/2021 (code CEIPSA-2021-TD-0018). Furthermore, I employed triangulation by cross-referencing the collected information with document analysis, encompassing academic literature, policy documents, reports, and news media. This approach ensured a comprehensive and reliable data collection process. Overall, it offers valuable insights into how fishers and traders perceive and rationalise their harmful and criminal actions.

Background to the European eel and their trade and traffic

The European eel has a long and complex life cycle (Aranburu et al., 2015). Spawning and eggs have never been observed in the wild (Dekker, 2000), but in 2022 was gathered the first direct evidence of the migration of the European eel back to their breeding place in the Sargasso Sea (Wright et al., 2022). The hatched larvae, called leptocephalus, use the Gulf Stream and the North Atlantic Current to reach the continental shelf of the Atlantic coast of Europe and North Africa. It is not until this point that they metamorphose into *glass eels* and enter continental waters. Once the glass (unpigmented) eel reaches the coast, it inhabits estuaries (where the freshwater and saltwater meet). Then, at the elver stage, they inhabit lagoons and lakes or migrate up the river basin before becoming yellow eels, the longest of their life stages (Bevacqua et al., 2015). Once they are in continental waters, they can settle or move across habitats, taking them up to 50 years to reach maturation and become silver eels (Bevacqua et al., 2015). Once they metamorphose into silver eels, they undertake a

5,000–7,000 km journey southwest using the Canary and Azores current systems back at the south-western region of the Sargasso Sea (Aarestrup, 2009; Miller et al., 2019), where they spawn and die (Bevacqua et al., 2015; Dekker, 2000). This complex life cycle, involving ocean and continental water habitats, makes them especially vulnerable to anthropogenic threats.

According to Dekker (2016), the European glass eel experienced a population decline of 90–99% since the 1980s. This is reflected in the extreme decrease in the levels of recruitment of glass eels in coastal areas (Miller et al., 2019). However, the decline of the species not only affects the European eel but also other temperate eels, such as the American and Japanese eel (Drouineau et al., 2018). The causes of the decline of eel populations involve several factors: climate change, pollution, habitat loss, the appearance of parasites, and overfishing (Bornarel et al., 2018; Dekker, 2003; Drouineau et al., 2018; Moriarty & Dekker, 1997).

To start, the process of eel migration is significantly affected by global warming and consequent changes in ocean conditions. This impact extends to various stages such as leptocephalus drift, survival, and the spawning migration of silver eels (Drouineau et al., 2018). Additionally, the lipid-rich nature of eels makes them susceptible to organic contaminants, heavy metals, and pesticides. *Habitat loss* due to the fragmentation caused by weirs and dams has been identified as one of the most important threats to the survival of the species (Drouineau et al., 2018; Pons-Hernandez, 2024). Furthermore, the introduction of the alien parasitoid *Anguillicola crassus* through aquaculture poses a widespread threat to European eels (Drouineau et al., 2018). Lastly, the intensive exploitation of eels at all life stages exacerbates the decline of the species; overfishing has been identified as a significant threat (Drouineau et al., 2018). Eels, valued as both a commercial and recreational "resource", are targeted throughout their continental life stages—ranging from glass eels and elvers to yellow eels and silver eels. Eel consumption, coupled with modernised fishing and commercial practices, has led to the overexploitation of freshwater species (Dekker, 2019). Because captive breeding of eels has not been achieved, all individuals come from the wild (Drouineau et al., 2018).

Population decline linked to all these anthropogenic threats, and the impossibility of captive breeding to meet demand led to the inclusion of the European eel in Appendix II of CITES and resulted in an export and import ban in the European Union—along with the adoption of other measures within the EU. Nowadays, it is estimated that up to 350 million glass eels are removed from European waters and trafficked to Asia every year (Europol, 2019). However, before the suspension of exports and imports in 2010, as Ringuet et al. (2002) report, three shipments of glass eels were sent every day by plane to China. Based on what has been described in the World Wildlife Crime Report (UNODC, 2020), the number of European eels legally fished for European consumption appears to be less than those illegally traded to Asia for farming. Following the ban, the demand has been supplied through poaching and the illegal market and by EU neighbouring countries with European eel populations, but without the EU ban, which increased the exports to Asia (e.g. Morocco, Algeria, Tunisia, or the United Kingdom) (Crook & Nakamura, 2013; Nijman, 2017). Bearing in mind the critical state of the European eel population, both trade and trafficking are equally harmful and impact the survival of the species (Gutierrez & Duffy, 2023). Thus, both legal and illegal fishing and trade are to blame for the decline of the European eel population. As discussed in the following pages, neutralisation techniques are used for those who *continue* exploiting the species despite being conscious of their delicate situation. For this reason, in this case, neutralisation techniques appear to be used to persist in harmful and criminal behaviour rather than leading to the first commitment of the crime (Maruna & Copes, 2005).

The denial of injury and the denial of the victim in the (illegal) glass eel trade

Denial of injury within the trade and traffic of European eels

The "denial of injury" was identified as a neutralisation technique since the original formulation of the theory (Copes & Maruna, 2017). Skyes and Matza (1957) defined it as a claim that no one was hurt or that harm was not intended. Instead, here it is defined as a denial of harm attached to eel fishing and trading. While it is widely accepted that both the trade and traffic of nonhuman species harm the species—threatening them with extinction—and the welfare of the individuals affected along the supply chain (Wyatt, 2022; Wyatt et al., 2021), glass eel fishers and traders deny that their actions injure the European eel and use the "denial of injury" as a neutralisation technique.

Before interviewing eel fishers and traders, I interviewed members of the scientific community to better understand the European eel as a species and the current discussions surrounding their imminent extinction. In multiple conversations, I have been told that both the legal and illegal trade, along with climate change and the loss of habitat, were the main threats to the species, but I also have been told that fishers and traders use the latter two to remove their responsibility. In the words of one of the participants, "What they [glass eel fishers and traders] will tell you is ¡No! ¡*The problem is the dams!* (…) For them, there is always someone else to blame". So, as expected, both glass eel fishers and traders resorted to the loss of habitat, the construction of dams to impede the migratory process, and other threats to European eels to neutralise their shared responsibility in pushing the species to the brink of extinction. For instance, Alex, a third-generation glass eel fisher, blamed the pollution of the water when claiming that:

> Overfishing is not a problem for the glass eel, ships are the real problem. Look, the river is the one that impacts the eel, and it impacts a lot. Having clear water during October signals that it is going to be a good fishing year, but with the ships…water is not clear anymore.
>
> *[My translation]*

Similarly, while one of the scientists interviewed referred to the strength and high adaptability of the eels to water habitats where the conditions were not optimal, an eel fisher claimed that "we are not the problem, the problem is pollution".

Nevertheless, the most recurrent threat to European eels, used as a neutralisation technique by members of the fishing industry to deny the harm to the species caused by overexploitation, is habitat loss due to fragmentation, commonly referred to as "hydropower mortality". In this vein, an eel trader stated:

> [the eel] has decreased a lot, but not as much because of overfishing, there is also climate change, pollution, and what scientists say is mainly the loss of habitat, the transformation of habitat (…) [the eel] has a cycle opposite to salmon, it grows in rivers and goes to saltwater to spawn and die. What happens? As we make cuts to the rivers to construct dams, we put obstacles to migration. People come from Zaragoza and tell us that they used to have eels. This affects *production*. There are countries like the Netherlands that are working on building systems to help eels cross. Instead, we are just blaming and restricting the fishery.
>
> *[My translation, emphasis added]*

Copes and Maruna (2017, p. 126) argue that "what is important is the function of neutralisation techniques (what they *do*), rather than the specific varieties in which they come". As found, fishers and traders deny the harm attached to their actions. As reflected in the quotations above, fishers and traders use the term "overfishing" instead of "fishing", turning a blind eye to the inherent "overexploitation outside sustainable limits" attached to the term *overfishing*. For this reason, the denial of injury helps them to overcome the guilt and continue harming the species through overexploitation. This technique of neutralisation can hinder compliance with the regulations adopted to avoid their extinction—limiting the fishing licences or reducing the fishing season—because the actors involved do not identify their actions (fishing) as harmful or particularly harmful.

Denial of the European eel as the victim of (wildlife) crime

Eel fishers and traders may accept *some* degree of responsibility for the declining numbers of European eels. When they accept the impacts of (over)exploitation on the European eel, they appeal to one superior neutralisation technique, the "denial of the victim". The denial of the victim has also been defined in the original formulation to argue that the offender may claim that some victims acted improperly, so they deserved punishment or retaliation. Or, as Carrabine and colleagues (2014, p. 81) describe, "the delinquent denies that the person injured or wronged is really a victim". In this vein, they use the sentence "the victim of a mugging was 'only queer'" to exemplify how the denial of the victim can be used as a neutralisation technique (Carrabine et al., 2014, p. 81). Instead, here I reject the premise that the victim needs to be human because the direct victims of wildlife crimes are not human but nonhumans (Wyatt, 2022). Failing to acknowledge nonhuman species as victims of crime is what I refer to here as the denial of the victim. Therefore, to exemplify what I refer to as "denial of the victim", I use this sentence, instead: "The victims of traffic were 'only eels'". In other words, in this research, the denial of the victim refers to the denial of victimhood of nonhuman species. In this sense, the denial of the victim refers to the denial of their self-worth and individuality.

In the first step of the supply chain, when eels are abducted (Sollund, 2019), they are constantly referred to by eel fishers in terms of private property. As claimed by an eel fisher, "Eels are a resource that we have here in the environment, it is my resource, so I live out of this" [my translation]. Accordingly, the value of the European eel is subjugated to the purpose they serve to the humans of a given territory. Moreover, moving to the next step of the supply chain, once pulled out of their habitat, European eels are framed as a commodity. For instance, in a conversation with a member of the eel industry—in which we were talking about the extent of the mortality of the eels due to exploitation of the species for consumption—, I was told that the industry was doing its part in reducing mortality by saying: "We've reduced the mortality in capture and transport to the farms. Now is less than 2% of deaths from catch to plate".

To this, I responded: "Well, in the plate there is 100% death".

To which the member of the eel industry replied: "Well, yeah. At the end that's 100%, but before it was 50% in catch and transport".

This excerpt from our conversation reflects that eels are largely identified as a commodity, so reducing mortality is only a tool to increase productivity and profitability. As seen, businesses may perceive themselves as the victims, as the ones affected by the population decline and the extinction of the species because they are the ones losing their "stock". Because European eels are not considered the victims, this crime is viewed by its perpetrators as victimless (Nurse, 2021). Framing the European eel as a commodity means framing the species in terms of economic value (Hutchinson, 2023). Thus, in their mind, in the same way that the victim of a robbery is not the phone, jewellery, or another material object, the victim of this crime is not the European eel. This denial of the victim, similar to the denial of the injury discussed above, perpetuates the overexploitation of the species not only because it frees them from guilt but also because it helps to diminish the extent of the harm and the worthiness of its victims.

Of particular interest when exploring glass eel trade and trafficking is that both neutralisation techniques appear to intertwine when looking into the harm to the individual eels—instead of to the harm to the species—because the denial of the victim as worthy appears to be intrinsically linked to the denial of the injury itself. This intricate connection becomes apparent as the act of dismissing the victim's worth is intricately linked to a simultaneous denial of the injury. The trade of glass eels and eels, from abduction to transportation, farming, slaughter, and even their reintroduction into the wild, is marked by instances of violence (Pons-Hernandez, 2024). However, despite the horrors to which they are subjected, fishers, traders, and even consumers deny the existence of injury, not because it does not exist, but because they deny that the individual eel can *feel* it.

For example, when a member of a local community was explaining to me how they kill the glass eels to prepare one of the traditional recipes, they said: "To cook glass eels the glass eel needs to be alive, so yes, it is a murder, but it does not feel anything, in two minutes it is dead, it dies in no time, it is very fast" [my translation]. By saying this, they neutralise their guilt by denying the torture to which they are subjected through their belief that they are not able to feel pain. So, this denial of injury is not rooted in the absence of injury but rather in the refusal to acknowledge the eel's capacity to experience it. Essentially, by rejecting the notion of injury, they are, in essence, rejecting the very sentience of the eels and the other nonhuman species. This dynamic brings forth that the denial of injury in the context of wildlife trafficking and other wildlife crimes is paired with the denial of the victim itself. In this context, the denial of the injury transcends a mere negation of physical suffering; it turns into a deliberate refusal to recognise the cognitive capacities of these sentient beings. The actors involved shield themselves from the moral implications of their actions by constructing a narrative that delegitimises the eel's ability to feel pain. So, by denying the injury, they deny their sentience. For this reason, the "denial of injury" seems to adopt a new dimension that overlaps with the "denial of the victim" when referring to wildlife trafficking and other wildlife crimes.

Conclusion

The "denial of injury" and the "denial of the victim" within the trade and traffic of European eels form a complex web of rationalisations that sustain the overexploitation of this endangered species. In the realm of "denial of injury", glass eel fishers and traders employ neutralisation techniques to absolve themselves of responsibility for the harm they inflicted on European eels. As seen, they blame factors like habitat loss, pollution, and climate change, framing their actions as innocuous within a broader context of threats. This denial not only allows them to evade guilt but also impedes compliance with regulations aimed at preventing extinction. However, the "denial of the victim" manifests how eel fishers and traders devalue the European eels by framing the species as a commodity or natural resource. In this vein, they overlook the intrinsic value of the species, perpetuating a view that sees their exploitation as victimless. This denial not only rationalises their actions but also downplays the severity of the crime, hindering efforts to save the European eel from extinction. Together, these denial mechanisms are found to enable glass eel trade and trafficking and to create a barrier to meaningful conservation, allowing the exploitation of European eels to persist unchecked. However, both neutralisation techniques seem to overlap when referring to the harm to the individual victims of trade and traffic instead of to the species. In this case, the denial of injury and the denial of the victim seem to overlap when the actors involved construct a narrative that delegitimises the eel's ability to feel pain, defending themselves against the moral implications. Consequently, the denial of injury extends to a denial of the victim's sentience, creating an overlap between both in the context of wildlife trafficking and other wildlife crimes.

References

Aarestrup, K., Økland, F., Hansen, M., Righton, D., Gargan, P., Castonguay, M., Bernatchez, L., Howey, P., Sparholt, H., Pedersen, M., & McKinley, R. (2009) Oceanic spawning migration of the European eel (Anguilla anguilla). Available at: https://science.sciencemag.org/content/325/5948/1660 (Accessed 25 November 2023).

Aranburu, A., Díaz, E., & Briand, C. (2015) Glass eel recruitment and exploitation in a South European estuary (Oria, Bay of Biscay). *ICES Journal of Marine Science*, 73(1), 111–121. https://doi.org/10.1093/icesjms/fsv116

Beirne, P. (2018) *Murdering Animals: Writings on Theriocide, Homicide and Nonspeciesist Criminology*. Cham: Palgrave Macmillan.

Bevacqua, D., Melià, P., Gatto, M., & de Leo, G. (2015) A global viability assessment of the European eel. *Global Change Biology*, *21*, 3323–3335. https://doi.org/10.1111/gcb.12972

Bornarel, V., Lambert, P., Briand, C., Antunes, C., Belpaire, C., Ciccotti, E., Diaz, E., Diserud, O., Doherty, D., Domingos, I., Evans, D., de Graaf, M., O'Leary, C., Pedersen, M., Poole, R., Walker, A., Wickström, H., Beaulaton, L., & Drouineau, H. (2018) Modelling the recruitment of European eel (*Anguilla nguilla*) throughout its European range. *ICES Journal of Marine Science*, *75*(2), 541–552. https://doi.org/10.1093/icesjms/fsx180

Brisman, A. (2014) Of theory and meaning in green criminology. *International Journal for Crime, Justice and Social Democracy*, *3*(2), 21–34.

Carrabine, E., Cox, P., Fussey, P., Hobbs, D., South, N., Thiel, D., & Turton, J. (2014) *Criminology: A Sociological Introduction*. Third Edition. Routledge.

Copes, H., & Maruna, S. (2017) Techniques of Neutralisation. In A. Brisman, E. Carrabine, & N. South (Eds.), *The Routledge Companion to Criminological Theory and Concepts* (pp. 125–129). Routledge.

Crook, V., & Nakamura, M. (2013) Glass eels: Assessing supply chain and market impacts of a CITES listing on *Anguilla* species. *TRAFFIC Bulletin*, *25*(1), 24–30.

Dekker, W. (2000) The fractal geometry of the European eel stock, *ICES Journal of Marine Science*, *52*, 109–121. https://doi.org/10.1006/jmsc.1999.0562

Dekker, W. (2003) Did lack of spawners cause the collapse of the European eel, *Anguilla anguilla*? *Fisheries Management and Ecology*, *10*, 365–376. https://doi.org/10.1111/j.1365-2400.2003.00352.x

Dekker, W. (2016) Management of the eel is slipping through our hands! Distribute control and orchestrate national protection. *ICES Journal of Marine Science*, *73*(10), 2442–2452.

Dekker, W. (2019). The history of commercial fisheries for European eel commenced only a century ago. *Fisheries Management and Ecology*, *26*, 6–19

Drouineau, H., Durif, C., Castonguay, M., Mateo, M., Rochard, E., Verreault, G., Yokouchi, K., & Lambert, P. (2018) Freshwater eels: A symbol of the effects of global change. *Fish and Fisheries*, *19*(5), 903–930. https://doi.org/10.1111/faf.12300

Eliason, S. L. (2003) Illegal hunting and angling: The neutralisation of wildlife law violations. *Society and Animals*, *11*(3), 225–243.

Eliason, S. L., & Dodder, R. A. (1999) Techniques of neutralisation used by deer poachers in Western United States: A research note. *Deviant Behavior: An Interdisciplinary Journal*, *20*, 233–252.

Enticott, G. (2011) Techniques of neutralising wildlife crime in rural England and Wales. *Journal of Rural Studies*, *27*(2), 200–208.

Europol. (2019) Over 5 tonnes of smuggled glass eels seized in Europe this year. Available at: https://www.europol.europa.eu/newsroom/news/over-5-tonnes-of-smuggled-glass-eels-seized-in-europe-year (Accessed 25 November 2023).

Europol. (2022) Environmental crime in the age of climate change: 2022 threat assessment. Available at: https://www.europol.europa.eu/publications-events/publications/environmental-crime-in-age-of-climate-change-2022-threat-assessment

Gutierrez, L., & Duffy, R. (2023) Harms and the illegal wildlife trade: Political ecology, green criminology and the European eel. *Critical Criminology, 32,* 61–76. https://doi.org/10.1007/s10612-023-09734-4

Hutchinson, A. (2023). Wildlife We Love to Harm: How charisma impacts conservation responses to the illegal wildlife trade in Europe. Available at: https://beastlybusiness.org/wp-content/uploads/2023/02/Charisma.pdf

Liddick, D. (2013) Techniques of neutralisation and animal rights activists. *Deviant Behavior, 34*(8), 618–634.

Mackenzie, S., & Yates, D. (2016) Collectors on illicit collecting: Higher loyalties and other techniques of neutralisation in the unlawful collecting of rare and precious orchids and antiquities. *Theoretical Criminology, 20*(3), 340–357.

Maruna, S., & Copes, H. (2005) What have we learned from five decades of neutralisation research? *Crime and Justice: A Review of Research, 37*(4), 442–460.

McKie, R. E. (2019) Climate change counter movement neutralisation techniques: A typology to examine the climate change counter movement. *Sociological Inquiry, 89*(2), 288–316.

Miller, M. J., Westerberg, H., Sparholt, H., Wysujack, K., Sørensen, S. R., Marohn, L., Jacobsen, M. W., Freese, M., Ayala, D. J., Pohlmann, J., Svendsen, J. C., Watanabe, S., Andersen, L., Møller, P. R., Tsukamoto, K., Munk, P., & Hanel, R. (2019) Spawning by the European eel across 2000 km of the Sargasso Sea. *Biology Letters, 15*(4), 20180835. https://doi.org/10.1098/rsbl.2018.0835

Mir Mohamad Tabar, S. A., Rizzolo, J. B., Mazlom Khorasani, M., & Noghani, M. (2022) Social learning, neutralisation, and environmental crimes: An empirical test of differential association and neutralisation theories in Iran. *Society & Natural Resources, 36*(2), 150–170. https://doi.org/10.1080/08941920.2022.2139442

Mir Mohamad Tabar, S. A., South, N., Brisman, A., Akbar Majdi, A. (2022) An empirical test of techniques of neutralisation regarding polluting behaviours in rural Iran. *Crime, Law and Social Change, 78*(1), 79–103. https://doi.org/10.1007/s10611-022-10016-3

Moriarty, C., & Dekker, W. (1997) Management of the European eel. *Fisheries Bulletin, 15,* 110.

Nijman, V. (2017) North Africa as a source for European eel following the 2010 EU CITES eel trade ban, *Marine* Policy, *86,* 133–137. https://doi.org/10.1016/j.marpol.2017.06.036

Nurse, A. (2017) Green criminology: Shining a critical lens on environmental harm. *Palgrave Communications, 3*(10), 1–4. https://doi.org/10.1057/s41599-017-0007-2

Nurse, A. (2021) Green criminological perspectives on dogfighting as organised masculinities-based animal harm. *Trends in Organized Crime, 24,* 447–476. https://doi.org/10.1007/s12117-021-09432-z

Nurse, A., & Wyatt, T. (2020) *Wildlife Criminology*. Bristol: Bristol University Press.

Pohja-Mykrä, M. (2017) Community power over conservation regimes: Techniques for neutralising the illegal killing of large carnivores in Finland. *Crime, Law and Social Change, 67,* 439–460. https://doi.org/10.1007/s10611-016-9666-y

Pons-Hernandez, M. (2024) "Missing the Trees for the Forest?" An analysis of the harms to European eels caused by their trafficking and trade. *Critical Criminology, 32*(1), 77–95.

Pons-Hernandez, M. (2023) Tráfico de Especies Silvestres en España: una exploración desde la criminología verde crítica no-especista. *Revista Española de Investigación Criminológica, 21*(2), e815. https://doi.org/10.46381/reic.v21i2.815

Pons-Hernandez, M. (2024). Habitat Loss. In H. Pontell (Ed.), *Oxford Research Encyclopedia of Criminology and Criminal Justice*. Oxford University Press. https://doi.org/10.1093/acrefore/9780190264079.013.764

Ringuet, S., Muto, F., & Raymakers, C. (2002) Eels: Their harvest and trade in Europe and Asia. *TRAFFIC Bulletin, 19*(2), 2–27.

Ritchie, J., Lewis, J., Elam, G., Tennant, R., & Rahim, N. (2014) Designing and Selecting Samples. In J. Ritchie, J. Lewis, C. Nicholls, & R. Ormston (Eds.), *Qualitative Research Practice: A Guide for Social Science Students and Researchers* (2nd Edn.). SAGE, 111–146.

Ruggiero, V., & South, N. (2010) Critical criminology and crimes against the environment. *Critical Criminology, 18*(4), 245–250.

Rytterstedt, E. (2016) 'I Don't See Myself as a Criminal': Motivation and Neutralisation of Illegal Hunting by Swedish Norrland Hunters. In G. R. Potter, A. Nurse, & M. Hall (Eds.), *The Geography of Environmental Crime: Conservation, Wildlife Crime and Environmental Activism* (pp. 217–239). Palgrave Macmillan.

Skyes, G. M., & Matza, D. (1957) Techniques of neutralisation: A theory of delinquency. *American Sociological Review, 22*, 664–670.

Sollund, R. (2013) Causes for Speciesism: Difference, Distance and Denial. In R. White (Ed.), *Transnational Environmental Crime* (1st ed.) (pp. 75–96). London: Routledge.

Sollund, R. (2019) *The Crimes of Wildlife Trafficking: Issues of Justice, Legality and Morality*. London: Routledge.

Sollund, R. (2020) Wildlife crime: A crime of hegemonic masculinity? *Social Sciences, 9*(6), 93. https://doi.org/10.3390/socsci9060093

UNODC. (2016) World Wildlife Crime Report: Trafficking in protected species. Available at: https://www.unodc.org/documents/data-and-analysis/wildlife/World_Wildlife_Crime_Report_2016_final.pdf

UNODC. (2020) World Wildlife Crime Report: Trafficking in protected species. Available at: https://www.unodc.org/documents/data-and-analysis/wildlife/2020/World_Wildlife_Report_2020_9July.pdf

White, R. (2011) *Transnational Environmental Crime: Toward an Eco-Global Criminology*. London: Routledge.

White, R., & Heckenberg, D. (2014) *Green Criminology: An Introduction to the Study of Environmental Harm*. London: Routledge.

Wright, R. M., Piper, A. T., Aarestrup, K., Azevedo, J. M. N., Cowan, G., Don, A., Gollock, M., Rodríguez Ramallo, S., Valterop, R., Walker, A., Westerberg, H., Righton, D. (2022) First direct evidence of adult European eels migrating to their breeding place in the Sargasso Sea. *Nature, 12*, 15362.

Wyatt, T. (2022) *Wildlife Trafficking: A Deconstruction of the Crime, Victims and Offenders*. Cham: Palgrave Macmillan.

Wyatt, T., & Brisman, A. (2016) The role of denial in the 'Theft of Nature': Comparing biopiracy and climate change. *Critical Criminology, 25*(3), 325–341.

Wyatt, T., Maher, J., Allen, D., Clarke, N., & Rook, D. (2021) The welfare of wildlife: An interdisciplinary analysis of harm in the legal and illegal wildlife trades and possible ways for.
Crime, Law and Social Change, 77, 69–89. https://doi.org/10.1007/s10611-021-09984-9

23

WILDLIFE TRAFFICKING VIA SOCIAL MEDIA

Implications for biodiversity conservation and public health in Brazil

Ophélia Miralles, Dener José Giovanini, Thiago Vargas and Raulff Zanferrari Lima

Introduction

The ownership of companion animals, including exotic pets, is globally trending. Indeed, 57% of global households already own at least one pet (Anderson, 2019). During times of unrest, companion animal ownership seems to increase (Wallace, 2020), as seen during the coronavirus pandemic (Pet Food Manufacturers' Association (PFMA), n.d.). Whereas dogs and cats are the most common companion animals, another side of the market is "exotic" pets (Hall, 2019). Brazil, as a biodiversity hotspot, grapples with extensive wildlife trafficking, driven by a demand for rare and unique companion animals. This chapter delves into the intricate web of wildlife trafficking facilitated by social media platforms, shedding light on the extent of the issue, its impact on biodiversity, economy, society, and public health, and the need for concerted efforts to combat it.

Historical context of wildlife trade in Brazil

Since the colonisation period, Brazil has been a hub for the collection and domestication of wild animals (parrots, monkeys, turtles, and snakes for pets), largely driven by Indigenous practices (Hagenbeck, 1910) and European fascination. The Portuguese and other Europeans not only carried out scientific expeditions throughout Brazil (Maurice de Nassau, Von Spix, Von Martius and Von Langsdorff) in order to identify, record, and collect the fauna, but also adopted the practice of keeping these species as pets in Brazil and in Europe (Hagenbeck, 1910). The Indigenous Peoples, for their part, understood quickly the European demand for wildlife and thus began the ongoing trade based on the commercialisation of wild animals and derived products (Hagenbeck, 1910).

The demand for rare and unique companion animals across the world is one of the main components of wildlife trafficking (Wyatt, 2022), and this holds true in Brazil (Giovanini, 2006). The attractiveness of rainforest species for the "exotic" companion animal (pet) market is undeniable, as it is a biodiversity hotspot. Yet, wildlife trafficking likely removes more

DOI: 10.4324/9781003333906-27

than 38 million animals from the wild in Brazil every year, posing a huge threat to regional and global biodiversity (RENCTAS, 2014). Sixty per cent of these animals are sold on the domestic pet market with the remainder illegally exported. The harm to smuggled animals is also serious. On average, of every ten animals taken, only one arrives at the final intended destination; the others die either during capture or in transport (RENCTAS, 2014). Further understanding of the pet market is thus needed and important. Wildlife trafficking in Brazil is a highly profitable illegal market. The national Brazilian non-governmental organisation RENCTAS—Rede Nacional de Combate ao Tráfico de Animais Silvestres—reports that Brazilian species such as the Lear's Macaw (*Anodorhynchus leari*) are traded for more than USD 100,000 each. The traffickers' profit is exorbitant, with wild birds being purchased from Indigenous Peoples' communities for USD 1 and resold for up to a thousand. Wildlife trafficking also involves the exploitation of Brazil's most vulnerable communities, including Indigenous Peoples and underprivileged youth (RENCTAS, 2014). In many poor communities, wildlife trafficking is an important source of income for local communities, and wildlife traffickers take advantage of this social vulnerability to expand their pool of animal collectors (RENCTAS, 2014).

The rise of social media in wildlife trafficking

In response to growing demand, wildlife trafficking has taken advantage of social media and private messaging platforms' global reach, speed, and anonymity (Broséus et al., 2016; Lee & Roberts, 2020; Martin, 2014; Di Minin et al., 2019). A basic social media search reveals a substantial global wildlife trade in the open, but the full extent of the online illegal trade remains challenging to quantify due to law enforcement's limited access to encrypted apps and private social media groups (Hawkins, 2021).

In recent years, the extent of wildlife trafficking on the internet has become a significant concern for local NGOs all over the world. In 2016, the Wildlife Justice Commission identified 51 Vietnamese traffickers responsible for dealing in more than $53 million worth of illegal wildlife products, including ivory, rhino horn, and tiger parts. Alarmingly, up to a third of this illicit trade unfolded on online platforms, including Facebook and WeChat (ACCO, 2023; Wildlife Justice Commission, 2016). The Project to End Great Ape Slavery also highlights a disturbing trend in the wildlife crime landscape. More than half of the global ape trade is estimated to occur on social media platforms, with Facebook and Instagram accounts becoming hotspots for the sale of young apes to collectors, primarily concentrated in the Gulf region (Dasgupta, 2018). The Cheetah Conservation Fund (CCF), a non-profit research and lobby organisation based in Namibia, adds to this concerning narrative, reporting that a staggering 70% of the illegal cheetah trade is conducted online, with Instagram emerging as a prominent hub for such activities, followed by 4Sale and YouTube (ACCO, 2023; Nuwer, 2023). The study by the CCF highlights that most online trades had links with Gulf countries, with the three best sellers based in Saudi Arabia (Ravenscroft, 2018).

Private messaging indeed provides a fast worldwide reach, ensures privacy and is flexible; if one group gets deleted, another can be created with a different name. On Facebook, some wildlife trafficking groups can count up to 40,000 members. These groups are generally private, and new participants usually have to answer questions to be admitted. Monitoring those private groups for research is thus an ethical challenge, as it may involve deception of participants in the data collection process.

Monitoring social media

Established in 1999, RENCTAS is a local Brazilian non-governmental organisation that has been collecting data and intelligence on wildlife trafficking with the goal of decreasing it. RENCTAS continuously monitors a spectrum of online platforms. This includes keeping a watchful eye on approximately 800 WhatsApp groups and multiple secretive hubs facilitating illicit wildlife trade on Facebook. RENCTAS' vigilance also extends to dedicated monitoring efforts on popular platforms such as Instagram and TikTok. Importantly, they broaden their scope to include e-commerce platforms, with Mercado Livre being a noteworthy example. These platforms often serve as channels through which wildlife traffickers operate. The influence of social networks extends beyond the illicit trade in live animals, encompassing the commerce of animal parts. These clandestine transactions predominantly cater to the fashion industry, where items like pelts, leather, bones, and teeth find buyers. Additionally, the trade in artisanal objects crafted from the remains of wild animals for ornamental purposes has achieved a concerning level of prevalence (RENCTAS, 2014).

The online monitoring conducted by RENCTAS is executed on a daily basis by a team of professionals who infiltrate virtual communities while systematically keeping an eye on social media for suspicious advertisements. This initiative encompasses several pivotal activities, including the identification of animal sales, the classification of involved species, comprehensive price documentation, and the systematic surveying of regions where animals are illicitly harvested and traded. One way that they have collected such data is to create 200 fake Facebook profiles to access private messaging groups on Facebook and WhatsApp. In their data collection, RENCTAS adheres to the Association of Internet Researchers' parameters and Brazilian law (Internet Civil Rights Act 12.965/2014). No personal information is shared publicly. Once access to the private messaging groups is gained, the relevant contents of the group chat are downloaded in the form of screenshots. The screenshots are securely stored, and the intention is to record the data from each screenshot. RENCTAS now has a repository of more than 4 million messages from 800 private groups collected in the last few years.

During the COVID-19 crisis, RENCTAS observed 15,000 online advertisements per day for rainforest wildlife-related trade. This number reflects RENCTAS' maximum capacity for message monitoring, meaning the actual number of daily advertisements for wildlife online is likely far greater. In Brazil, all restricted species of wild animals can only be legally purchased at a store through breeders certified by the local environment agency IBAMA (Instituto Brasileiro do Meio Ambiente e dos Recursos Naturais Renováveis no date) (see Brazilian Federal law 5.197/67, 9.605/98). Hence, any sale of wildlife online and through groups on private messaging platforms such as those monitored by RENCTAS is illegal.

A pilot study recently published in the journal Biological Conservation (Wyatt et al., 2022) explores RENCTAS' anonymised database of screenshots advertising "exotic" pets in Brazil. Ethical approval was obtained via Northumbria University's Department of Social Sciences (#26787) for the recording of the data from a random sample of 500 screenshots collected between August and November 2020 into a spreadsheet for analysis. When it was possible, the species traded and their origin, the type of transaction (sell or buy order), the price, and the specific social media and group used for this advertisement were recorded. The number of individual animals was assumed to be one if the number for sale was not specified. Additional information such as the animal class, scientific name, was added if not provided in the original post. Conservation status using different lists (ICMBio (Instituto Chico Mendes de Conservação da Biodiversidade—the national organisation overseeing

Brazilian national protection), IUCN (International Union for the Conservation of Nature) Red List, Convention on International Trade in Endangered Species of Wild Fauna and Flora (CITES)) was used to provide insights into level of protection for species being traded and possible conservation implications. By exploring the contents of private messaging group advertisements on WhatsApp and Facebook for the first time, this study aims to raise awareness about the extensive use of social media and messaging platforms for illegal activities that threaten Brazilian ecosystems, economy, society, and public health.

Patterns and trends in wildlife trafficking

This section delves into the findings of the pilot study (Wyatt et al., 2022). The messages contained mainly "for sale" advertisements for wildlife from 105 different private groups, revealing an estimated minimum of 1,682 individual animals from 260 distinct species. Reptiles and birds dominated the advertisements, accounting for 84% of the total. Mammals, arachnids, and amphibians constituted the remaining 16% (see Table 23.1). The ads included species from different continents, with the majority originating from Brazil (81%). This pilot contains data from only 500 of 4 million messages, indicating that there are

Table 23.1 Breakdown of advertisements by species' taxonomic class (Wyatt et al., 2022, p. 3)

Species' taxonomic class	*# of ads*	*% of ads*
Amphibian	29	1.72
Arachnid	115	6.82
Bird	670	39.72
Mammal	133	7.88
Reptile	740	43.86
Grand total	1,687	100.00

Table 23.2 Species who account for most of the advertisements (Wyatt et al., 2022, p. 3)

Species in Latin	*Species common name*	*# of ads*	*% of ads*
Boa constrictor	Boa constrictor	123	7.29
Chelonoidis sp.	Turtle	110	6.52
Iguana iguana	Green iguana	102	6.05
Pantherophis guttatus	Corn snake	93	5.51
Ara ararauna	Blue and yellow macaw	89	5.28
Saltator similis	Green-winged saltator	59	3.50
Sapajus sp.	Capuchin	50	2.96
Sporophila caerulescens	Double collared seedeater	49	2.90
Callithrix sp.	Marmoset	31	1.84
Sicalis flaveola	Saffron finch	30	1.78
Amazona aestiva	Turquoise-fronted parrot	28	1.66
Python sp.	Python	28	1.66
Tupinambis sp.	Tegu (lizard)	21	1.24
Tyto furcata	Barn owl	21	1.24

profound implications for conservation and biodiversity loss stemming from the demand for Brazilian species as companion animals. Species originating in the United States accounted for 6.25% of individuals, followed by just under 4% of species from Africa. Only two individuals were species from Europe (Russia and Spain/Portugal).

Certain species more popular in the pet trade appear more than other species. In 13 ads, the information about the species was unidentifiable. One hundred and fourteen species only appeared once in one advertisement. One hundred and six species appeared in 2–9 ads (accounting for 418 ads), and only 23 species appeared in 10–19 ads (for a total of 308 ads). The remaining 834 ads contained only 14 species, accounting for approximately 49% of all ads. The list of these species is in Table 23.2.

Price disparities and conservation implications

Data available for prices were limited: prices might be missing because it is common knowledge among these private groups' members, or because negotiating is expected and thus it is purposefully omitted in the messages. The total value of all individuals was 780,207 Brazilian real or USD 148,239. However, there were 673 instances (~40% of the total number of individuals) which had no price information for the animal for sale. From the information that was available, we suggest that there is a clear unevenness in the value of each species category, with mammals and birds each accounting for approximately USD 95,330 (or 64% of the total USD value) and reptiles accounting for USD 51,726 (35% of USD value). Mammals had the highest value per advertisement on average (USD 792), followed by birds (USD 166 per ad) and reptiles (USD 130 per ad). The average higher price for mammals may be a reflection of lower population numbers overall, fewer mammals being available for sale, or the greater perceived value. The economic incentive for trafficking mammals indicates the need for focused conservation efforts.

The messages were analysed according to different lists of conservation status, namely the CITES Appendices, the IUCN Red List categories, and the Conservation Categories from Brazil's Ministry of the Environment (ICMBio). Approximately one-third of ads were for species for which some level of trade under CITES is allowed (Appendix II). Almost 4% of ads, however, are for species in Appendix I (which in essence means no international trade is allowed). Ads for CITES Appendix I listed species were split evenly between birds, mammals, and reptiles. However, the distribution changes for Appendix II listed species with reptiles accounting for almost 70% and birds and mammals accounting for approximately 15% each (see Table 23.3). Few ads, only 30, were for species in the Critically Endangered and Endangered categories of the IUCN Red List. However, approximately 3% of identifiable ads were for species designated as Near Threatened (5.30% are near threatened as defined by ICMBio), meaning their trade, in addition to other factors, could make them at risk for qualifying for a more threatened category in the future. The majority of ads (56%) were designated as being in the Least Concern category (see Table 23.4). Similarly, for the advertisements by ICMBio category, a majority of the species were Least Concern (Table 23.5). Critically Endangered, Endangered, Near Threatened, and Vulnerable species were less than 9% combined. Unevaluated species (those that have not been scientifically assessed) and unidentified species accounted for over 30% of the total. Despite the majority of species being categorised as "Least Concern," trade in near-threatened species and species with undetermined conservation statuses raises concerns about their vulnerability to endangerment due to the wildlife trade's cumulative impact.

Table 23.3 The distribution of advertisements across taxonomic class within each CITES Appendix (Wyatt et al., 2022, p. 3)

Cites	*# of ads*	*% of ads*
Appendix I	54	3.20
Bird	16	29.63
Mammal	16	29.63
Reptile	22	40.74
Appendix II	605	35.86
Amphibian	4	0.66
Arachnid	26	4.30
Bird	72	11.90
Mammal	76	12.56
Reptile	427	70.58
Appendix III	15	0.89
Bird	3	20.00
Mammal	3	20.00
Reptile	9	60.00
Total	674	

Table 23.4 Advertisements by IUCN Red List category (Wyatt et al., 2022, p. 4)

IUCN category	*# of ads*	*% of ads*
Critically endangered	19	1.13
Endangered	11	0.65
Near threatened	45	2.67
Vulnerable	79	4.68
Least concern	941	55.78
Not evaluated	337	19.98
Unidentifiable (NP)	255	15.12

Table 23.5 Advertisements by ICMBio category (Wyatt et al., 2022, p. 4)

ICMBio	*# of ads*	*% of ads*
Critically endangered	16	0.95
Endangered	7	0.41
Near threatened	64	3.79
Vulnerable	54	3.20
Least concern	996	59.04
Not evaluated	362	21.46
Data deficient	10	0.59
Unidentifiable (NP)	178	10.55

Illegal trade and public health risks

The illicit wildlife trade is exacerbated by its online nature, enabling easy access and anonymity for traffickers. Illegally traded animals do not undergo any health inspections and are more likely to transmit diseases, including unknown ones, to domestic animals and human beings. This poses public health risks due to zoonotic disease transmission, which is particularly poignant given the coronavirus pandemic that continues to cause human death and suffering around the world. COVID-19 is the latest disease to be transmitted from animals to humans, likely through consumption (Kimbrough, 2020), as has happened in previous zoonotic outbreaks (e.g. Severe Acute Respiratory Syndrome (SARS), HIV, Ebola—Swift et al. (2007)). Wildlife trade and the related markets have been identified as possible points of transmission and mutation of potential viral pathogens and thus risk factors for zoonotic spill over (Aguirre et al., 2020; IPBES, 2020; World Animal Protection, 2020). The diversity of species brought to highly populated cities is thought to contribute to the potential for disease emergence (Swift et al., 2007), as are the cramped and stressful conditions for wildlife, insufficient diet, and exchange of excrement and viruses (Kimbrough, 2020). The welfare of wildlife in these spaces—particularly the conditions in which they are caught, killed, transported and kept—has direct and fatal consequences for human health and safety (One Welfare, 2019; World Health Organisation, 2017).

The COVID-19 pandemic's origins from animals underscore the urgency of addressing zoonotic spillover risks associated with wildlife trade, including the companion animal market. However, in the debates around wildlife trafficking and zoonotic diseases, it appears that food consumption is the focus, and the companion animal market has been largely overlooked. Yet many of the species in demand for "exotic" pets also carry the risk of transmitting diseases to humans. For instance, in one small study in Brazil, 55.8% of animals (24 out of 43 total—41.7% and 58.3% of the carnivores and nonhuman primates, respectively) were found to have at least one zoonotic parasite species (Lima et al., 2021). Bezerra-Santos et al. (2021) cite instances of documented cases of transmission of viruses, bacteria, and parasites from "exotic" pets such as small mammals, reptiles, amphibians, fish, and birds to humans. Thus, in Brazil, in addition to the conservation implications, there are real concerns for public health arising from the "exotic" pet trade.

Need for legislative reform and digital platform accountability

RENCTAS' data collection efforts are systematically transformed into practical insights, which are then shared with Brazil's environmental control and monitoring authorities. This collaborative approach allows for swift and informed responses to mitigate the impact of wildlife trafficking on our planet's ecosystems and biodiversity. Despite RENCTAS' ongoing awareness-raising initiatives regarding the threat of wildlife trafficking, legal actions against online trafficking have struggled to keep pace with its rapid expansion. What appears to be lacking is a sincere commitment from digital platforms to actively assist megadiverse countries in addressing this criminal activity, which poses a significant threat to their biodiversity.

In 2023, RENCTAS embarked on a significant endeavour to combat online wildlife trafficking. Over a four-month period, the organisation monitored YouTube to identify videos depicting illegal hunting of wild animals. Thousands of such videos were discovered, and the data was compiled into a comprehensive dossier. This dossier formed the foundation for a legal action taken by RENCTAS, aiming to have all illicit poaching videos removed from

YouTube. While RENCTAS initially achieved a notable victory in court, with an order for YouTube to promptly delete these illegal hunting videos under threat of a substantial daily fine of R$10,000, YouTube contested this decision and filed an appeal. Presently, the case is pending in a higher court, and regrettably, videos depicting illegal hunting of wild animals continue to proliferate on the platform, despite such activities being expressly prohibited in Brazil.

RENCTAS' outreach extends beyond YouTube and includes several attempts to engage with other social networks, notably Instagram, which have yet to yield positive outcomes. Even when Facebook eventually reached out to RENCTAS, their commitment to taking necessary measures to combat wildlife trafficking on their platform appeared to be lacking. It is worth noting that addressing this criminal activity on social networks does not present insurmountable technological challenges, as these platforms already possess the tools necessary to combat numerous other illicit activities, such as child exploitation, human trafficking, or terrorism. Unfortunately, there seems to be a lack of interest from these social networks in effectively addressing animal trafficking, as evidenced by the persistence of reported pages engaged in illegal activities, which often continue to operate with apparent impunity. The primary argument put forth by digital platforms centres on the principle of freedom of expression. However, the current threat to worldwide biodiversity underscores the urgent need for legal actions. Without appropriate regulation of social media content, these platforms are likely to persist in their disregard for the ecological damage caused by wildlife trafficking. Such reform is thus crucial to hold digital platforms directly accountable for the content they facilitate and distribute (Stiles, 2019).

Conclusion

Due to its relatively recent emergence and the multitude of possible execution methods, online animal trafficking has evolved into a significant challenge for governments, researchers, and organised civil society. The utilisation of social networks is contributing to a daily rise in the number of trafficked animals, consequently placing a growing number of species at risk. The 1,682 advertisements analysed in Wyatt et al. (2022) provide insights into the exotic and illegal pet trade in Brazil. Reptiles and birds appear to be a majority of the volume of traded species in Brazil's online illegal wildlife trade, whereas mammals likely earn sellers more money per animal. A vast majority of the trade is in Brazilian species, and it does not appear that many of the species are endangered under the three conservation status lists we consulted. However, there is a significant number of advertisements where the status of the species cannot be determined, so there may be more trade of endangered species than our figures indicate. Moreover, when scaled up to 4 million ads and over time, the trade could underpin large losses to specific species populations if unchecked, thereby threatening their conservation status.

Regardless of conservation status, the recorded trades are illegal under Brazilian law. As mentioned, only certified breeders are permitted to advertise and sell wildlife in Brazil. Despite this legal limitation, the scale of Brazilian trade seems to be enormous. Each message contains, on average, 3.3 ads for wildlife. Extrapolated to the entire RENCTAS' dataset of 4 million messages, this means 13,200,000 ads, and there are more out there from other groups not being monitored. This is just ads; this is not the number of individual animals, which is undoubtedly more. This is from only a 12-month period of monitoring by RENCTAS. A vast majority of the wildlife are Brazilian, so the scale has direct implications for conservation and the sustainability of species populations in Brazil. Furthermore, it is not possible to inspect uncertified breeders to check that they are maintaining welfare and hygiene standards. This could mean increased risk of zoonotic viruses, bacteria, and parasites from the illegal trade of companion animals within Brazil and beyond its borders.

Beyond the need to analyse the full extent of online illegal wildlife trade in Brazil, further research needs to be done to track where these wildlife are going to, but it appears that there is a substantial domestic market in Brazil (which means CITES is irrelevant as CITES does not regulate national level trade and does not always reflect the national list of protected species). Thus, demand reduction campaigns in Brazil are likely necessary. This means further research is needed into who is selling and buying wildlife, who are organising these private messaging groups, how the online interaction links to physical interactions (i.e. how wildlife is delivered), and the level of knowledge of people involved in the online wildlife pet market in Brazil of its unlawfulness. Given the recent and well-founded concern about zoonotic disease spread from the companion animal trade, further research might also seek to understand the level of risk posed by commonly traded species in Brazil and identify measures to reduce the risk of zoonotic spill over to human populations. There is a substantial illegal trade taking place via private messaging groups, which raises unique challenges for conservation and for law enforcement. Although efforts are being made by technology companies to address illegal trade online (see the Coalition to End Wildlife Trafficking Online, for instance), encrypted private messaging poses particular hurdles. Further research is needed to support law enforcement in their approach to online illegal trade in wildlife in Brazil and all over the world for both species' conservation and public health purposes.

The findings highlight the need for comprehensive research to understand the dynamics of wildlife trafficking, including sellers' and buyers' profiles, online-to-physical interactions, and levels of awareness of the trade's illegality. This chapter emphasises the imperative of demand reduction campaigns within Brazil, given the significant domestic market for these illicitly traded species. It also underscores the challenges posed by private messaging groups, calling for technological solutions and law enforcement collaboration to curb wildlife trafficking online. Addressing this issue requires collaborative action at local, national, and international levels. By addressing the wildlife trafficking crisis in Brazil and specifically the online marketplaces, we can contribute to safeguarding species, ecosystems, and human well-being on a global scale.

References

ACCO. (2023). *Wildlife in Peril: Animals are Traded, Tortured on Social Media*. https://www.counteringcrime.org/s/WildlifeFactSheetFINAL-f4xm.pdf

Aguirre, A. A., Catherina, R., Frye, H., & Shelley, L. (2020). Illicit Wildlife Trade, Wet Markets, and COVID-19: Preventing Future Pandemics. *World Medical & Health Policy*, *12*(3), 256–265. https://doi.org/10.1002/wmh3.348

Anderson, J. (2019, March 7). *Pet statistics around the world*. Pet Search. https://www.petsearch.com.au/pet-statistics-around-the-world/

Bezerra-Santos, M. A., Mendoza-Roldan, J. A., Thompson, R. C. A., Dantas-Torres, F., & Otranto, D. (2021). Illegal Wildlife Trade: A Gateway to Zoonotic Infectious Diseases. *Trends in Parasitology*, *37*(3), 181–184. https://doi.org/10.1016/j.pt.2020.12.005

Broséus, J., Rhumorbarbe, D., Mireault, C., Ouellette, V., Crispino, F., & Décary-Hétu, D. (2016). Studying Illicit Drug Trafficking on Darknet Markets: Structure and Organisation from a Canadian Perspective. *Forensic Science International*, *264*, 7–14. https://doi.org/10.1016/j.forsciint.2016.02.045

Dasgupta, S. (2018, October 22). *Illegal cheetah trade continues through Instagram, 4sale, YouTube*. Mongabay Environmental News. https://news.mongabay.com/2018/10/illegal-cheetah-trade-continues-through-instagram-4sale-youtube/

Di Minin, E., Fink, C., Hiippala, T., & Tenkanen, H. (2019). A Framework for Investigating Illegal Wildlife Trade on Social Media with Machine Learning. *Conservation Biology*, *33*(1), 210–213. https://doi.org/10.1111/cobi.13104

Giovanini, D. (2006). Taking Animal Trafficking Out of the Shadows: RENCTAS Uses the Internet to Combat a Multi-Billion Dollar Trade. *Innovations: Technology, Governance, Globalization*, *1*(2), 25–35. https://doi.org/10.1162/itgg.2006.1.2.25

Hagenbeck, C. (1910). *Animales y hombres: Recuerdos y experiencias…* hijos de Carlos Hagenbeck. https://books.google.ch/books?id=pnnhHAAACAAJ

Hall, J. (2019, February 20). *Exotic pet trade, explained*. National Geographic. https://www.nationalgeographic.com/animals/article/exotic-pet-trade

Hawkins, S. (2021, April 27). *Social Media Platforms Hotbeds for Wildlife Trafficking*. https://www.courthousenews.com/social-media-platforms-hotbeds-for-wildlife-trafficking/

Instituto Brasileiro do Meio Ambiente e dos Recursos Naturais Renováveis. (n.d.). *IBAMA*. Retrieved August 31, 2023, from https://www.gov.br/ibama/pt-br/pagina-inicial

IPBES. (2020). *Workshop Report on Biodiversity and Pandemics of the Intergovernmental Platform on Biodiversity and Ecosystem Services (IPBES)* (1.4). Zenodo. https://doi.org/10.5281/ZENODO.4147317

Kimbrough, L. (2020, March 13). *Conservationists set the record straight on COVID-19's wildlife links*. Mongabay Environmental News. https://news.mongabay.com/2020/03/conservationists-set-the-record-straight-on-covid-19s-wildlife-links/

Lee, T. E., & Roberts, D. L. (2020). Moving beyond Simple Descriptive Statistics in the Analysis of Online Wildlife Trade: An Example from Clustering and Ordination. *Tropical Conservation Science*, *13*, 194008292095840. https://doi.org/10.1177/1940082920958401

Lima, V. F. S., Ramos, R. A. N., Giannelli, A., Schettino, S. C., Galina, A. B., Oliveira, J. C. P. D., Meira-Santos, P. O., & Alves, L. C. (2021). Zoonotic Parasites in Wild Animals Such as Carnivores and Primates That Are Traded Illegally in Brazil. *Brazilian Journal of Veterinary Medicine*, *43*, e113720. https://doi.org/10.29374/2527-2179.bjvm113720

Martin, J. (2014). Lost on the *Silk Road*: Online Drug Distribution and the 'Cryptomarket.' *Criminology & Criminal Justice*, *14*(3), 351–367. https://doi.org/10.1177/1748895813505234

Nuwer, R. (2023, May 9). *Baby apes are being stolen for pets—And little is being done to stop it*. Project to End Great Ape Slavery (PEGAS). https://freetheapes.org/tag/trafficking/

One Welfare. (2019). *About One Welfare*. https://www.onewelfareworld.org/about.html

Pet Food Manufacturers' Association (PFMA). (n.d.). *PFMA Confirms Dramatic Rise in Pet acquisition among Millennials*. Retrieved August 31, 2023, from https://www.ukpetfood.org/resource/pfma-confirms-dramatic-rise-in-pet-acquisition-among-millennials.html

Ravenscroft, H. (2018, September 27). Cheetah conservation fund data analysis confirms social media role in advertising illegal wildlife trade, including trafficking of cheetahs for illegal pet trade. *Cheetah Conservation Fund*. https://cheetah.org/press-releases/cheetah-conservation-fund-data-analysis-confirms-social-media-role-in-advertising-illegal-wildlife-trade-including-trafficking-of-cheetahs-for-illegal-pet-trade/

RENCTAS. (2014). *1st National Report on the Traffic of Wild Animals*. https://www.renctas.org.br/wp-content/uploads/2014/02/RELATORIO-INGLES_final.pdf

Stiles, D. (2019, October 25). *Holding social media companies accountable for facilitating illegal wildlife trade (commentary)*. Mongabay Environmental News. https://news.mongabay.com/2019/10/holding-social-media-companies-accountable-for-facilitating-illegal-wildlife-trade-commentary/

Swift, L., Hunter, P. R., Lees, A. C., & Bell, D. J. (2007). Wildlife Trade and the Emergence of Infectious Diseases. *EcoHealth*, *4*(1), 25. https://doi.org/10.1007/s10393-006-0076-y

Wallace, A. (2020, October 16). *Demand for pets is soaring during the pandemic. This company is cashing in | CNN Business*. CNN. https://www.cnn.com/2020/10/16/business/west-paw-dog-toys-boom/index.html

Wildlife Justice Commission. (2016, November 17). *Independent Panel Confirms that Immediate Action by the Vietnamese Government is Required to Shut Down Wildlife Trafficking Networks in Vietnam*. https://wildlifejustice.org/independent-panel-confirms-immediate-action-vietnamese-government-required-shut-wildlife-trafficking-networks-viet-nam/

World Animal Protection. (2020). *To end wild animal exploitation forever, we must focus on the global wildlife trade*. https://www.worldanimalprotection.org/latest/news/end-wild-animal-exploitation-forever-we-must-focus-global-wildlife-trade/

World Health Organisation. (2017). *One Health*. https://www.who.int/news-room/questions-and-answers/item/one-health

Wyatt, T. (2022). *Wildlife Trafficking: A Deconstruction of the Crime, Victims and Offenders*. Springer International Publishing. https://doi.org/10.1007/978-3-030-83753-2

Wyatt, T., Miralles, O., Massé, F., Lima, R., Costa, T. V. da, & Giovanini, D. (2022). Wildlife Trafficking via Social Media in Brazil. *Biological Conservation*, *265*, 109420. https://doi.org/10.1016/j.biocon.2021.109420

24

ILLEGAL WILDLIFE TRADE IN NON-TIMBER ORNAMENTAL PLANTS

Jared Margulies and Amy Hinsley

Introduction

Illegal wildlife trade in non-timber ornamental plants is global in scope and affects a wide array of plant taxa. The impacts of these trades on the conservation status of thousands of wild plant species are significant, and more than 20 species are now thought to be globally extinct or extinct in the wild as a direct result of overexploitation, ranging from cycads to orchids to cacti (Hinsley et al., 2023; Margulies et al., 2019; Wyatt and Hutchinson, 2023). Although in the past few years a variety of major newspaper and media outlets drew heightened attention to a variety of illegal trades in ornamental plants based on the supposition that the COVID-19 pandemic, which ushered in a contemporary houseplant 'fever', might have also caused a spike in illegal plant trade (Burke et al., 2022; Hübschle & Gore, 2024), international illegal trade involving ornamental plants is nothing new, while the overexploitation of wild plants for the ornamental trade dates back centuries (Maunder, 2022).

While evidence is lacking to support clear causality between the growth in popularity of ornamental plants with a growth in *illegal* trade, the legal trade is clearly globally significant and growing, and keeping houseplants is unambiguously popular; in 2022 the export value of live plants and bulbs was valued at $13 billion (Burke et al., 2022; CITES Secretariat, 2022; International Trade Centre, 2023). While some illegal trades in ornamental plants seem doomed to persist, either because affected species may prove technically challenging to cultivate in meaningful quantities, or because wild collection is more profitable than cultivating plants (largely due to the labour and infrastructural costs of the latter), most wild plants currently exploited through illegal wildlife trade could, theoretically, be substituted for sustainably cultivated plants of the same species. There is good reason to imagine then it is possible to reduce overexploitation of wild species through their substitution with cultivated plants, yet surprisingly, empirical evidence is largely lacking to support this kind of substitution effect (Liu et al., 2019; Williams et al., 2014; Kay et al., 2011). At the same time, concerns of overexploitation of wild plants for the ornamental trade date back decades, so it is important to consider the socioecological and economic reasons why illegal trades persist when, for so many impacted species, sustainable alternatives theoretically could substitute for them.

DOI: 10.4324/9781003333906-28

As of 2024, there are 34,310 plant species included on one of three appendices of the Convention on International Trade in Endangered Species of Wild Fauna and Flora (CITES), which regulates international trade in wild species. While many CITES-listed plant species are traded for non-ornamental value (e.g. plants with nutritive, medicinal, fibre, or other human uses), the majority are traded primarily as ornamental plants. Most CITES-plants are in the Orchidaceae (orchid) family, with all ca 29,000 species listed. The entire Cactacae (cactus) family of ca 1,500 species is also listed in an appendix of CITES, as are all Cycadales (cycads). This already accounts for over 85% of all species listed on CITES (recognising that some orchids are also traded as medicines, cosmetics, or for their consumptive value, such as the production of salep, a flour made from *Orchis* tubers and used commonly in beverages in West Asia). The impact of wildlife trade on ornamental plants is therefore significant, yet wildlife trade in plants remains significantly understudied compared with wildlife trade in animals (Margulies et al., 2019).

This chapter reviews and discusses the diversity of affected species, geographic distribution of the trade and illicit supply chains, and evidence about motivations for involvement in this form of wildlife crime. This chapter concludes with recommendations and opportunities to both further research on illegal ornamental plant trade and reduce overexploitation of wild species.

Diversity of affected species

Ornamental plants are plants valued for their decorative and aesthetic attributes by humans. They often possess showy and charismatic features but may also be prized for faring well in either the indoor biome or the garden under human care, frequently far from their native habitat range (Maunder, 2022). They are sought by people around the world to be enjoyed for their aesthetic qualities, improvement to indoor and outdoor home, work, or play environments, as well as for their collecting value (this especially applies to rarer plants). The category of non-timber ornamental plants is non-specific and likely includes thousands of plant species, yet also includes more than 100,000 additional cultivars, hybrids, and subspecies forms (De, 2017). Today, many of the world's most popular ornamental plants have global, mass-market appeal (cacti and monstera, for example). Other ornamental plants, either due to their growing requirements or due to cultural and historical value, are more geographically specific in their popularity or suitability as ornamental plants (e.g. cycads).

According to Royal Botanic Gardens, Kew's 2023 State of the World's Plants and Fungi, there are 350,386 known vascular plant species in the world (Antonelli et al., 2023). And yet, between 2020 and 2023, close to 9,000 new species of plants were named by scientists. This suggests there remain major gaps in basic knowledge of plants globally. These gaps are both taxonomic and geographic; certain kinds of plant groups are historically understudied, while certain geographies also remain challenging to access for researchers for a variety of reasons, such as armed conflict or extreme remoteness, for comprehensive biodiversity surveys (Antonelli et al., 2023). While there is a clear need for continued emphasis on basic botanical research to identify, describe, and catalogue species, the world of plants (both known and unknown to science) simultaneously faces numerous intersecting threats. Of all assessed plant species in the International Union for the Conservation of Nature's Red List of threatened species, 38.5% are assessed as threatened with extinction. However, only 17% of all plants have been assessed through the IUCN Red List process to date (IUCN, 2024). This means that for most of the world's known vascular plants, their formal conservation status is unknown.

Amidst this gap in conservation knowledge of plants, within the domain of ornamental plants, a far more specific list of taxa and categories of plants emerges as plants whose conservation status is impacted by illegal trade. These include orchids, succulents (inclusive of a range of plant taxa), cacti, carnivorous plants, geophytes (bulb plants), arums (aroids), ferns, and a select number of water plants (e.g. water lilies). In the extreme, Sprenger's tulip *Tulipa sprengeri* and the Chilean blue crocus *Tecophilaea cyanocrocus* are reported as extinct in

Table 24.1 Key ornamental plant taxa impacted by illegal wildlife trade

Plant taxa	*Specific genera/taxa*	*Regions impacted by wild collection*	*Key references**
Cactaceae	Many, including *Copiapoa, Ariocarpus, Aztekium, Discocactus, Lophophora, Melocactus, Sclerocactus, Turbinocarpus*, and *Uebellmania*	US Southwest, Mexico, Central America, South America	Arroyo-Quiroz and Wyatt (2019); Goettsch et al. (2015); Lavorgna and Sajeva (2021); Margulies et al. (2023); Novoa et al. (2017);
Orchidaceae	Many, including *Cypripedium, Mexipedium, Paphiopedilum, Phragmipedium*, and *Selenipedium* (slipper orchids)	Southeast Asia, China, South America	Phelps and Webb (2015); Hinsley et al. (2018); Ticktin et al. (2020).
Carnivorous plants	*Nepenthes, Sarracenia, Dioneae*	Southeast Asia, Southern US	Cross et al. (2020); Margulies et al. (2024); Phelps and Webb (2015); Shirey et al. (2013).
Other Succulent plants and geophytes (bulbs)	Many, including *Avonia, Cycadales, Conophytum, Dudleya, Galanthus* (snowdrops), *Lithops, Othonna, Pachypodium, Tulipa* (tulips), and many Southern and East African geophytes.	Southern Africa, East Africa, Europe, Mediterranean region	Cousins and Witkowski (2017); Krigas et al. (2021); Read (1989); Margulies (2023); *review articles across succulent taxa are lacking.

Table 24.1 summarises key categories of plants known to be impacted by wildlife crime for their ornamental value. Although there are more locally or sub-nationally known instances of arums (Araceae) such as Alocasia being wild-collected and traded for their ornamental value (e.g. Beligolo & Buenavista, 2023), there are no Araceae currently listed on CITES (several were removed from the CITES Appendices in the 1990s). This is in part due to the relative ease in vegetative propagation of arums, enabling their widespread distribution when compared to cacti, bulbs, and many succulents. It would therefore be seen as administratively prohibitive and inefficient to regulate these species via CITES given that most of the international trade is in artificially propagated plants. However, the lack of regulation of arums by CITES also means there is less readily available international trade data on seizures of plants (for instance, plants missing phytosanitary permits, or wild-collected plants seized by customs and border officials because of national biodiversity laws). Like illegal orchid trade, it is also possible that trade in wild-harvested arums may occur undetected, laundered within much larger volumes of trade in cultivated specimens (Hinsley et al., 2017).

the wild due to overexploitation for the bulb trade (Maunder et al., 2001), and five species of cycads are thought to be extinct in the wild from harvesting for the wildlife trade (Hinsley et al., 2023). Many species of orchids and cacti are thought to be locally extinct because of the ornamental trade, including slipper orchids, alongside dozens of cacti (Goettsch et al., 2015; Hinsley et al., 2023).

Hinsley et al. (2023) recently reviewed seizure data from 2013 to 2022 from the TRAFFIC Wildlife Trade Data Portal. As they note, seizure data only reflects a limited view into illegal trades, as the data is prone to biases in confiscation efforts, the challenge of intercepting all forms of illegal plant trade (e.g. seeds), and its relatively low priority in customs and border enforcement. Still, such data can provide some evidence of extant illegal plant trades, and they found 104 incidents during that period linked to a seizure or illegal harvesting for ornamental plants totalling over 100,000 plants or seeds (Hinsley et al., 2023). The majority of incidents involved seized cacti and orchids, followed by a diversity of succulents, carnivorous plants (pitcher plants *Napenthes* spp. and Venus flytraps), cycads, and aquatic plants.

Geographic diversity

As shown in Table 24.1, there are a few regions of the world whose ornamental plants are not affected by illegal wildlife trade. Although there are clear hotspots in terms of plant origin (Latin America, Southeast Asia, Southern and East Africa), for ornamental plants impacted by illegal wildlife trade today, the entire supply chain, inclusive of consumers, is truly global. Like global wildlife trade writ large, much of the international flow of plant products moves from Global South countries overlapping with areas of species richness (countries of supply) to Global North countries (countries of consumption), but many countries like China, the United States, and many European countries are both major importers of wildlife products (inclusive of plants) as well as major exporters (Andersson et al., 2021; Fukushima et al., 2020; Scheffers et al., 2019). Because of the rapid pace of global and regional plant trends, geographic hotspots are prone to sudden shifts and the rise of alarming overexploitation. This is seen today, for instance, in the relatively recent emergence in demand for *Dudleya* succulents from California and Mexico (Margulies, 2020), as well as the ongoing overexploitation of a variety of bulb and succulent species from Southern Africa (Hübschle and Margulies, 2024; SANBI, 2022).

International scale and scope of illegal trade

A unidirectional flow of plants from the Global South to the Global North does not fully capture the complexity and diversity of the global illegal wildlife trade in ornamental plants. Although Global North countries are the dominant consumers of illegally traded ornamental plants, they are also affected by the illegal sourcing of plants. For instance, the United States is likewise affected by international illegal trade in cacti genera including *Lophophora* and *Ariocarpus* (Muneta, 2020; Robbins, 2003), carnivorous plants such as *Sarracenia* and the Venus flytrap (Cross et al., 2020; Margulies et al., 2024), as well as a national illegal trade in Saguaro (*Carnegia gigantica*), a species protected under Arizona state law (Robbins, 2003).

Ornamental plant collection is popular across many cultural and geographic contexts, leading to both domestic illegal trades (see Figure 24.1) as well as intra-regional international trade (e.g. cactus trade within Latin America, orchid trade throughout Southeast Asia) (Phelps & Webb, 2015). For instance, studies in China and Thailand have detailed the importance of international regional trade in orchids across East and Southeast Asia (Phelps & Webb, 2015; Gale et al., 2019). Similarly, wild collection of bulbs from Greece is believed to largely support a European market (Krigas et al., 2021).

National scale and scope of illegal trade

Many illegal plant trades are likely under-reported or absent from international trade databases because the trade takes place at sub-national scales. For example, the Philippines is known to have robust local trade in wild-harvested plants affecting arums, orchids, and *Nepenthes* pitcher plants, among other species. This trading often takes place at open-air regional plant markets (Phelps & Webb, 2015). While some of this trade is illegal because it enters the international sphere and comes under the jurisdiction of CITES, much of this trade is believed to occur sub-nationally yet remains illegal under Philippines law (Protecting Wildflowers and Plants in the Philippines, Act No. 3983, 1932). Collection of ornamental plants is popular in the Philippines, and so domestic consumption is understood as significantly impacting native flora conservation (Beligolo & Buenavista, 2023). Similar illegal domestic trade dynamics are recorded in China (Gale et al., 2019) and in the United States (Shirey et al., 2013).

Supply chain characteristics

Wildlife crime involving ornamental plants is global in reach but is also multi-nodal in structure. The flow of plants through illegal supply chains often involves international intermediary transit locations, compared with trades with only clear source and destination locations (Hinsley et al., 2016; Lavorgna & Sajeva, 2021; Margulies, 2020). For instance, the recent problem of illegal trade in *Dudleya* and *Conophytum* from California and South Africa, respectively, principally involved Korean harvesters and an initial destination of South Korea, yet the final consumer base proved to be global in scope, involving consumers in China, Japan, Europe, and beyond (Hinsley et al., 2023; Margulies, 2020 2023).

Figure 24.1 shows four (non-exhaustive) models through which ornamental plants move via illegal supply chains. These generalised models are based on synthesising results from different kinds of ornamental plant trades, including cacti and succulents, orchids, and carnivorous plant species, based on both our own research and depictions of supply chain structures in published literature (e.g. Phelps and Webb, 2015). As Figure 24.1 suggests, there are a variety of structures of illegal supply chains in ornamental plants, which are influenced by geography, socio-cultural characteristics, economics, horticultural technologies, and plant physiology.

Individual collector model

The individual collector model is perhaps the oldest form of ornamental plant wildlife crime, though in earlier eras, such as the Victorian period, (especially wealthy) collectors were more likely to hire someone to collect plants on their behalf, to be sent back to them for their personal enjoyment (Maunder, 2022; Voskuil, 2017). The practice of collectors travelling to different parts of the world to obtain wild-harvested plants for their collection historically occurred before the implementation of CITES in 1975 (Sanders, 2017), meaning that a great deal of this collecting occurred either legally or in a grey area depending on the laws of the countries (or earlier, colonies) where plants were being collected. The practice (and enjoyment) of wild collecting by individual collectors is therefore also tied for many collectors to a cultural history and practice of seeking out plants as a mode of exploration, adventure, and discovery. Unsurprisingly, such cultural histories and practices are closely linked to eras and ideations of empire and colonialism (Batsaki et al., 2016; Subramaniam, 2024; Weber, 2021). Today, although there is evidence that ornamental plant collectors within some surveyed communities do not approve of wild-collecting plants, the practice persists (Hinsley et al., 2017; Margulies et al., 2023). As Figure 24.1(1a) details, in this model, the individual collector (often travelling in a small group, but not always) controls the entire supply chain, seeking out plants, extracting them, and coordinating and managing their shipment back to the collector's place of origin. This may occur either through shipping live plants or plant parts (or seeds) via postal services (usually falsifying contents to avoid detection or the need for phytosanitary permits or export paperwork), or depending on the taxa, shipping plants within their personal luggage or even directly on their persons (this is especially easy for seeds or plant parts but is also possible for many smaller species). Plants and seeds are not easy to detect via standard airport security systems, and this offers collectors a lower risk of apprehension unless they are searched at customs at their port of entry. The individual collector model is found across nearly all illegal ornamental plant trades.

Multi-nodal international model

The most complex, and perhaps, most worrying form of wildlife crime impacting ornamental plants is described in Figure 24.1(1b). This is described here as a multi-nodal international model because it involves the coordinated efforts of actors in multiple, often very distant, locations. At its most direct, it may involve just two actors or sets of actors – those on the demand side and those on the sourcing side. However, there are often additional multiple intermediary/transit stage actors, including transport actors (actors who physically move plants from the site of extraction or harvesting by vehicle) and export actors (actors responsible for packaging and preparation of plants for export). Export actors are typically the go-betweens for harvesting actors and international buyers, whether the latter are individuals looking to purchase plants for themselves, or more likely, wholesalers abroad with their own lists of clients.

This multi-nodal model is, within the world of ornamental plant trade, the model which most closely approximates an organised crime model. However, this is usually better characterised as a coordinated yet *dis*organised model of wildlife crime (Wyatt et al., 2020). This is because, with some noted exceptions, it is unusual for formally organised syndicates or crime organisations (e.g. cartels, violent crime groups or gangs) to be directly engaged in ornamental plant crimes, though media articles often depict these groups otherwise. However, due to the coordinated nature of these crimes, this model is also the most likely to develop into more hierarchical crime networks or introduce more serious criminal actors into the trade. Such networks often become of greater concern to law enforcement and security agencies, which in turn can lead to the potential for more violent and serious crime actors to become engaged in these crimes to facilitate thwarting or subverting heightened surveillance and enforcement (Duffy, 2022). The emergence of the multi-nodal international model is, without question, facilitated by e-commerce and the internet, as well as encrypted peer-to-peer communication systems, which allow for rapid and coordinated communication between harvesters, buyers, and intermediaries in real time. Social media and e-commerce platforms have also developed as important digital communities and locations through which buyers, intermediaries, and sellers can connect, communicate, and coordinate wildlife crimes (Lavorgna & Sajeva, 2021; Olmos-Lau and Mandujano, 2016). The multi-nodal model is worrying in part because its increased complexity is suggestive of increased awareness by engaged actors of its illegality, and its persistence and at times expansion suggest a lack of concern for the conservation impacts of these trades. An example of this model being operationalised is seen in ongoing illegal succulent trades from Southern Africa (Hübschle & Gore, 2024; Hübschle & Margulies, 2024; SANBI, 2022).

Regional hub model

In many ways, the regional hub model described in Figure 24.1(1c) is like the multi-nodal international model, except this model often relies on: (a) more geographically proximate supply chains and actors; (b) reduced law or conservation enforcement oversight; and (c) opportunities to sell illegally obtained material in otherwise licit settings. The regional hub model is frequently observed to be international, relying on porous (usually terrestrial) borders with a lack of enforcement (or presence of corruption) to facilitate the movement of plant material from source countries to consumer countries. In the regional hub

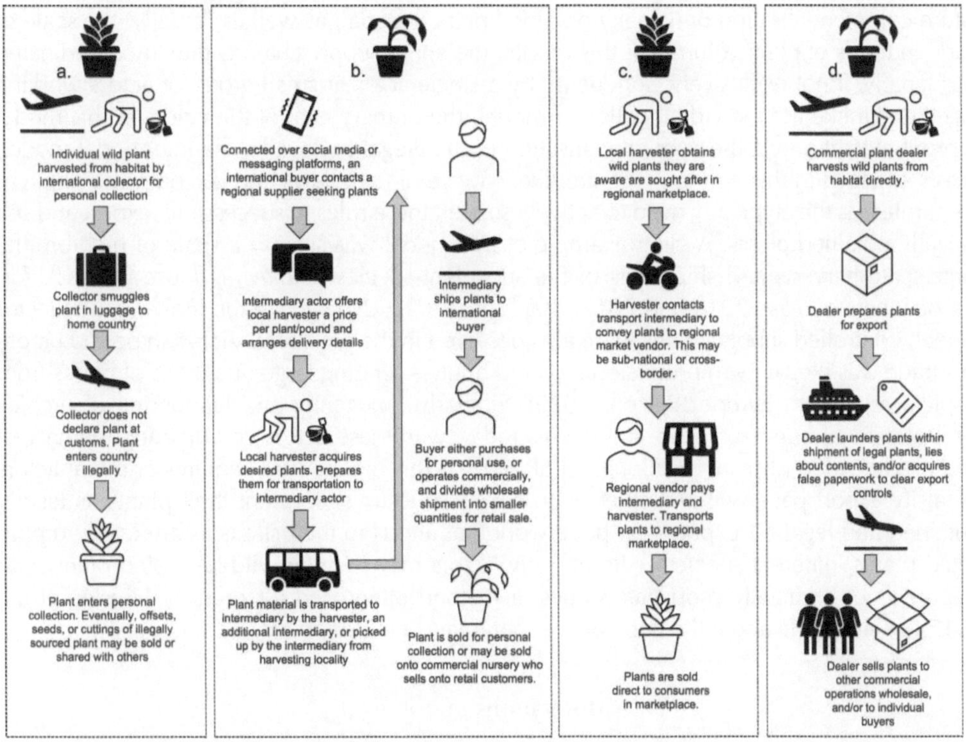

Figure 24.1 Four generalised models of illegal trade in wild ornamental plants: (a) individual collector model; (b) multi-nodal international model; (c) regional hub model; and (d) controlled supply chain model. These models are not meant to exhaust all examples or possibilities of illegal plant trade supply chains, but characterise general patterns observed across a range of wildlife crime studies involving ornamental plants.

model, consumers often reside in the countries where the sale of plants takes place (i.e. the collectors are domestic, rather than international). These trades typically feature source, intermediary, and demand actors, but unlike the multi-nodal international model, these relationships may be ongoing for long periods of time as opposed to more ephemeral interactions between intermediary actors (most typically in the country of plant origin) and demand actors (who may be connected to these intermediary actors via online social media groups or peer-to-peer groups). This model is well evidenced within the regional market for many ornamental plants ranging from ferns to arums to orchids in Southeast Asia, where plants obtained in Laos or Myanmar may move across borders to be sold in legal markets in Thailand, where the same markets may also sell both legally and illegally obtained plants domestically (Phelps & Webb, 2015). Historically, the Venus flytrap's illegal trade is understood to have operated similarly in North Carolina (albeit in a sub-national context), with harvesters paid by legal commercial nurseries for wild-collected plants which were then laundered into the cultivated and licit trade (Margulies et al., 2024).

Controlled supply chain model

In structure, the controlled supply chain model described in Figure 24.1(1d) is very similar to the individual collector model (Figure 24.1, 1a), yet with key differences regarding the future commercialisation of illegally obtained plant material, as well as (usually) the scale of trade in terms of plant volume. In this model, the entire supply chain is directly coordinated and largely, if not exclusively, conducted by a single actor or small group of actors working together. Unlike the individual collector model, the primary goal of the actors in this model is profit extraction via the commercialisation of the illegal trade. Like the inter-nodal model, actors engaged in this type of trade are likely aware of trade laws and customs and enforcement rules, as they typically need to actively subvert these rules to successfully export and sell illegally obtained plants. A clear example of this model was seen in a spate of high-profile, large-scale harvests and shipments of the succulent species *Dudleya farinose* from California between (at least) 2015 and 2017 (Margulies, 2020, 2023). Like the individual collector model, controlled supply chains were also observed in the pre-CITES Victorian period before this trade was illegal, with professional orchid hunters sending large quantities of plants from tropical regions to Europe (Bilston, 2025). Similarly, smuggling and laundering of orchids by professional nursery owners still occurs today, with these actors coordinating the harvest and illegal export of plants (Bullough et al., 2021). This model often requires criminal actors to falsify export paperwork or other documentation to (mis)represent their plants as legally obtained and legal for export (e.g. paperwork that attests to their status as artificially propagated plants, different species). Alternatively, actors may launder wild-harvested plants into shipments of artificially propagated ones, a method often used for orchids (Hinsley et al., 2017), or simply lie about the contents of containers being shipped.

Motivations of crime

In the case of many of the trade models described in Figure 24.1(b–d), profit extraction from the harvesting and sale of illegally obtained, traded, or sold plants is the primary motivation for wildlife crime involving ornamental plants. However, there are often multiple factors incentivising or driving such trades, and especially in the case of the individual collector model (Figure 24.1, 1a), profit may not be a motivating factor at all in the persistence of this form of wildlife crime. In both the inter-nodal and controlled supply chain models, for instance, in addition to profit, another motivating factor may be obtaining sufficient plant 'parent material' to produce a viable artificial propagation programme abroad (Hinsley and Roberts, 2018). Although legally obtainable genetic material may be available (e.g. in the form of seed), stealing large quantities of reproducing and mature plants may rapidly accelerate the ability for a nursery operation to scale plant propagation and out-compete other operations in bringing mature, higher-value plants to market more quickly. This may be especially important in markets where fast 'fads' in plants are observed; this can be exacerbated in the case where plants move in and out of fashion quickly, but the plants themselves are slower to grow. This was observed, for instance, in what at least in part drove the theft of *Dudleya farinosa* in the South Korean succulent market (Margulies, 2023). In other cases, newly described species, which are often commercially valuable, may take many years to raise from seed, incentivising breeders to speed up the process using illegal material (Hinsley & Roberts, 2018).

Consumer desires are also an important motivating factor in driving illegal wildlife trade in ornamental plants, something which has been observed across all ornamental plant trades. Even in the trade models where profit and commercialisation are driving factors, ultimately these trades only exist because there is a sufficient collector and consumer demand for plants. Where in the individual collector model it is apparent that collector desires are a primary motivating factor for obtaining illegally sourced plants or plant material, it is important to remember that consumer demand is a core feature of what drives the emergence of other models of trade (Margulies, 2023).

'Consumer desire' is a rather vague, catch-all term that indexes a wide array of consumer motivations for wildlife trade products. As Thomas-Walters et al. (2021) describe, motivations for the use of wildlife trade products across the tree of life are diverse and range from experiential and social to functional, financial, and spiritual. The same holds true within the ornamental plant trade; even if ornamental plants are sought out for their aesthetic values, motivations for engaging in ornamental plant caretaking include recreational and sensory experiences (collecting plants with beautiful flowers or scents, the pleasure of caring for plants in greenhouses, etc.) as well as important social functions (Thomas-Walters et al., 2021). Alongside these motivational forces, specific qualities or traits of plants may incentivise certain kinds of collector demands, as summarised in Table 24.2, drawing on the insights from a wide range of sources.

Motivational factors may also be related to broader social contexts in which illegal harvesting and trade occur, as well as behavioural factors (Eliason, 1999; Skidmore, 2023).

Table 24.2 Plant traits that frequently influence collector demands across taxa

Trait	*Examples*
Natural rarity	Species that grows across a wide habitat or region but found infrequently.
Limited geographic extent (geographic rarity)	Species is abundant where found but limited in geographic extent.
Endemism	Species is unique to one country, region, or geographic area.
Rare form/colour/mutation	Cristate cacti or other unusual morphology, atypical flower colour, etc.
Unique provenance	Plant may come from a historical type locality or other notable location (e.g. a population that no longer exists).
Unique features compared to other closely related species or genera	Exceptional flower colours, forms, shapes, etc.

Many plant collectors are part of both in-person and online social communities organised around ornamental plant collection; obtaining new, rare, or highly fashionable plants can therefore be motivated by reputational, relational, or social pressure drivers (Thomas-Walters et al., 2021). Although evidence suggests collectors in the cactus and succulent trades in the United States and the United Kingdom generally disapprove of wild-harvesting plants, for example, those actors who still engage in this behaviour may do so for a variety of reasons (Margulies et al., 2024). Like other forms of poaching, one reason is that the illegal taking of wildlife remains commonly seen by many in plant collector communities as a 'folk crime' (Eliason, 1999; Muth, 1998). This means that while it is recognised as illegal, it is seen as less serious compared to other crimes and therefore may be considered more socially acceptable. For others, the thrill in engaging in wrongdoing or criminal behaviour with relatively low risk of serious consequences is itself a form of pleasure and enticement (Margulies, 2024). Still others may engage in illegal trade on normative or moral-ethical grounds. This can include desires to either 'save' species from obscurity by removing species perceived as threatened from their habitat, or because they believe taking plants otherwise represents a means to protect or conserve them (Mackenzie & Yates, 2016 Margulies, 2023). Still others may even engage in illegal wildlife trade as an act of political or social defiance, as seen in fish and animal wildlife trades (Bell et al., 2007; Hübschle, 2017).

The same holds true for financial motivations: even if profit is a motivational driver of their illegal trade, it may be immediate or future profit that drives the trade; this is especially true for ornamental plants that are long-living and may acquire more value as they grow and become more mature plants (Thomas-Walters et al., 2021). Conversely, it is important to recognise that 'profit' indexes different needs or wants; the opportunity for foreign collectors to make money by selling illegally harvested plants acquired abroad is not the same motivating force as domestic actors who may turn to illegal harvesting of plants out of necessity for their basic survival, as has recently been observed in South Africa's Succulent Karoo (Hübschle & Margulies, 2024). Unpacking the underlying motivations of ornamental plant trades, therefore, requires intensive and well-designed social research that can provide insights into ornamental plant consumer cultures. In addition to conscious decisions to take part in illegal plant trade, some actors may also take part without knowledge of the legality of the trade. Unwitting illegal consumers, either at a street market, an international plant show, or on an e-commerce website, may buy a plant without knowing that it is wild-harvested or traded in contravention of CITES rules. This may be especially true online, where buyers may not be specialists or may have little understanding of the required paperwork to import plants (Lavorgna & Sajeva, 2021; Margulies et al., 2023; Novoa et al., 2017).

Conclusion: opportunities to develop sustainable alternatives in illegal ornamental plant trade

Only about 10% of all known vascular plants are considered currently or potentially threatened by international trade to such a degree that they are listed on CITES, and not all listed plants are valued for their ornamental traits (for instance, there are ~800 tree species valued as timber listed on CITES). In the full scope of the plant kingdom, relatively few plants are valued by humans for their ornamental qualities, and fewer still are threatened by wildlife crime. But those taxa that are threatened by illegal wildlife trade can quickly become imperiled or even made extinct or locally extirpated through rapid overexploitation. This is especially true in the past few decades in which the internet has become a critical platform through which international plant trades occur (Lavorgna et al., 2018, 2020; Lavorgna &

Sajeva, 2021). Global demand for selective taxa of charismatic, showy, and rare plants is persistent (Hall et al., 2008); even where fads and tastes may change relatively quickly for species or genera (e.g. some species of succulents pass in and out of fashion with impressive speed), for naturally rare and slow growing plants, such global demand can quickly lead to local extinctions and even total extinction of wild species.

Despite increasing attention in recent years to illegal wildlife trade in ornamental plants, plants still receive far less attention from granting agencies concerned with wildlife conservation and/or illegal wildlife trade (Margulies et al., 2019). This hinders the ability of conservation practitioners to implement practical interventions and for researchers to gather robust bodies of evidence that can inform equitable and socially and environmentally just responses to this historically overlooked form of wildlife crime. The diversity and number of species of plants impacted by illegal trade for their ornamental value are significant, and because of the demand for new wildlife commodities to collect and possess, novel species are often especially threatened from the horticultural trade, as evidenced, for example, by the rapid local extinctions of many newly described cactus species.

Across professional and amateur conservation, trade, and collector circles, it is a popular notion that 'flooding the market' with artificially propagated plants is the best way to thwart illegal collection and trade in highly sought-after ornamental plant species. In fact, CITES has recognised the implicit wisdom of this approach. At the CITES CoP15 in 2010, resolution 9.19 was passed, which stated:

> RECOGNIZING that, by making specimens readily available, the artificial propagation of specimens of species included in Appendix I reduces the collecting pressure on wild populations and, thus, has a positive effect on their conservation status.

The full resolution introduced the ability for nurseries to artificially propagate specimens of Appendix I plant species for legal export, something which was not previously permitted. The goal of this programme was to have vetted nurseries registered and approved by individual countries' CITES Management Authorities to artificially propagate Appendix I plants as a mechanism to reduce pressure on wild plant populations. Yet despite passing this resolution, very few countries have approved nurseries to do so, and as of this writing, only 12 countries have CITES-approved plant nurseries for Appendix I plants. At the same time, there is scant empirical evidence that the approach of 'flooding the market' curbs the wild collection of plants. The few examples of published studies that exist present mixed responses, with some suggesting that access to artificially propagated does not meaningfully reduce pressure on wild plants, while others suggest they can in some cases (Phelps et al., 2015; Gale et al., 2019). Furthermore, there is evidence that cultivation can shift the benefits to different actors, usually those with the resources and land rights to establish nurseries (Williams et al., 2014). Clearly, more empirically driven studies are needed to parse how, and under what conditions, artificially propagated plants do reduce pressure on wild plant populations, and when they do not (e.g. Entwistle et al., 2002).

There are other reasons, however, that the seemingly obvious choice to make desirable plants widely available through legal channels is not always forthcoming. As Yota Batsaki and Peter Crane (2024) recently wrote:

> Collecting is increasingly regulated at local, national, and international levels, with many benefits but also unintended consequences. The law is a blunt instrument. It used to be said "if you have a rare plant give it away", but this simple advice, focused

> on reducing the risk of ex situ extinction, is increasingly thwarted by multiple legal layers that seek to pursue other valid aims, such as enhancing "biosecurity" by reducing the spread of pests and disease, or eliminating the incentives for collecting plants that are valuable, rare, or endangered in the wild (p. 25).

There are other reasons the 'flooding the market' approach as a sustainable response to illegal wildlife trade is less straightforward than it may seem. For instance, Access and Benefit Sharing agreements as mandated by the Nagoya Protocol of the Convention on Biological Diversity, a tool meant to ensure that individual countries financially benefit from the access, trade, and sale of their biodiversity and genetic resources, can lead some countries to restrict the trade and sharing of plant materials. To date, few ornamental plants have been the focus of access and benefit sharing agreements, and more work is needed to make future agreements beneficial as well as feasible to implement (Blackhall-Miles, 2021). More research is needed to understand the long-term consequences of such decisions for plant conservation, and how they may help or hinder the conservation of ornamental plants weighed against the intended social benefits.

With these caveats aside, the artificial propagation of plants remains an important consideration for designing demand reduction programmes and socially and environmentally just responses to this form of wildlife crime. Still, important questions remain; for instance, who directly benefits from the sale of artificially propagated plants intended to offset pressure on wild plant populations? Who is entitled to a financial stake in plant profits framed within a model of environmental justice? What do sustainable global plant supply chains look like? The answers to these questions will be geographically and culturally specific. But following key interventions within the emerging scholarly area of southern green criminology (Goyes, 2018, 2019, 2021), which draws attention to and critiques the reproduction global inequalities and colonial modes of policing in addressing green crimes, it is more important than ever to consider how addressing matters of both poverty *and* inequality can feature in designing responses to wildlife crime (Lunstrum et al., 2023). Designing illegal wildlife trade interventions focused on reducing socioecological harms and amplifying community empowerment and social justice is not an easy matter (Hübschle & Margulies, 2024); yet working to do so remains a pressing imperative to achieve truly sustainable results with lasting impact for human and nonhuman communities alike.

References

Andersson, A. A., Tilley, H. B., Lau, W., Dudgeon, D., Bonebrake, T. C., & Dingle, C. (2021). CITES and beyond: Illuminating 20 years of global, legal wildlife trade. *Global Ecology and Conservation, 26*, e01455.

Antonelli, A., et al. (2023). *State of the World's Plants and Fungi 2023*. Royal Botanic Gardens, Kew.

Arroyo-Quiroz, I., & Wyatt, T. (2019). Wildlife trafficking between the European Union and Mexico. *International Journal for Crime, Justice and Social Democracy, 8*(3), 23–37.

Batsaki, Y., Cahalan, S. B., & Tchikine, A. (Eds.). (2016). *The Botany of Empire in the Long Eighteenth Century*. Dumbarton Oaks Research Library and Collection.

Batsaki, Y., & Crane, P. (2024). Collections, extinction, abundance. *Arnoldia, 81*(2), 20–25.

Beligolo, F. A., & Buenavista, D. P. (2023). Is popularity good for plant conservation? Impacts of "plantito and plantita culture" in Valencia city, Bukidnon, Philippines. *Acta Scientifica Naturalis, 4*(2), 126–143.

Bell, S., Hampshire, K., & Topalidou, S. (2007). The political culture of poaching: A case study from northern Greece. *Biodiversity and Conservation, 16*, 399–418.

Bilston, S. (2025). *The Lost Orchid: A Story of Victorian Plunder and Obsession*. Cambridge: Harvard University Press.
Blackhall-Miles, R. (2021). *Plant Hunting for Commercial Horticulture and Access and Benefit Sharing: Complementary or Competitive Uses of Wild Novel Plants?* (Unpublished master's thesis, University of Kent [United Kingdom]).
Bullough, L. A., Nguyễn, N., Drury, R., & Hinsley, A., (2021). Orchid obscurity: Understanding domestic trade in wild-harvested orchids in Viet Nam. *Frontiers in Ecology and Evolution, 9*, 631–795.
Burke, R., Sherwood, O. L., Clune, S., Carroll, R., McCabe, P. F., Kane, A., & Kacprzyk, J. (2022). Botanical boom: A new opportunity to promote the public appreciation of botany. *Plants, People, Planet, 4*(4), 326–334.
CITES Secretariat. (2022). *World Wildlife Trade Report 2022*. Geneva, Switzerland.
Cross, A. T., Krueger, T. A., Gonella, P. M., Robinson, A. S., & Fleischmann, A. S. (2020). Conservation of carnivorous plants in the age of extinction. *Global Ecology and Conservation, 24*, e01272. https://doi.org/10.1016/j.gecco.2020.e01272
Cousins, S. R., & Witkowski, E. T. F. (2017). African cycad ecology, ethnobotany and conservation: A synthesis. *The Botanical Review, 83*, 152–194.
De, L. C. (2017). Improvement of ornamental plants - a review. *International Journal of Horticulture, 7*(22), 180–204.
Duffy, R. (2022). *Security and Conservation: The Politics of the Illegal Wildlife Trade*. Yale University Press.
Eliason, S. L. (1999). The illegal taking of wildlife: Toward a theoretical understanding of poaching. *Human Dimensions of Wildlife, 4*(2), 27–39.
Entwistle, A., Atay, S., Byfield, A., & Oldfield, S. (2002). Alternatives for the bulb trade from Turkey: A case study of indigenous bulb propagation. *Oryx, 36*(4), 333–341.
Fukuṣhima, C. S., Mammola, S., & Cardoso, P. (2020). Global wildlife trade permeates the tree of life. *Biological Conservation, 247*, 108503.
Gale, S. W., Kumar, P., Hinsley, A., Cheuk, M. L., Gao, J., Liu, H., Liu, Z. L., & Williams, S. J. (2019). Quantifying the trade in wild-collected ornamental orchids in South China: Diversity, volume and value gradients underscore the primacy of supply. *Biological Conservation, 238*, 108204.
Goettsch, B., Hilton-Taylor, C., Cruz-Piñón, G., Duffy, J. P., Frances, A., Hernández, H. M., Inger, R., Pollock, C., Schipper, J., Superina, M., Taylor, N. P., Tognelli, M., Abba, A. M., Arias, S., Arreola-Nava, H. J., Baker, M. A., Bárcenas, R. T., Barrios, D., Braun, P., … Gaston, K. J. (2015). High proportion of cactus species threatened with extinction. *Nature Plants, 1*(10), 15142. https://doi.org/10.1038/nplants.2015.142
Goyes, D. R. (2018). Green criminology as decolonial tool: A stereoscope of environmental harm. In K. Carrington, R. Hogg, J. Scott, & M. Sozzo (Eds.), *The Palgrave Handbook of Criminology and the Global South* (pp. 323–346). Springer International Publishing. https://doi.org/10.1007/978-3-319-65021-0_17
Goyes, D. R. (2019). *Southern Green Criminology: A Science to End Ecological Discrimination*. Emerald Publishing Limited. https://www.emerald.com/insight/content/doi/10.1108/978-1-78769-229-920191015/full/html
Goyes, D. R. (2021). Environmental crime in Latin America and southern green criminology. In *Oxford Research Encyclopedia of Criminology and Criminal Justice*. https://oxfordre.com/criminology/display/10.1093/acrefore/9780190264079.001.0001/acrefore-9780190264079-e-588
Hall, R. J., Milner-Gulland, E. J., & Courchamp, F. (2008). Endangering the endangered: The effects of perceived rarity on species exploitation. *Conservation Letters, 1*(2), 75–81.
Hinsley, A., Hughes, A. C., Van Vankelburg, J., Stark, T., Van Delft, J., Sutherland, W. J., & Petrovan, S. O. (2023). The true cost of the global ornamental plant trade. https://doi.org/10.31219/osf.io/qujm8
Hinsley, A., Nuno, A., Ridout, M., John, F. A. S., & Roberts, D. L. (2017). Estimating the extent of CITES noncompliance among traders and end-consumers, lessons from the global orchid trade. *Conservation Letters, 10*(5), 602–609.
Hinsley, A., Lee, T. E., Harrison, J. R., & Roberts, D. L. (2016). Estimating the extent and structure of trade in horticultural orchids via social media. *Conservation Biology, 30*(5), 1038–1047.
Hinsley, A., & Roberts, D. L. (2018). The wild origin dilemma. *Biological Conservation, 217*, 203–206.
Hinsley, A., Willis, J., Dent, A. R., Oyanedel, R., Kubo, T., & Challender, DWS. (2023). *Trading species to extinction: Evidence of extinction linked to the wildlife trade. Cambridge Prisms: Extinction, 1*, e10, 1–9. https://doi.org/10.1017/ext.2023.7

Hübschle, A. M. (2017). The social economy of rhino poaching: Of economic freedom fighters, professional hunters and marginalized local people. *Current Sociology, 65*(3), 427–447.
Hübschle, A., & Gore, M. L. (2024). Lessons in resilience from the illegal wildlife trade during COVID-19 lockdowns. *Science of the Total Environment, 916*, 170365.
Hübschle, A., & Margulies, J. D. (2024). The need for a socioecological harm reduction approach to reduce illegal wildlife trade. *Conservation Biology, 38*(5), e14335. https://doi.org/10.1111/cobi.14335
International Trade Centre. (2023). Trade Map: Trade statistics for international business development. International Trade Centre. https://www.trademap.org/. Accessed June 2024.
IUCN. (2024). Summary Statistics. Table 1a: Number of species evaluated in relation to the overall number of described species, and numbers of threatened species by major groups of organisms. Updated 27th June 2024: https://www.iucnredlist.org/resources/summary-statistics
Kay, J., Strader, A. A., Murphy, V., Nghiem-Phu, L., Calonje, M., & Griffith, M. P. (2011). Palma corcho: A case study in botanic garden conservation horticulture and economics. *HortTechnology, 21*(4), 474–481.
Krigas, N., Lykas, C., Ipsilantis, I., Matsi, T., Weststrand, S., Havström, M., & Tsoktouridis, G. (2021). Greek tulips: Worldwide electronic trade over the internet, global ex situ conservation and current sustainable exploitation challenges. *Plants, 10*(3), 580.
Lavorgna, A., Middleton, S. E., Pickering, B., & Neumann, G. (2020). FloraGuard: Tackling the online illegal trade in endangered plants through a cross-disciplinary ICT-enabled methodology. *Journal of Contemporary Criminal Justice, 36*(3), 428–450. https://doi.org/10.1177/1043986220910297
Lavorgna, A., Rutherford, C., Vaglica, V., Smith, M. J., & Sajeva, M. (2018). CITES, wild plants, and opportunities for crime. *European Journal on Criminal Policy and Research, 24*(3), 269–288. https://doi.org/10.1007/s10610-017-9354-1
Lavorgna, A., & Sajeva, M. (2021). Studying illegal online trades in plants: Market characteristics, organisational and behavioural aspects, and policing challenges European. *Journal on Criminal Policy and Research, 27*(4), 451–470. https://doi.org/10.1007/s10610-020-09447-2
Liu, H., Gale, S. W., Cheuk, M. L., & Fischer, G. A. (2019). Conservation impacts of commercial cultivation of endangered and overharvested plants. *Conservation Biology, 33*(2), 288–299. https://doi.org/10.1111/cobi.13216
Lunstrum, E., Givá, N., Massé, F., Mate, F., & Jose, P. L. (2023). The rhino horn trade and radical inequality as environmental conflict. *The Journal of Peasant Studies, 50*(3), 1085–1105.
Mackenzie, S., & Yates, D. (2016). Collectors on illicit collecting: Higher loyalties and other techniques of neutralization in the unlawful collecting of rare and precious orchids and antiquities. *Theoretical Criminology, 20*(3), 340–357.
Margulies, J. D., Trost, B., Hamon, L., Kerr, N. Z., Kunz, M., Randall, J., Shew, R., Shew, D., Starke, L., Suiter, D., West, Z. (2024). Expert assessment of illegal collecting impacts on Venus flytraps and priorities for research on illegal trade. *Conservation Biology, 38*(5), e14320.
Margulies, J. D. (2020). Korean 'housewives' and 'hipsters' are not driving a new illicit plant trade: Complicating consumer motivations behind an emergent wildlife trade in Dudleya farinosa. *Frontiers in Ecology and Evolution, 367*, 604921.
Margulies, J. D. (2023). *The Cactus Hunters: Desire and Extinction in the Illicit Succulent Trade*. U of Minnesota Press.
Margulies, J. D., Bullough, L., Hinsley, A., Ingram, D. J., Cowell, C., Goettsch, B., Klitgård, B. B., Lavorgna, A., Sinovas, P., & Phelps, J. (2019). Illegal wildlife trade and the persistence of "plant blindness". PLANTS, *PEOPLE, PLANET, 1*(3), 173–182. https://doi.org/10.1002/ppp3.10053
Margulies, J. D., Moorman, F. R., Goettsch, B., Axmacher, J. C., & Hinsley, A. (2023). Prevalence and perspectives of illegal trade in cacti and succulent plants in the collector community. *Conservation Biology, 37*(3), e14030. https://doi.org/10.1111/cobi.14030
Maunder, M. (2022). *House Plants*. Reaktion Books.
Maunder, M., Higgens, S., & Culham, A. (2001). The effectiveness of botanic garden collections in supporting plant conservation: A European case study. *Biodiversity & Conservation, 10*, 383–401.
Muneta, J. D. (2020). Peyote crisis confronting modern indigenous peoples: The declining peyote population and a demand for conservation. *American Indian Law Journal, 9*, 135.
Muth, R. M. (1998). The persistence of poaching in advanced industrial society: Meanings and motivations--An introductory comment. *Society & Natural Resources, 11*, 5–7.

Novoa, A., Le Roux, J. J., Richardson, D. M., & Wilson, J. R. (2017). Level of environmental threat posed by horticultural trade in Cactaceae. *Conservation Biology*, *31*(5), 1066–1075.

Olmos-Lau, V. R., & Mandujano, M. C. (2016). An open door for illegal trade: Online sale of Strombocactus disciformis (Cactaceae). *Nature Conservation*, *15*, 1–9.

Phelps, J., & Webb, E. L. (2015). "Invisible" wildlife trades: Southeast Asia's undocumented illegal trade in wild ornamental plants. *Biological Conservation*, *186*, 296–305.

Read, M. (1989). The bulb trade—A threat to wild plant populations. *Oryx*, *23*(3), 127–134.

Robbins, C. S., & Luna, R. T. B. (2003). *Prickly Trade: Trade and Conservation of Chihuahuan Desert Cacti*. Traffic North America.

Scheffers, B. R., Oliveira, B. F., Lamb, I., & Edwards, D. P. (2019). Global wildlife trade across the tree of life. *Science*, *366*(6461), 71–76.

Skidmore, A. (2023). Exploring the motivations associated with the poaching and trafficking of Amur tigers in the Russian Far East. *Deviant Behavior*, *44*(3), 331–358.

Shirey, P. D., Kunycky, B. N., Chaloner, D. T., Brueseke, M. A., & Lamberti, G. A. (2013). Commercial trade of federally listed threatened and endangered plants in the United States. *Conservation Letters*, *6*(5), 300–316.

South African National Biodiversity Institute. (2022). National response strategy and action plan to address the illegal trade in South African succulent flora. Department of forestry, fisheries, and the environment. Republic of South Africa. February.

Subramaniam, B. (2024). *Botany of Empire: Plant Worlds and the Scientific Legacies of Colonialism*. University of Washington Press.

Thomas-Walters, L., Hinsley, A., Bergin, D., Burgess, G., Doughty, H., Eppel, S., MacFarlane, D., Meijer, W., Lee, T. M., Phelps, J., Smith, R. J., Wan, A. K. Y., & Veríssimo, D. (2021). Motivations for the use and consumption of wildlife products. *Conservation Biology*, *35*(2), 483–491. https://doi.org/10.1111/cobi.13578

Ticktin, T., Mondragón, D., Lopez-Toledo, L., Dutra-Elliott, D., Aguirre-León, E., & Hernández-Apolinar, M. (2020). Synthesis of wild orchid trade and demography provides new insight on conservation strategies. *Conservation Letters*, *13*(2), e12697

Voskuil, L. (2017). *Victorian Orchids and the forms of Ecological Society*. Strange science: Investigating the limits of knowledge in the victorian age, 19–39.

Weber, A. (2021). Natural history collections and empire. In Andrew Goss (Ed.), *The Routledge Handbook of Science and Empire* (pp. 80–86). Routledge.

Williams, S. J., Jones, J. P. G., Annewandter, R., & Gibbons, J. M. (2014). Cultivation can increase harvesting pressure on overexploited plant populations. *Ecological Applications*, *24*(8), 2050–2062. https://doi.org/10.1890/13-2264.1

Wyatt, T., & Hutchinson, A. (2023). Plant Blindness and the law on international trade in wildlife. *Erasmus Law Review*, *16*, 14.

Wyatt, T., Van Uhm, D., & Nurse, A. (2020). Differentiating criminal networks in the illegal wildlife trade: Organized, corporate and disorganized crime. *Trends in Organised Crime*, *23*(4), 350–366. https://doi.org/10.1007/s12117-020-09385-9

25
CONCLUSION
Reflections on researching and understanding wildlife crime

Rebecca W.Y. Wong, Francis Massé, Laura Gutiérrez, Annette Hübschle, and Tanya Wyatt

Reflections on researching and understanding wildlife crime

The 24 chapters in this book provide readers with an all-rounded view of issues related to wildlife crime, which has become a global concern. These contributions illustrate that crimes against wildlife are not confined to specific regions, species, or environments, yet have lasting global implications.

In Part 1, contributions investigated the complex interfaces and interconnections between wildlife crime with gender, cultural, and social norms. These intersections could result in a distortion between what is legal and illegal and, in turn, complicate enforcement efforts and turn attention away from harm and sustainability. The importance of culturally embedded gender norms, particularly masculinity, indirectly facilitates hunting practices and influences how harm against wildlife is perceived. Trading in wildlife, such as the lion body parts trade in South Africa, demonstrates the fact that wildlife crime is, to a large extent, a product of social and cultural constructs. The importance of traditional cultural practices is further exemplified in the chapter on camel racing, which reveals the underlying political and cultural complexities of this ongoing practice, a practice that also blurs the line about what is 'wildlife'. In addition, the involvement of business corporations in wildlife crime cannot be overlooked. Economic motives and commercial decisions all play a part in the exploitation of the environment and wildlife.

The chapters included in Part 2 highlight the gap between researching the real-life practice and the theoretical understanding of wildlife crime and related harms. The chapters also tackle the methodological challenges of researching wildlife crime head-on. Bringing together insights from multiple disciplinary backgrounds, the contributions of this section thus offer plausible methodological and research approaches to address these challenges and potential inconsistencies. In the chapters, a research framework noting the cultural and economic differences in understanding Asian consumer demand for high-value wildlife products is developed, while a set of reliable and ethical quantitative methods for estimating wildlife crime prevalence is also put forward. We are also reminded of the role of socioeconomic factors in escalating environmental tensions and conflict within the context of commercial poaching and green militarisation, and how this requires taking a broader lens of environmental conflict.

 DOI: 10.4324/9781003333906-29

The chapters in Part 3 zoomed in on the range of alternative and emerging responses to wildlife crime. The collection of work in this section shows that effective responses do not necessarily require legal redress but can take the form of non-punitive and non-legalistic approaches, as demonstrated in tackling illegal, unreported, and unregulated fishing in Argentina and developing community-based approaches to poaching in southern Africa. Clearly, regulations and legislation, though not free of challenges and difficulties in enforcement, are equally important in curbing the exploitation of wildlife. This is illustrated through the potential for regulating the legal trade in rhino horn and by how legislation and regulations are enforced in the illegal wildlife trade in Hong Kong and the prescription of rhino horn in Traditional Chinese Medicine in mainland China. These contributions critically point to the many regulatory, policy, and ontological possibilities for holistic, just, and sustainable pathways to addressing harms involving and against wildlife.

The final section of this handbook reported on ongoing issues in wildlife crime with contributions on the illegal captive breeding of wildlife, the shark fin trade in Hong Kong, online wildlife trafficking in Brazil, and the concepts of 'denial of injury' and 'denial of the victim' discussed in the context of the European eel trade. The work on the unlawful trade in non-timber ornamental plants illustrates the broader scope of wildlife crimes beyond nonhuman animals. Despite the diverse contextual backgrounds on which these chapters are based, the message conveyed by these contributions is clear: addressing wildlife crime requires multi-dimensional approaches.

As with any handbook and edited collection, there are, of course, things that are missing. For example, missing from this collection is emerging research on the use of nonhuman animals in bloodsport, the sexual assault of nonhumans, the expanding trade in fungi, financial crimes associated with wildlife crime, and the infectious disease and public health concerns of wildlife trade and consumption. Hence, this handbook is not meant to be an exhaustive overview of wildlife crime, efforts to address it, or critical approaches to researching it. It is a snapshot of some of the emerging research at a moment in time that highlights the need, potential, and promise of critical research and approaches for understanding and effectively addressing harms against and using wildlife.

As we conclude this handbook, the trends in wildlife crime continue to change. For example, the 2024 World Wildlife Crime Report notes a decline in the standard seizure index of nonhuman animals, but a large increase in the seizures of plants, resulting in an overall increase in seizures. This points to changing preferences and dynamics of the species being traded. We are also witnessing gradual advancements in addressing wildlife crime with the aid of technologies such as drones for detecting deforestation, GPS tracking to locate protected wildlife, and the gradual integration of AI to monitor and investigate online platforms for illegal trading in protected wildlife. While these approaches and technologies are often being harnessed to strengthen an enforcement-first approach, they can also be mobilised to underpin community-based, holistic approaches to curbing wildlife crime.

These and many other advancements in addressing wildlife crime, as well as the expansion of wildlife crime and related harms into new spaces, species, and practices, only point to the continued need for critical approaches in understanding these dynamics. It is in these critical approaches that a holistic, multifaceted focus on socioecological harms, well-being, and sustainability can productively be centred. While methods are constantly evolving and changing, we hope this handbook can provide some continued inspiration to safeguard the planet's wildlife and related socioecological well-being.

INDEX

Note: **Bold** page numbers refer to tables; *italic* page numbers refer to figures and page numbers followed by "n" denote endnotes.